Library of Ethics and Applied Philosophy

Volume 44

The Library of Ethics and Applied Philosophy addresses a broad range of topical issues emerging in practical philosophy, such as ethics, social philosophy, political philosophy, philosophy of action, It focuses on the role of scientific research and emerging technologies, and combines case studies with conceptual and methodological debates. We are facing a global crisis, which raises a plethora of normative issues, but also poses a challenge to existing conceptual and methodological resources of academic philosophy. The series aims to contribute to a philosophical diagnostic of the present by exploring the impact of emerging techno-scientific developments on zeitgeist, collective self-image, and worldview. Concrete and urgent emerging issues will be addressed in depth and from a continental philosophical perspective, which may include dialectics,hermeneutics, phenomenology, post-phenomenology, psychoanalysis, critical theory and similar approaches.

Rasmus Sandnes Haukedal

Agency and Organisation: The Dialectics of Nature and Life

Springer

Rasmus Sandnes Haukedal
East China Normal University
Shanghai, China

ISSN 1387-6678 ISSN 2215-0323 (electronic)
Library of Ethics and Applied Philosophy
ISBN 978-981-95-1923-1 ISBN 978-981-95-1924-8 (eBook)
https://doi.org/10.1007/978-981-95-1924-8

This Springer imprint is published by the registered company Springer Nature Singapore Pte Ltd.
The registered company address is: 152 Beach Road, #21-01/04 Gateway East, Singapore 189721, Singapore

The original version of this book has been revised. A correction to this book can be found at https://doi.org/10.1007/978-981-95-1924-8_9

Acknowledgements

During the work on this book, I have been spoiled by the company of multiple great colleagues and comrades. This work, as any dialectical whole, is not self-contained. I am not its sole author since I could not have done it without the help and insights of many others. But the responsibility for all its flaws and mistakes is, of course, mine. First among those that require my thanks is my Ph.D. supervisor, the unrelenting Professor Gerald Moore, whose call to 'always gloss' I have tried to heed. Without your patience and encouragement, this work would not have been possible or even conceivable. Then, the (in)famous Geraldians, whose friendship has been invaluable in these interesting times. I am forever grateful to all of you. I extend this to the other friends I made in Durham. You are present in this work in one way or another, but I can only mention Mauro Senatore and Philippe Lynes here. Kerstin Oloff at the Centre for Culture and Ecology deserves special mention, as do the people at the Institut de recherche et d'innovation in Paris and Bernard Stiegler, who left us too soon. The influence of Giuseppe Longo and Maël Montévil has been considerable, as the direction of this book changed extensively after being introduced to them and their work. I also want to thank my current supervisor, Tong Shijun, at East China Normal University, for all his encouragement and for giving me the opportunity to work patiently on this project. My friends and colleagues at ECNU, and the talented baristas at the Coffee Station, cannot go unmentioned. They have provided me with an excellent context for finishing this project. I should also thank former colleagues and friends at my alma mater, the Norwegian University of Science and Technology – Thomas Netland and my previous supervisor Helge Høibraaten in particular. I would be remiss if I forgot Øystein Tennfjord, who in some ways inspired this book, and his better half, Elida. I also thank the staff at Ustinov and Josephine Butler colleges in Durham, and the school for Modern Languages and Culture, for all their assistance. The same goes for the staff at ECNU, Hong Cheng in particular. My family and friends in Norway deserve my sincerest gratitude for their unwavering support and for putting up with my strange habitus. I single out my nieces Tuva and Sunniva and my nephews Magnus and Peter, but I am indebted to you all. Finally, there is one person whose importance I cannot even begin to describe. I dedicate this book to you, my love.

Contents

Chapter 1
Introduction: The Return of Life and Dialectics

> Whenever inner and outer, cause and effect, end and means, subjectivity and objectivity, etc., are one and the same, there is life. (Hegel 2004: 274)

For a long time, I did not know how to structure this volume. Sometimes, it seemed that the biologists could speak for themselves without the help of philosophers. While there is some truth to this, I soon realised that scientists were calling for philosophy to leave its isolation. It also seemed that I could not make sense of the recent turn in biology except by looking at the philosophical tradition that it rejuvenates, tied to an organic conception of nature, and opposed to the mechanistic understanding upon which modern science is founded. This neglected tradition promulgated a dialectical biology that placed the organism at the centre of evolution. But modern biology abandoned *organicism*, reified nature and thereby forgot about life. Both these trends are evident in the Modern Synthesis (MS), the tradition I aim to reconstruct in this work.

This book is first and foremost an attempt at philosophy of nature, not philosophy of science. I am not content discussing the meaning and interpretations of science, past and present, but seek to criticise it from within. This not only demands that we do not assume anything as given but question all presuppositions but also entails that philosophy and science must share subject matter, nature itself (Unger and Smolin 2015). One of the tasks then, is to show how philosophy of science is also, implicitly or explicitly, philosophy of nature.

Arguably, the neglect of such issues has led to a crisis within philosophy (Mohan 2021). To combat these matters, I aim to show how philosophy is itself science, but not of particulars, like the natural sciences. Instead, philosophy is a science of totality and thus cannot have a clearly defined and specific subject matter (Illetterati 2021). In this view, anything can and should become the subject of philosophical critique.

This view has been anathema to modern philosophy—which was characterised by an epistemological bias and two prevalent views on philosophy's relation to science.

R. Sandnes Haukedal, *Agency and Organisation: The Dialectics of Nature and Life*, Library of Ethics and Applied Philosophy 44,
https://doi.org/10.1007/978-981-95-1924-8_1

First, the hermeneutical view considers philosophy a discourse outside and independent from science. It thus insists on the autonomy of philosophy. The analytical approach, on the other hand, moves in the opposite direction. It emulates science, adopts its notion of 'scientificity' uncritically, and scorns anything deemed metaphysical (Illetterati 2022). The irony is that both sides have tended to reduce the natural sciences to positivism, the view that all knowledge derives from the 'positive data' provided by experience that is immediate. But this understanding of the natural sciences is superficial at best and obfuscating at worst. In what follows, I will not spend time on these traditions but propose a view that transcends both.

By contrast to these positions, G.W.F. Hegel says philosophy is a science precisely by breaking with the scientific ideal of modern science, and instead proposes a science of totality that may complement the particular sciences. Philosophy cannot impose categories on science from without but should strive to 'situate the sciences within their broader non/extra-scientific contexts' (Johnston 2019: 55) and show how they contain more metaphysics than they are aware of, Hegel opines. In this view, the a priori conditions of experience are not 'something separate from the *a posteriori*. The *a priori* is, if anything, the process of gradual conceptual justification of that world of experience which constitutes the starting point of the cognitive endeavour and of the need for philosophy' (Illetterati 2022: 66).

Science, as a denial of pure immediacy, is an integral part of the same process as philosophy since it also partakes in 'the rational justification of experience' (Illetterati 2022). I seek to show how developments within science, like its failure to generalise and integrate findings (Johnston 2019), point toward its self-subversion—a trend especially evident in evolutionary biology. While philosophy has lagged behind the developments in natural science, scientists are 'prisoners of philosophy' (Engels 1934). The task is therefore twofold: Both mapping the assumptions of science *and* contributing to its further conceptualisation by making conjectures about nature itself, in a speculative move. Importantly, when I extrapolate, I try to stay but a few steps ahead of the sciences.

For these reasons, I would not call my approach exegetical or historical. I do not, for instance, go into the philosophical background that shaped Hegel's view, even if I use facets of Kant's philosophy as an argumentative prop at several junctures. And although I spend a fair amount of time on the history of biology, my presentation is selective, serving as a stepping stone for the theories I discuss. Maybe it is more accurate to call my method systematic, as I am searching for an encompassing framework that can grasp complex systems without distorting simplification. I am more interested in *applying* philosophical insights outside their usual context than unearthing their 'true' meaning. Specifically, I aim to show how the re-formulation of the neglected tradition of organicism is anticipated both by a dialectical tradition within science and by Hegel's philosophy—and how these theories may inform the ongoing shift within evolutionary biology called the Extended Evolutionary Synthesis (EES).

The central notion that connects the theories under scrutiny is *organisation*, a notion that preoccupied the organicists. Theirs was a view of nature first formulated by German Idealism. Within biology, however, it was predominantly a British phenomenon. And it was already in its heyday, the start of the nineteenth century, formulated as a critique of the mechanical worldview (Beiser 2005). Later, this perspective was accommodated to mechanism by way of quantitative analysis (Merchant 1989).[1]

I argue that the current rejuvenation of the organic perspective calls for an updated version of organicism informed by a new understanding of Hegel's philosophy. The move away from the standard textbook version builds on aspects of Hegel's philosophy that have been neglected or marginalised in the literature. These aspects are not minor aberrations of his overall system but applied to recast his whole philosophy. They also imply a renewed commitment to science. It is, therefore, incumbent upon us to reveal the convergences between Hegel's philosophy and modern science and how it may inform its theory and practice.

Part of this shift is viewing Hegel's notion of organic form as more than an analogy (Beiser 2005). We perceive this in his discussion of Newton, and his emphasis on the *systemic* interdependencies between celestial bodies—a systemic dimension overlooked by Kant (Westphal 2020). All models of development in Hegel's philosophy, including his logic, are based on this model of organismic growth, which is not linear. If we assume that there is a primordial unity at the outset that precedes opposing elements, individual development simply unfolds a given potential. This is a deductive model based on first principles (Westphal 2008). Hegel's view of science is not deductive but informed by a deep involvement with the sciences and confidence in their ability to disclose natural processes and forces.

Philosophy must rely on and organise the results of empirical science, even if it cannot complete this task. Hegel views the sciences as cognitively valid but limited by unjustified assumptions, and engages with their findings to disclose these assumptions. He seeks a reflexive science that applies its results to its own practice, that accounts for its own precondition scientifically. This is how form and content may coincide (Radnik 2024). He finds that mechanical and chemical processes do not suffice if we want to understand the self-relational process of differentiation and reunification that makes life, cognition, and ultimately science itself possible. The mechanistic understanding of nature, by contrast, reduces the organism to a mere object of evolution, not an active subject. It removed its agency as it could not account for its existence except in abstract and derivative terms.

'Dialectical biologists' Richard Lewontin and Richard Levins (1985, 2007) seek to undermine this view. One of the core facets of their dialectical theory is to consider the organism an active agent in its own evolution. They understand the relation between organism and environment as reciprocal and symmetric and emphasises the dialectical interplay between parts and whole and between actuality and potentiality. In their view, organisation is a crucial feature of biological systems, where top-down processes constrain bottom-up processes while at the same time depending on them. This interplay, I argue, is ubiquitous in nature and society, and the proper dialectical

[1] Throughout this study, I apply the term 'mechanism' instead of 'mechanicism' to the mechanistic perspective. While it might be misleading—as 'mechanism' also applies to the subject matter of the mechanistic worldview—it follows the usage in much of the literature I survey.

understanding of it must account for how possibilities are continuously produced in and through the way they are actualised.

I am fortunate to write this work during a wave of reinterpretations of Hegel's naturalism, a veritable 'Hegel renaissance' (Beiser 2008; Ferrini 2014), to which I am heavily indebted. My modest contribution lies in establishing connections to current science, not in offering a unique take on Hegel. Less a reading of Hegel than a Hegel-inspired understanding of nature and science, I will not contrast the different interpretations of Hegel's philosophy in any systematic way. Instead, I make use of what I find most relevant for my purpose. This approach, which mostly concerns the application of Hegel's thoughts, and my attempt at utilising them further, follows from an understanding of his theory as profoundly historical. I argue that it is forever incomplete or aporetic because it is a theory that is concerned with the unfinished process of nature cannot itself be complete. There are always limits to our capacity to understand nature, limits immanent to nature itself (Padui 2010: 252). Therefore, there cannot be a final sublation. In this view, nature is not unknowable to us but inexhaustible.

Against interpretations that depict Hegel's approach as strictly a priori, and his dealing with science to be non-systematic—simply used to confirm a priori principles—I hold that Hegel saw the empirical sciences as the material through which philosophy of nature must work to avoid being a self-enclosed logic.[2] Crucially for this investigation, it also

> seems to suggest that a distinctly Hegelian philosophy of nature would remain perpetually open to thinking through new developments in the empirical sciences, situating them in terms of conceptuality and the architecture of the speculative framework. (Furlotte 2018: 21)

Echoing this stance, Slavoj Žižek says that to remain faithful to Hegel we must repeat him: Apply him to an unfamiliar context by reading him against modern philosophy and science. Catherine Malabou, on her part, emphasises how reading Hegel's corpus is an act of construction since his system is open to radical changes. It cannot be possessed, only repeated. We should understand his theory as a permanently unfinished research programme, not as a doctrine (Zwart 2022). Without striving for closure, we must seek to complement Hegel's theory. At times, this takes the form of diagnostics, simply registering that scientists are practising a Hegelian approach; at other times, I argue that a Hegelian interpretation could elucidate their findings; at still others, it is natural science that helps elucidate while philosophy takes the back seat.

To establish a fertile exchange between science and philosophy, I examine the theories of Žižek and Malabou, as well as Donna Haraway. In this company, Haraway might seem the odd one out, as she does not deal much with Hegel. Nonetheless, I believe there are important lessons to be learned from her work, extending from her dissertation on organicism to her recent publications. I highlight her profoundly

[2] The other reading is empirical, and says that Hegel is acting as a scientist. From this view, his philosophy of nature is dismissed on the grounds that science has changed a lot since Hegel's times. As Westphal (2008) argues, both these readings miss the holistic character of Hegel's philosophy, where there is identity-in-difference between science and philosophy.

ecological way of thinking about the entanglement between organism and environment. I continue this line of reasoning through Žižek's philosophy. He provides a version of Hegel that emphasises contingency over necessity and the notion that nature is unfinished. This ontological incompleteness suggests that subjectivity is inscribed in the order of being as a possibility. Nature is underdetermined, and it is this failure of determination which enables something like consciousness to emerge.

Perhaps it is correct to say, like Ruda (2022), that nature has returned to centre stage because it can no longer be ignored—because 'it started to become a problem' (114). I aim to show that nature was always a problem, but it was not recognised as such. Philosophers made compromises that left nature in the background as a silent premise, never thematised, and their theories consequently suffered from ambiguity. Malabou's philosophy of plasticity, by contrast, is re-reading Hegel's philosophy of nature through the lens of modern science. Plasticity, a concept that Malabou finds in the periphery of Hegel's thinking and uses as a way of unlocking his whole thinking, is both the reception and the construction of form. In her dealing with neuroscience, she applies this dialectical relationship between the neuronal and mental to get at their unity-in-difference. Plasticity operates in the leeway left open by nature. This indicates an overlap between the perspectives at hand.

The way I deal with these thinkers might seem opportunistic, but I aim to show both how they may contribute to contemporary biology and how they may learn from it. I question the rigid distinctions between the sciences and between science and philosophy that have led to a Balkanisation of science both internally and externally, making scientific integration more challenging. I combat this tendency by showing convergent trends not only within natural science but between natural science and natural philosophy. Here, the so-called *organisational approach*, a modern restoration of organicism, is central.

The book culminates in a call for a new kind of science based on the organic view of nature. It must leave the shackles of mechanistic and analytic science behind and acknowledge the centrality of history and ecology. And it demands expanding the notion of agency from being about efficient causes alone. An evolutionary agent is not merely a force acting on organisms on the outside in a mechanistic manner; rather, agency is constituted through the interaction of organism and environment.

The book is divided into six chapters and a conclusion: I start with two chapters concerned with the MS and the EES. These chapters set up the discussion and function as introductions to the biological concepts that are under debate, so that we have a clear sense of the history of the MS and the challenge it is currently facing. The first chapter takes us through the establishment of the MS—to render its fragility as a research programme and how it has been challenged and reformed several times. I argue that the MS is no longer operative as a complete theory but that most of its underlying assumptions are nonetheless intact. I end the chapter by arguing that the MS favours Mendel's views more than Darwin's and that the current calls for extension constitutes a return to some of Darwin's original theses.

In the next chapter, I present some of the core facets of the EES, outline the main aspects of the organisational approach, and how it relates to the older tradition of organicism. They both aim at reintroducing the organism to centre stage as an active subject in its own evolution, a view that is also promoted by Lewontin and Levins. Next, I sketch the basics of niche construction theory, the topic of Chap. 7. I also discuss dynamical systems theory and how it is limited when applied to historical systems, where you cannot predefine the possibility space. This is a fundamental point that will follow us throughout the study. I end by asking what success would constitute for the EES.

I should note that the whole work is interspersed with discussions of topics from evolutionary biology, even if I spend the most time on them at the beginning and end of the thesis. I further develop the notions that we introduce in the first two chapters throughout the work, but I do not aim at a final definition or settling biological debates. I simply aim to provide a reading that makes evident the conceptual overlaps between disciplines that are often considered irrelevant to each other.

After the chapters on biology, I turn to Donna Haraway (Chap. 4). More specifically, I deepen the debate on organicism by presenting her PhD thesis on the subject. I then go on to critique the notion of autopoiesis, both based on the organisational approach and based on Haraway's adoption of the term *sympoiesis*, which is tied to the notion *holobiont*—the organism plus its persistent symbionts—a challenge to simplistic notions of biological individuals. In this view, individual organisms are not enclosed from their outside world but consist of multiple genomes—a consortium of microorganisms. I argue that sympoiesis encapsulates the contradiction between self-organisation and symbiosis which is only implicit in autopoiesis. This can be further elucidated by the notion of closure in the organisational approach, and inform how we consider the relationship between organisms that mutually enable each other's maintenance.

Having discussed how the dialectical perspective has influenced biology, I move to Žižek and Hegel, to get at a version of the latter that takes nature and science seriously. I have divided this chapter in two, starting with an introduction of central terms in Hegel's dialectics. I then present his important thesis on the weakness (*Ohnmacht*) of nature, which simply put concerns its contingency or lack of rigid structure.[3] I claim that this explains why Žižek has taken an interest in quantum physics. I also contend that his views dovetail with how the quantum realm is theorised in the organisational approach. Here, the related notion of *bio-resonance* recasts the interplay between different levels of organisation and leads us to the question of whether the universe is canalised or not and the implications for evolution.

The second part of the chapter deals extensively with Hegel's notion of necessity and contingency as seen in his analysis of the interplay of possibility and actuality and the different modalities involved. I then return to a discussion of emergence and argue for an ontological understanding of it through Stuart Kauffman's theory, informed by Hegel's views. I go on to claim that Hegel's views are already implicit in the organisational approach, which is not surprising when we look at its genealogy.

[3] I should note that this is not all, as it also speaks to a lack of power in nature: 'For Hegel himself, nature simply does not have, as it were, the power as a whole to live up to all its other powers, to be what it is supposed to be or at least what would follow from its being external to (logical) thought' (Pinkard 2024: 266).

The two final chapters before the conclusion take us to plasticity and niche construction theory, respectively. The first begins with a general discussion of plasticity and canalisation, before reconstructing Malabou's theory of plasticity. I argue that we should understand plasticity as an ecological phenomenon and therefore acknowledge the interplay of different forms of plasticity. I also argue that plasticity, as any dialectical concept, contains its apparent opposite, canalisation. Then, we move on to the notion of habit formation as a process of *in-habiting* the niche and try to relate her concept of explosive plasticity to similar notions in the organisational perspective, arguing for their convergence. I argue that explosive plasticity is a generic potential found in every organisation and denotes its capacity to be overturned. Moreover, I try to show that Malabou's notion of plasticity can be informed by the organisational approach and inform it in return.

Building on the previous chapter, we turn to niche construction and its possible extension in Chap. 7. This extension involves acknowledging that construction is not only about material changes but also an epistemic process, and that mediation has vital importance for how we imagine the interplay between organism and environment. This shift, I argue, takes us in the direction of a more dialectical understanding of niche construction, aligned with Lewontin's (2000) proposal. In this view, the environment underdetermines the actions of the organism—thereby opening a space for it to retroact on it. I relate this to Hegel's understanding of the concept as embodied in purposive action and nature as the idea in its externality (Ng 2020) and argue that this constitutes an organisation of the environment that accommodates it to the organism. The determination of the environment is simultaneously a kind of self-determination since the organism is not externally related to what it is not but has integrated it into its functioning.

In the conclusion, I tie together some of the insights we have encountered and then move to a more general discussion of science to ask which scientific ideal we should opt for. I do not aim for closure but to suggest a way forward. I argue that the radicality of the EES might be undermined by the scientific ideal it propounds. Specifically, I claim that we should move towards a more dialectical understanding of causality, found in the notion of enablement proposed by the organisational view. Finally, I outline a kind of scientific unification without subsumption or eradication of differences. I propose this not only because I believe this is the most theoretically fruitful way forward, but also as a practical way for the sciences—and philosophy—to regain their vitality. This follows from Malabou's notion that disciplines who isolate themselves from others are destined to stagnate, and that this lack might explain our current predicament.

By these means, I aim to demonstrate that the current ramifications of Hegel's philosophy of nature can be appreciated more fully through Haraway, Žižek, and Malabou on one side and the modern organisational approach on the other. This is not an exposition of modern biology; nor is it a comprehensive reconstruction of Hegel's philosophy. It is a bit of both and neither. I mostly avoid philosophical quibbles unless they shine a light on the matter at hand, and I will cut many discussions short. Some arguments will appear apodictic, and Hegel scholars might be annoyed by a lack of detail and rigour in my discussions—as will scholars of Haraway, Žižek, or

Malabou. Biologists, and historians of science and biology, will equally see gaps and biases in my exposition, if not outright mistakes. But if scholars from these schools of thought and disciplines may nonetheless perceive something worthwhile in what I am doing, I have succeeded.

I further believe my contribution lies in the philosophical examination of evolutionary theory and in establishing points of connection that should be developed further. It is an attempt at *philosophising* nature—of applying the tools provided by philosophy and science to gain a foothold on nature and understand how it applies to science in general. Hopefully, we may hereby overcome the forgetfulness of nature and life within biology and philosophy.

Bibliography

Beiser, Frederick. 2005. *Hegel*. New York and London: Routledge. https://doi.org/10.4324/9780203087053.

Beiser, Frederick. 2008. Introduction: The Puzzling Hegel Renaissance. In *The Cambridge Companion to Hegel and Nineteenth-Century Philosophy. Cambridge Companions to Philosophy*, ed. Beiser F. C., 1–14. Cambridge University Press.

Engels, Friedrich. 1934. *Dialectics of Nature*. Moscow: Progress Publishers.

Ferrini, Cinzia. 2014. From Disparagement to Appreciation: Shifting Paradigms and Interdisciplinary Openings in Interpreting Hegel's Philosophy of Nature. *Esercizi Filosofici* 9 (1): 1–13.

Furlotte, W. 2018. *The Problem of Nature in Hegel's Final System*. Edinburgh: Edinburgh University Press. https://doi.org/10.3366/edinburgh/9781474435536.001.0001.

Hegel, Georg Wilhelm Friedrich. 2004. *Philosophy of Nature: Part Two of the Encyclopaedia of the Philosophical Sciences*, With Foreword by J. N. Findlay, translated by A. V. Miller. Oxford, UK: Oxford University Press.

Illetterati, L. 2021. Philosophy as the Science of Freedom. In *Hegel's Encyclopedic System*, ed. S. Stein and J. Wretzel. New York and London: Routledge.

Illetterati, Luca. 2022. Science as Experience of Freedom: Hegel on the Scientific Nature of Philosophy. In *The Relevance of Hegel's Concept of Philosophy: From Classical German Philosophy to Contemporary Metaphilosophy*, ed. Luca Illetterati and Giovanna Miolli, 55–74. London and New York: Bloomsbury Academic.

Johnston, Adrian. 2019. *Prolegomena to Any Future Materialism, Volume Two: A Weak Nature Alone*. Evanston, IL: Northwestern University Press.

Levins, Richard, and Richard Lewontin. 1985. *The Dialectical Biologist*. Cambridge, MA: Harvard University Press.

Lewontin, R. 2000. *The Triple Helix: Gene, Organism, and Environment*. Cambridge, MA: Harvard University Press.

Lewontin, Richard, and Richard Levins. 2007. *Biology Under the Influence: Dialectical Essays on Ecology, Agriculture, and Health*. New York, NY: Monthly Review Press.

Merchant, Carolyn. 1989. *The Death of Nature: Women, Ecology and the Scientific Revolution*. New York, NY: HarperOne.

Mohan, Shaj. 2021. On the Bastard Family of Deconstruction. *Philosophy World Democracy*. https://www.philosophy-world-democracy.org/other-beginning/on-the-bastard-family-of-deconstruction.

Ng, K. 2020. *Hegel's Concept of Life: Self-Consciousness, Freedom, Logic*. New York, NY: Oxford University Press. https://doi.org/10.1093/oso/9780190947613.001.0001.

Padui, Raoni. 2010. The Necessity of Contingency and the Powerlessness of Nature. *Idealistic Studies* 40 (3): 243–255. https://doi.org/10.5840/idstudies201040316.

Pinkard, T. (2024). The Prospects for an Idealist Natural Philosophy: Logic and Nature. In *Hegel's Philosophy of Nature: A Critical Guide*, ed. M. F. Bykova, 259–272. Cambridge: Cambridge University Press. https://doi.org/10.1017/9781009023160.018.

Radnik, Borna. (2024). *Freedom, in Context: Time, History, and Necessity in Hegel*. London, UK: Bloomsbury Academic.

Unger, Roberto Mangabeira, and Lee Smolin. 2015. *The Singular Universe and the Reality of Time*. Cambridge, UK: Cambridge University Press.

Westphal, Kenneth R. 2008. Philosophizing about Nature: Hegel's Philosophical Project. In *The Cambridge Companion to Hegel and Nineteenth-Century Philosophy*, ed. Frederick Beiser, 281–310. Cambridge, UK: Cambridge University Press. https://doi.org/10.1017/CCOL9780521831673.012.

Westphal, Kenneth. 2020. Causality, Natural Systems, and Hegel's Organicism. In *The Palgrave Hegel Handbook*, ed. Marina F. Bykova and Kenneth R. Westphal, 219–39. Cham: Springer International Publishing. https://doi.org/10.1007/978-3-030-26597-7_12.

Žižek S, F. Ruda, and A. Hamza. 2022. *Reading Hegel*. Cambridge, UK: Polity.

Zwart, Hub. 2022. *Continental Philosophy of Technoscience*. Cham: Springer.

Chapter 2
Setting the Stage: The Evolution of Evolution

Abstract This chapter explores the conceptual foundations and historical trajectory of the Modern Synthesis (MS), the framework that united Darwinian natural selection with Mendelian genetics to form the cornerstone of modern evolutionary theory. I critically examine the MS's theoretical commitments, highlighting its ambiguities and inner tensions, particularly the balance between mechanistic and teleological explanations in biology. Key themes include the role of adaptationism, the debate over whether natural selection is creative or eliminative, and the unresolved question of variation's origin. The chapter also interrogates the MS's narrative coherence, challenging the notion of a seamless Darwin-Mendel partnership and exposing the compromises that shaped the synthesis. By engaging with critiques from mutationists, neutralists, and developmental biologists, the chapter sets the stage for later discussions on extending or replacing the MS, emphasising the need for a more inclusive evolutionary theory that accounts for organismal agency, plasticity, and process-oriented ontologies.

Introduction

> The aim of science is to seek the simplest explanations of complex facts. We are apt to fall into the error of thinking that the facts are simple because simplicity is the goal of our quest. The guiding motto in the life of every natural philosopher should be, Seek simplicity and distrust it. (Whitehead 2015: 104)

What follows is preparation for the philosophical debates later—an introduction to core facets of the Modern Synthesis (MS), with a view to overlaps between biology and philosophy. As a non-historian and non-biologist, I concentrate on conceptual matters and leave rigorous historical and empirical works to others. My aim is to examine the conceptual commitments of the MS, not to intervene directly in debates that go beyond my expertise.

R. Sandnes Haukedal, *Agency and Organisation: The Dialectics of Nature and Life*, Library of Ethics and Applied Philosophy 44,
https://doi.org/10.1007/978-981-95-1924-8_2

To get at these commitments, we must first outline the overarching tendencies of the enterprise. Not pretending to be comprehensive, I focus on the most characteristic—yet often controversial—aspects of this history, to illustrate how current attacks and alternatives have been articulated, and how they were accommodated.

In these chapters on biology, and the book as a whole, I seek to render evident the claim that biology is a 'profoundly historical and philosophical discipline' (Smocovitis 1996: 61). If what Huneman and Walsh (2017a) write is correct, the debate about the right path for biology is more about conceptual than empirical matters (Jaeger et al. 2012). For sure, the debate is historical in nature, making matters increasingly complex (Smocovitis 2023). O'Malley and Dupré (2005) encourage philosophers to engage with what is happening in biology—not only with issues concerning bioethics or personal identity but metaphysics proper. This could not only strengthen the metaphysical underpinnings of biology but also impact philosophy, by revealing which ontology affords us the analytic tools to explain evolution adequately. I argue that dialectics provides us with the most useful tools for this, but I cannot justify this claim until later.

The MS usually refers to the marriage between Charles Darwin's theory of natural selection and the genetic theory of Gregor Mendel, and marks the founding of modern evolutionary theory. Unfortunately, the chaotic history of its ascent is often forgotten, as is the role played by dialectical materialism in moulding the approach of the scientists. This leads to the question of whether a substance or a process ontology is most adequate for science, about which I can only offer preliminary remarks.

In substance ontology, natural entities are premised on an underlying substrate or essence which is static. Everything that exists derives from combining these basic properties. A process-oriented approach, on the other hand, is dynamic since the fundamental constitution of reality is changing. Processes and relations are ongoing reconfigurations of the world, and substances are a mere abstraction of such processes, and thus only *relatively* invariant. From this view, it is impossible to pin down the exact meaning of concepts a priori as there are no eternal essences outside history. As we shall see, the process view is aligned with dialectics. It contrasts with metaphysics as it is commonly understood because it does not pretend to describe the cosmos in an ahistorical manner but instead is permanently open to revision.

In anticipation of what we will discuss later, from processes emerge new levels of organisation which are irreducible to the activity of their parts, as they display capabilities that the parts do not exhibit in isolation. This is the dialectical approach in a nutshell. Searching for an assumed ultimate level of causation is futile since no level is isolated or unconstrained.[1] As we shall see, this is akin to the perspective of the father of epigenetics, C.H. Waddington, and Theodor Dobzhansky, one of the central architects of the MS. The process view does not imply that we cannot speak of entities, nor that research treating processes as things is falsified as such;

[1] To anticipate a bit, let us take an example: While the genome encodes the sequence of amino acids in a protein, the physical process of protein folding—essential for biological function—is not genetically specified. Instead, folding depends on the biochemical and physical conditions of the cellular environment, such as temperature, pH, molecular chaperones, and solvent properties (see Jaeger 2024).

it only means that we must rethink entities as coagulated processes, that the objects of science are synchronic snapshots of diachronic processes (Juarrero 1999). Since some processes are more stable than others, they take on the form of entities, but this obfuscates their ontological status as processes (Lewontin and Levins 2007).

I return to such questions, tied to individuation and interlevel causation, in due course. For now, the most pressing matter is figuring out what the MS was and its current status. This could be the topic of this whole volume, but I will try to condense it into two chapters, where the first aim to articulate the main facets of the MS, while the latter moves beyond the MS towards its alternative.

I begin with some general remarks on terminology and how we should view the theoretical status of the MS. I claim that the MS was not a paradigm but more of a research tradition, which implies that it had a looser character. This makes it challenging to pin it down conclusively. In fact, part of the task is understanding the theoretical status of the MS.[2] After this, I move to depict the history of the MS more in detail, to show how it was a rather delicate compromise. After this groundwork, I turn to a central premise of the MS, *adaptationism*, the view that organisms are passively adapted to the environments due to natural selection acting on genetic variants. Here, the development and activities of the organism are irrelevant. I juxtapose this view with Dobzhansky's understanding of adaptation, which differed from the MS view. It epitomises a forgotten alternative adopted by his student, Richard Lewontin, whose dialectical perspective is explored throughout this study.

Next, we move to the question of whether natural selection is creative or not, whether it produces the variation it acts on or depends on other processes—a topic which is often forgotten in descriptions of the MS. I argue that the constitution of variation for natural selection to work on is still an unsolved problem. I relate this to one of the main challenges the MS has faced, which came from the so-called mutationists, who claimed that natural selection was a negative sieve that depended on the variation provided by genetic mutations, and not—as Darwin and the MS architects believed—a creative factor (Beatty 2016). I argue that this reveals that the marriage between Darwin and Mendel was less harmonious than it is portrayed as, and that we should ask who got the upper hand in the synthesis. I thus challenge the notion that the MS is a simple combination of Darwin and Mendel's views and argue that it is something qualitatively different even if it was based on their perspectives.

Some of the parts are a bit dense, as I try to get to the core rapidly and elaborate later. I will indicate this along the way. Also, I will not sketch the alternatives to the MS until the next chapter. There, I focus on niche construction theory (NCT) and dynamical systems theory, but topics such as plasticity, ecological development biology (eco-devo), endosymbiosis, multilevel selection, etc. are never far away, even if I cannot thematise them. Together, they form an interdependent cluster, aiming for a *repetition* of evolutionary theory that acknowledges the relevance of neglected perspectives. As Odling-Smee et al. (2003) note, such a reorientation would make

---

[2] As Smocovitis (2023) notes, the disputes of the 80s attracted philosophers to study biology, partly because it' 'defied any model of a scientific theory; they variously described [it] as a "supratheoretical framework," a "hypertheory," and a "metatheory"' (39).

biology more open to insights from other disciplines. I shed light on these openings to demonstrate how philosophy may contribute.

Remarks on Narrative Coherence

> A critical history is one that makes explicit the norm at work in practices of articulation and delivers it for the purposes of assessment and critique. It clarifies the commitment that the norm expresses together with the insurmountable risk internal to the commitment in question: that it is an institution to be maintained and can be debunked, that it can sustain precisely the opposite of what it purports to represent. (Zambrana 2015: 139)

The above quote articulates the ideal for my exposition: To render the instability and ambiguity of the normative commitments of the MS through immanent critique. As we shall see, the notion that an institution can sustain the opposite view of what it purports is particularly pertinent. In *Organism and Environment* (2015), Sonia Sultan sums up how evolutionary biology was led to its current state, with a narrow focus on genes and their selection. Varying only one factor experimentally led to a kind of self-fulfilling prophecy, as different outcomes will be attributable to the factor in question. Yet,

> this does not mean that other factors are irrelevant under natural conditions, where they too vary. Concluding that genes alone determine phenotypes based on these allelic variation studies is circular reasoning, to be sure. Yet, by a process of conceptual shortcuts, what began as a useful research program for basic transmission genetics has resulted in a deeply entrenched, unique status for genes as quintessential biological information. (Sultan 2015: 1)

Sultan emphasises the fixation on genetic factors, and how the circularity of the underlying reasoning led to a neglect of context and development in evolutionary research. Hence, it disregards the concrete organism and its historical constitution. We should note her use of the term ‘research program’, introduced by Imre Lakatos (1978). In contrast to Kuhn’s paradigms, it is a less rigid structure, allowing for reorganisation without revolution.[3] Unlike paradigms, research traditions change in a ‘continuous fashion and allow for amendments and processes of updating’ (Fábregas-Tejeda and Vergara-Silva 2018: 172). It captures the history of evolutionary theory, a discipline wherein the Darwinian view is accepted as largely correct, and where the point is not to uproot it but to improve it (Pigliucci et al. 2006).

Depew (2011) states that the MS ‘research tradition has never been, is not now, and never will be smoothly cumulative. It has gone through reformulations in which conceptual change has played as prominent a role as empirical discovery’ (94). The MS is less a coherent paradigm than a ‘moving target’—‘a cohesive if somewhat

[3] David J. Depew and Bruce Weber (1995) similarly apply this term, or the ‘research tradition’, when characterising the MS, which encapsulates how the MS was never a neatly integrated theory (see Malafouris 2015).

diffuse body of theory' (Huneman and Walsh 2017b: 2). It is narrated as something that was broad and inclusive hardening into a narrow research programme (Gould 1983), exhibiting 'overly strong reliance on the mechanism of natural selection' (Smocovitis 1996: 32). While successful, its opponents say the time has come to overthrow it, or at the very least extend it by including more factors into the evolutionary framework (see Noble 2006).

I do not disagree with this narrative or the need to extend or replace the MS, but it might understate the inner tension of the MS and the degree to which it has already changed. It might also obfuscate how some of its architects were keen on depicting it as more coherent than it was. We must question the 'Synthesis Historiography' formulated by the architects themselves (Stoltzfus 2017). Also, since calls to extend the MS were contemporaneous with the formulation of the MS, we must ask whether such calls are not part of its very functioning—and thus only seemingly challenged its core assumptions.

On first approximation, the MS was founded on the combination of Darwin's theory of natural selection with Mendel's genetic research on beans (Noble 2011).[4] However, as Huneman and Walsh (2017b) note, '[t]here is probably no definition of the MS' (2). Since the integrity of the MS might be overstated we run the risk of attacking a straw man (see Buskell and Currie 2017). Alan C. Love (2017) speaks of *theory presentations* as opposed to theories: In contrast to theories, theory presentations are partial, involving idealisations and black boxing. It follows that the conflict between the MS and its alternatives might be 'a kind of pseudo-conflict' (Love 2017: 196)—and there are cases in which there is only a difference in emphasis—where the MS acknowledges mechanisms outside the majority view but deems them marginal (or *proximate*). One example is the case of multilevel selection and questions about units of selection, with heretical figures such as Richard Goldschmidt, who held that this unit was the cell and not the gene (Gilbert 1988). Another example concerns the role played by development in introducing novelties into evolution (Gilbert and Epel 2015). We return to these topics below.

The predominant narrative centres around figures like Theodosius Dobzhansky and Ernst Mayr. It is a story of a pluralistic research programme gradually constricted into an orthodoxy around the 1950s, by becoming epistemically fixated on genes and their selection (Gould 1980).[5] But it seems that both sides describe the MS as more coherent than it was and is, for opposite reasons. One camp describes the MS as a cumulative and unified research tradition to reinforce their view of evolution, whereas the critics of the MS overstates its unity to exaggerate the radicality of their alternative. Depew (2017) accuses Gould of the latter. The MS traditionalists

---

[4] The MS disregarded other mechanisms, such as sexual selection, as an autonomous process and thus subsumed it under natural selection (Hosken and House 2011). Noble and Noble (2021) claim that the exclusion of sexual selection was one of the moves that led to how organisms were denied agency, as sexual selection involved the active choices of organisms.

[5] The exact periodisation is even subject to dispute: Mayr held that it emerged from 1936 to 1947, whereas Provine held that it was from 1920 to 1950. Since no 'standard historical narrative' exists, we have to be critical also of how the alternatives are formulated, as we see later (cf. Smocovitis 2023).

portray the MS as an ongoing phenomenon, 'the sum total of everything we know about evolution' (Laland et al. 2014: 161), which allowed them to recast criticisms of the MS as a rejection of evolutionary biology *per se* (Gawne et al. 2018). This is hyperbolic, as the critiques are not aiming for a return to pre-Darwinian biology. And even if some of them might run against facets of Darwin's theory, so did the founders of the MS, as we shall see (Stoltzfus and Cable 2014).

The main line of defence for the traditionalists is that the MS has already accommodated calls to extension; that the phenomena proposed to extend it have already been 'assimilated into evolutionary thinking without [...] major adjustment' (Huneman and Walsh 2017b: 11). This is not false but seems too simple. On the other hand, while there has not been a lack of criticisms of the MS, no comprehensive alternative has been proposed yet (Müller 2017). Stotz (2014) argues that the task might be to grasp the full implications of assumptions that have already been integrated into mainstream biology. An obvious candidate is the implicit teleology that we explore below.

Questions of autonomy and independence are at the centre of the discussion, and explanatory autonomy is entwined with causal autonomy. But let us pause to acknowledge how the MS was criticised even in its early development. In proceedings from the Pavia symposium in 1953, J.B.S. Haldane wrote: 'The current instar of evolutionary theory may be defined by such books as those of Huxley, Simpson, Dobzhansky, Mayr, and Stebbins. We are certainly not ready for a new moult, but signs of new organs are perhaps visible' (Haldane 1953: xviii–xix). He anticipated an extension, pointing to 'a broader synthesis in the future'. Considering the changes in biology since and how many of the techniques now applies were unknown to the founders of the MS, it might not be 'meaningful at all to speak of the Modern Synthesis as a continuing phenomenon, given the seismic shifts that have taken place since the 1950s' (Sepkoski 2019: 19). Indeed, Futuyma (2017) says it has been 'almost continuously extended' (1) since its formulation, even if its principal tenets have been salvaged. Many of the original architects modified their views, like Simpson, who emphasised the role of behaviour as a cause of evolution, while others came to promote a more hardened version of the MS, like Dawkins. If the MS is not a coherent theory—and might be said to be extended already—there are nevertheless predominant research agendas (Depew and Weber 2011). I focus on these agendas while remaining sceptical of the cumulative narrative.

'A Delicate Balance': Mechanistic Yet Purposive?

Due to lack of space and to not be held hostage to historical details, I simplify extensively. This section serves as a condensed outline of what I expound on below. Smocovitis seems correct in calling the MS 'one of the most vexing problems in the history of biology' (Smocovitis 1996: xii). My hypothesis is that the MS is too restrictive to function as a comprehensive approach to evolution. But we must also ask whether the MS is still predominant. Denis Noble says that 'as a *complete* theory, it [the MS] has already been falsified. We now need to admit processes outside its remit, so that it needs to be extended' (Noble 2011: 1008). Regardless of whether this description is correct, the MS seems to be an unstable disciplinary engagement instead of a coherent theoretical movement—a 'delicate balance', as Smocovitis (1996) puts it (see also Sepkoski 2019).[6]

If we look at the present status of evolutionary theory, it is vexing to learn that in the MS's nascency, between the two world wars, Darwinism was on the brink of theoretical extinction. It was 'the eclipse of Darwinism', the period before natural selection was acknowledged as the primary motor of evolution (Bowler 1992). According to this story, the rebirth relied on population genetics and was influenced by the positivistic ideal of unifying the sciences (Smocovitis 1996).[7] Population genetical and its *mathematisation* of Darwinism provided the scientific rigour evolutionary theory had lacked (Depew and Weber 2011). But as not to be subsumed by physics and/or chemistry, a balance between 'the unity of science and the autonomy of biology' had to be struck. If global determinism prevailed, there would be no need for a separate science to explain evolution.

In other words, the unity of science ideal would in the last instance make biology superfluous as a separate discipline. But if the teleological view won the day, there would not be any mechanistic or explanatory science of evolution in the first place. At least this was the opinion of the architects. While modelling itself on physics, evolutionary biology had to distance itself from it. It had to unify itself based on Newtonian ideals without succumbing to its global determinism. Hence, the delicate balance alluded to was between mechanism and teleology, reductionism and emergence. While the genocentric and mechanistic view won out, it had to be considered *emergent* on the global determinism of physical laws, and thus irreducible to them: 'If biology were to preserve its independent existence as a science, it could in some measure by dependent *on*, but could not be fully reducible *to*, physics and chemistry' (Smocovitis 1996: 104). More precisely, even if biological systems could be explained completely by reference to the structure and interactions of their components, the *function* of these parts was irreducible to physics or chemistry (Nicholson 2012). This led to a balancing act between vitalism and mechanistic materialism.

---

[6] Arguably, the MS was not so much a research tradition but more adequately understood as a philosophy of nature. Joseph Rouse (2024) interprets Smocovitis in this manner, and says that it was more a 'an organizing philosophy of nature that connected those domains heteronomically' than a 'systematic theoretical unification' (fn22, 18). I cannot entertain this notion now, but I return to it in the concluding chapter. For now, it is useful to maintain it as a rather loose research program.

[7] The story of decline might be overstated, partly due to Mayr's triumphalist self-depiction of the rise of the MS, and partly because of historians of science like Peter Bowler in *The Eclipse of Darwinism: Anti-Darwinian Evolution Theories in the Decades Around 1900* (1983). See Stoltzfus and Cable (2014) for a critique. Also, as Depew and Weber (1995) note, Smocovitis might have overstated the influence of positivism in the formation of the MS. Smocovitis' narrative might describe the British tradition but that the American tradition was more influenced by pragmatism than positivism.

The gene—reminiscent of the elementary particles of Newton's theory—was integral to the emulation of physics.[8] It offered several things. First, unity: Through genetics, evolutionary biology could surpass charges about not being explanatory since it could measure the effect of natural selection by looking at the distribution of genetic alleles within a population. All hereditary information from one generation to the other was stored in a discrete and stable manner. They hoped that this 'particulate theory of inheritance might do for biology what the atomic theory had done for chemistry' (Keller 2010: 13), by offering a straightforward way to study how heritable information is transmitted. Another advantage was explanatory and causal autonomy from other sciences. But this aspect also led the MS into a paradoxical position. It seemed to imply a progressive historical trajectory that the theory did not allow for—at least not without also permitting some purposiveness (or agency) in nature.

The residuum of purposiveness in the formulation of the MS is found in the idea that microparticles have agency over and beyond chemical/physical processes. More reductive approaches resurfaced with the rise of molecular biology and the discovery of the triple helix (DNA) in 1963. Crick and Watson opted for a reductionist reading of their findings, regarding 'life as "nothing but" chemistry' (Depew and Weber 2011: 93). Against this, the architects emphasised the hierarchical ordering of biological systems and emphasised the emergence of properties on different levels, irreducible to chemistry. Mayr, for example, defended the idea that biology was far more complex and 'had its *own* emergent properties unlike *any* of the physical sciences' (Smocovitis 1992: 59, emphasis original). Even if there were no emergence *within* biology, there had to be *between* biology and the physical sciences.

Dobzhansky's dictum: 'Nothing in biology makes sense except in the light of evolution' (Dobzhansky 1973) is a principle both of intelligibility[9]—evolution as the 'unifying principle' of biology (Woodger 2010)—and integration, by which biology could finally cut its bonds with botany and zoology and gain disciplinary autonomy and identity (Smocovitis 1992). While the rhetoric of freedom and progress was intact at the MS's inception, variation was increasingly considered a random process of changes in allele (gene variant) frequencies in populations. Genes were given agency above the level of physics and chemistry, an emergent level of causality. The relative fit between the organism and the environment was a product of the interplay

[8] Here, I say 'gene' instead of DNA because the structure of DNA would not be discovered until later. Genes are the hereditary content that DNA transmits, and the term was applied before DNA was discovered, as a statistical shorthand for anything that could not be explained by other factors. It was an abstraction, a hypothetical entity, 'based on the statistical quantification of parent-offspring resemblance' (Danchin et al. 2019: 2). Although the molecular mechanisms behind this resemblance (the proportion of phenotypic variation transmitted) were conjectures, 'none of this mattered. The gene could be anything that had the properties of transmittal with infrequent change' (Gilbert 2000: 179). For lack of space, I cannot elaborate on the status of genes before and after the discovery of DNA. But the short explanation is that DNA is simply the mode of transmission of genetic information. This information, not the DNA itself, is selected (Danchin 2013).

[9] As Depew 2011 points out, a strong reading of this dictum does not only say that phenomena are incommensurable without evolution; it also means that the phenomena are *invisible* without the evolutionary lens.

between genes and natural selection. In this way, the purposive activity of organisms did not explain adaptations. Instead, individual changes in alleles were expected to scale up to a population-genetic level linearly, warranting the neglect of development as an evolutionary factor (Depew 2011).

But as Depew (2017) writes: 'If population genetics explains [...] it is only in the weak sense of parsimoniously redescribing a confusing array of particular facts by using statistical summaries of events whose proper causes are ontogenetic' (56). It confuses bookkeeping with causal efficacy (Gould 2002). It 'explains' in an abstract manner, which cannot make sense of '*how* phenotypic variation is generated, *how* particular phenotypes come to be successful while others fail, and *how* offspring come to resemble their parents' (Lala et al. 2024: 38, emphasis original). These circumstances were ignored so that a simple model could be maintained. Within this framework, narrow genetic mutations evinced the mechanisms needed for microevolution and natural selection, leaving what is called the developmental or phenotypic gene—which is *not* random with regard to fitness since it is internally selected by the developmental process—in the dark (Gilbert 2007; Jablonka and Lamb 2020). Since microevolution was assumed to scale up to macroevolution anyway, it did not need to bother with how genes were *expressed* through the concrete development of organisms. In short, it black-boxed the *process* of actualisation, which involves nongenetic factors, as argued below.

Here, development is considered conservative and a matter of fine-tuning. Not seen as a source of adaptive bias on variation, it was pushed into the periphery. Development was reduced to the sheer unrolling of a programme inscribed in genes. One mode of evolution was thus supposed to be both *necessary* and *sufficient*, defined as 'smooth adaptive shifts via shifting gene frequencies' (Stoltzfus 2017: 6). In this view, genes are the only thing that is inherited by the next generation. Additionally, natural selection, both eliminative and creative, is considered the main driver of evolution.

Thus, there were three main points of agreement in the MS: (1) the primacy of natural selection, (2) the gradual rate of change 'operating at the level of small individual differences', and (3) a 'continuum between microevolution and macroevolution' (Smocovitis 1992: 47)—the only difference between them being their timeframes, as macroevolution is just microevolution over longer timescales (Lewontin and Levins 2007; Dobzhansky 1951). Population genetics provided the mathematical tools to quantify the success of any species, depending on a single variable: fitness, measured in terms of reproductive success. While it was evident that natural selection, variation, adaptation, and inheritance are entangled processes, they were treated as separate—considered independent conditions for natural selection to work on. Walsh (2015) calls this a *fractionation* of evolution: considering evolutionary processes discrete and relatively independent of each other—i.e. only externally related (see also Huneman and Walsh 2017a).

Theodosius Dobzhansky's Version of Adaptation

There is nothing inherently Darwinian about the view outlined above. Darwin did not consider development a mere product of transmission across generations but instead held that 'transmission was a product of development' (Winther 2000: 450). The inheritance of variation (of any kind) was the rule, non-inheritance anomalous. We might call the view described above 'Spencerian', as he was one of the first proponents of gene selectionism but he nonetheless emphasised how organisms integrate the materials provided into wholes and thus did not consider the accumulation and combination of genes as the sole cause of adaptation. According to his view, '[i]t is this harmony [integration of traits] that makes for the survival of the fittest—and for adaptedness' (Depew 2017: 39). In this, Spencer agreed with Darwin: The whole organism is the agent of adaptation (Walsh 2015).

By contrast, the atomistic version of natural selection stated that traits and their correlated genes are selected discretely one by one; adaptation does not concern whole organisms but independent traits within populations. Organisms are mere aggregate and carriers of such traits. According to Gould, this was one of the major characteristics of the hardening of the MS into what he and Lewontin called *the adaptationist programme* (Gould and Lewontin 1979). Depew (2011) states that before this hardening, organisms were acknowledged as developmental systems, causally on par with genetic factors. Mayr, for instance, first held the view that DNA acted as a complete program for development, but later abandoned this view and was therefore opposed to Ronald Fisher and Richard Dawkins' more atomistic view of selection (see Depew 2017; Longo 2020).[10]

The adaptationist viewpoint entailed that as soon as the blind processes of mutation give rise to traits which are deemed beneficial by selection and maintained by it, they would function in a deterministic fashion. Thus, evolution is about the interplay between randomness and determinism; everything in-between, such as organisms, is inconsequential. The central role given to genes does not mean that there is no autonomy involved but that it is displayed at the level of genes and emergent on physical and chemical microlevels (Martínez and Esposito 2014).

Dobzhansky's views the illustrate inner tensions of the MS. He was the first to formulate the view that evolution is 'a change in the frequency of alleles within a population'. In his seminal *Genetics and the Origin of Species* (1937), he argued that three levels could explain evolutionary changes: (1) genetic mutations provided the raw material, leading to changes in (2) populations as new variants spread and combined and (3) the fixation of such changes through reproductive isolation, which explains the formation of new species. Yet, he also considers natural selection constructive, producing the materials on which it acts—even if it depends on the raw

---

[10] This illustrates the difference between the American tradition, more pluralistic and less adaptationist, and the British, influenced by Fisher's restricted view of evolution. Neither Dobzhansky nor Mayr disregarded genetic drift, the relativity of fitness, and genetic connectivity—all facets that the British tradition neglected. The latter could then be said to be hardened all along (Depew and Weber 1995).

material of genes—and emphasises the organism's capacity for plastic responses to environmental changes. This fits a process-oriented approach to adaptation: 'the accumulation of germinal changes in the population genotypes is, in the long run, a necessity if the species is to preserve its evolutionary plasticity' (Dobzhansky 1951: 127).

Yet, as we see from the quotation, the plastic responses made by organisms were simply the effects of other processes which the organism may use to fine-tune its developmental trajectory. Plasticity is not itself a cause of evolutionary change. For an organism to solve the problems *posed from without*, the product of genetic change, the phenotype, must fit its environment. On the one hand, according to the adaptationist programme, an 'adaptive trait' is a discrete property which is decidable only after the fact, yet 'being adapted' implies being designed for something, as if serving a preordained purpose. This 'for' is decided by natural selection alone, according to adaptationists. No genetic mutation has a given function before natural selection works on it, favouring or discarding its related phenotype(s), which is what produces adaptations. In this view, an adaptation is the result of selection in the past. This leads to an infinite regress, as anything could be regarded as an adaptation by the simple fact that it exists, thus being 'just so stories' (Gould and Lewontin 1979). Lenny Moss calls this *the phylogenetic turn*:

> The intent of this phrase is to highlight the idea that as the gene and genetic program became understood to be the principal means by which adapted form is acquired, the theatre of adaptation changed from that of individual life histories, that is, ontogenies, to that of populations over multiple generations, that is, phylogenies. As the genetic program moved to the explanatory center stage, the individual organism, with its own adaptive capacities, began to recede from view. (Moss 2003: 4)

Against the populational outlook, adaptation also concerns an organism's capacities and the *process* through which it is fitted to the environment. Instead of depriving the organism of agency, Dobzhansky's view distinguishes between (1) adaptation as an evolutionary process, (2) adaptiveness as the selected state of fitting to one's environment, and (3) an adaptive trait as a specific aspect which increases the probability of survival (Depew 2011). Consequently, he casts 'populations as states of relative adaptedness' (Müller 2017: 3)—relative to dynamically changing environments.

Dobzhansky is reluctant to speak of 'an adaptation' as a substantial thing, implying design and taking for granted that the environment remains stable. He calls this view 'Cartesian' and ties it to ideas of natural selection as a process of optimisation, as if done by a rational agent (see Okasha 2018). He is thus mindful that adaptations are only adaptive insofar as the selective environment remains constant, which it rarely does. Adaptiveness is therefore not a consequence of selective design acting on phenotypes. It is rather the transient result of an ongoing process of adaptation which involves the actions of the whole organism. As we shall see, Dobzhansky's approach also anticipates current critiques of evolution as optimisation.

Even if he is a precursor to the developmental and teleological turn found in proposals to revise the MS, Dobzhansky remains one of the architects of the orthodox view, as seen in how he regards the plasticity and adaptive responsiveness of organisms as a secondary effect of past natural selection of genes. While he thus marginalised the activity of organisms, we should note the ambiguity of his position. There are aspects of it that do not fit the prevailing 'gene's eye view' of evolution, in which natural selection works on discrete traits to design increasingly better phenotypes.

For Dobzhansky's student, Lewontin, his teacher was on the right track but did not go far enough (Levins and Lewontin 1985). Dobzhansky acknowledged the plastic responses of the organism as a source of evolutionary variation but did not pursue this line of reasoning to the end—did not consider how organisms actively shape the environment to make adaptiveness possible. Depew writes:

> On Dobzhansky's view, problems are posed to organisms by environments. They do not solve these problems directly [...]. Instead, solutions to the problems organisms face arise indirectly by way of transgenerational responses to environmental contingencies by the populations to which organisms belong. Over time, the environment selectively amplifies fitter variants until a population's state of adaptedness has been optimally, if transiently, adjusted. It can be argued that this orthodoxy still assigns the lead role to external forces and in so doing portrays organisms as too passive to capture the agency that they exert in responding to and making their own environments. (2017: 55)

In other words, the plasticity an organism has—and its importance in development—only explains the *proximate* causes of how an adaptation comes about. Plasticity is *ultimately* the product of former cycles of natural selection. Dobzhansky is committed to this notion, even if he does not adopt Fisher's model, which is restricted mainly to additive genetic variation and selection alone and presupposes—but does not account for—stable environments (Depew and Weber 1995). In fact, Fisher's conceptualisation does the opposite of what we may expect, revealing another ambiguity. It presumes that natural selection is the dominant factor *when the environment is stable* but ignores how natural selection itself constantly changes the environment. Natural selection undermines its own centrality since it causes environmental instability that deprives it of primacy (Okasha 2006).

Dobzhansky's model is less linear and predictive than Fisher's. It includes the isolation mechanisms proposed by Sewall Wright, namely population size, drift, and migration,[11] and is therefore a more complex theory of adaptation than the strict adaptationist programme (Depew 2011).[12] While Dobzhansky came to downplay Wright's mechanisms in favour of natural selection, they share an emphasis on the

[11] Drift is a process where random sampling potentially undermines the diversity of variation within the population, counteracts selection, and produces statistical associations between loci (independently of natural selection). In contrast to selection, drift has no direction but is only governed by the mathematics of chance. It changes genotypic frequencies irrespective of adaptive value. Two examples are (1) the bottleneck effect, which is when a chance event, like a natural disaster, induces changes in variation because some organisms were lucky enough not to be affected, and (2) the founder effect, where a new population establishes a new niche, diverges from the original population from whence it came. It is thus a random sample from the original population. The effects of drift are higher in small populations than in large ones, as small populations have less variation.

[12] Fisher did not deny that genetic drift existed but limited it to small, interbreeding populations. Insofar as populations are large and panmictic (randomly mating), genetic drift is not considered a widespread evolutionary agent (see Depew and Weber 1995).

nonlinearity of evolutionary processes. In short, Dobzhansky includes cases of adaptation that are not the product of selection acting on specific traits within a population but involves the developmental capacities of whole organisms—in contrast to theories wherein the organism is a mere bundle of traits deprived of causal potency (Juarrero 1999).

This exegesis illustrates an important distinction within the MS: adaptation as a product and as a process. The former drives a wedge between evolution and development and does not concern itself with *adaptability*, organisms' ability to adapt to changing circumstances as an ongoing process (see Depew and Weber 1995). If hard inheritance is all that matters in evolution, if *random* genetic mutations ultimately explain where change and constancy stem from, then there is no role left for the organism, except as a result of other processes. Evolutionary theory combined inheritance and variation in one master molecule and thus neglected adaptation as a process involving organismic agency. The allegation of epistemic closure within evolutionary theory therefore seems pertinent (Turner 2017). We see this in how the analytic method that follows this view only works by assuming 'stable components and relations among them' (Marques and Brito 2014: 110), excluding a range of processes that does not fit this definition.

The Primacy (and Agency) of Natural Selection

> Neither mutation nor selection alone is creative of anything important in evolution; but the two in conjunction are creative. (Huxley 2010: 28)

Against Dobzhansky's views, the prevailing approach became the one in which discrete adaptations were promoted by selection because of their past capacities, 'each one individually adapted to performing a function in the way best suited for the organism's survival' (Müller 2017: 3). This analytic approach does not regard the whole organism as a locus of selection, and is unable to account for maladaptive traits. Organisms are perceived as aggregates, not integrated systems. This suggests the machine metaphor of the organism—treating it as an artefact, 'assemblies of independently adapted parts and traits' (Depew and Weber 2011: 97). According to this view, the organism cannot be an *intrinsically* purposeful agent (Longo and Mossio 2020). Instead, natural selection bestows purpose on the organism from without.

As indicated, selectionism is the flip side of adaptationism, in that it holds that adaptations are the product of previous cycles of selection. But is it creative or eliminative: does it produce its own variation, or simply choose between options that are presented to it?

One version of *agential thinking* depicts natural selection as an intelligent designer, improving its work bit by bit (see Okasha 2018). This line of reasoning dominates the popular imagination and is tangible e.g. in ideas of improving the intelligence of one's progeny by gene therapy.[13] It comes in two versions:

> Agential thinking is intimately linked with the idea of fitness-maximization in biology. This idea has two variants, both controversial. The first is that evolved organisms will exhibit traits that are adaptive, hence maximize their fitness relative to some set of alternative traits. [...] The second is that the process of natural selection itself involves maximization, in the sense of continually changing a population's composition so as to achieve higher fitness. (Okasha 2018: 3)

I can only highlight the second view, that natural selection is directed towards a rational goal when it acts on organisms and ecosystems, in a process of optimisation. Here, purpose is not eliminated but explained away, reduced to a mechanism (Buskell and Currie 2017), *viz.* acting on the organism from the outside, like a cue striking a billiard ball. Still, this mechanism is an expression of natural agency (see Moss 2012).

Another dominant idea is that natural selection is a one-sided process which favours the fit and culls away the ill-fitted. This is the sieve metaphor of natural selection, proposed by Hugo de Vries. Spencer, who coined the term 'survival of the fittest', shared his view, and saw natural selection as purely eliminative. One problem which arose, however, was accounting for the abundance of variation and why it was not 'concentrated on the good and bad ends of the distribution curve' (Depew and Weber 2011: 92). Since it was much more diverse, another explanation was needed. To explain how variation is produced, Darwin and the architects of the MS held that natural selection is creative. Darwin's evidence for the efficacy of natural selection was based on an analogy with artificial selection (Gayon 1997). But to justify the power of natural selection, it could not explain fitness by differential survival alone; it also had to explain how adapted species came about in the first place. Natural selection had to account for changes in the mean and the range of variation (Beatty 2016).

But while one species' superiority over another will be '*expressed* as differential survival [...] it is not *defined* by it' (Gould 1991: 42). In other words, accounting for sheer numbers does not explain how the qualitative changes are brought on by evolution, it only registers the product of this process quantitatively. Mechanisms that induce variation are needed to explain the constitution of fit organisms. But since the genesis of variation was seen as blind in terms of adaptative value, such mechanisms could largely be black-boxed by the MS. Danchin and Pocheville (2017) says that this feature—the epistemically economical way in which adaptations come about, in which variation 'literally come from nowhere, [...] is probably where the core of the neo-Darwinian paradigm lies' (131). We have thus identified a problem that is not explained so much as explained away.[14]

---

[13] The popular uptake of Dawkins' views might be caused by the fact that it fits the idea that Darwinism concerns selfishness and competition 'that the public has long associated with Darwinism' (Depew and Weber 2011: 94).

[14] While I include this quotation in which the term is applied, I should note that I avoid the term 'neo-Darwinian' in what follows, as it is often used in a non-analytical and derogatory sense. It

Irrespective of whether the material provided for natural selection are produced blindly, natural selection is not a random process. It introduces biases in genetic populations by channelling populations towards certain phenotypes, even if this does not mean that the variations it acts on is biased. Evolution is commonly described as a 'two-stage process', with genetic variation coming first and natural selection then acting on genotypes. Darwin entertained this possibility before eventually discarding it (in the fifth edition of the *Origin of Species*). Initially, he thought that new variants might drive evolutionary changes by themselves. But he was led to the view the variants required for natural selection are always present and that the role of directing evolution is done by natural selection as an architect assembling material for a building (Beatty 2016). It allowed Darwin to fend off religious arguments about the origin of variation. We might thus understand why the notion of natural selection as a creative force is part and parcel of Darwinism (Gould 1991). Gould points to the metaphors applied by some of the proponents of the MS: an architect (Darwin), a composer (Dobzhansky), a sculptor (Mayr), and a poet (Simpson), and states that

> [t]he essence of Darwinism lies in its claim that *natural selection creates the fit*. Variation is ubiquitous and random in direction. It supplies the raw material only. Natural selection directs the course of evolutionary change. It preserves favourable variants and builds fitness gradually. (1991: 44, my emphasis)

It might come as a surprise since natural selection is often presented as getting rid of the unfit but not inducing the fit. By contrast, Darwin's final position was that natural selection initiates evolutionary changes and is also responsible for producing the variation it acts upon. It moulds it by filtering and reinforcing genetic or epigenetic changes. As such, natural selection could produce a fit phenotype independently of other mechanisms. (This view undergirds the fractionation mentioned above.) If this were not the case, natural selection would rely upon the variation produced by other processes and thus have less importance:

> The idea that selection merely eliminates or preserves the variation provided, suggests that there can be no evolution in any particular direction until variation in that direction is made available; selection must wait for variation to act upon. On the contrary, as Darwinian defenders of the creativity of natural selection have argued, it is natural selection, not the production of variation, that initiates evolutionary change. Moreover, natural selection is in an important sense responsible for the variation that it acts upon. One way in which this occurs is when selection in a particular direction results in the production of further variation in the same direction. (Beatty 2016: 2)

It rests on three assumptions: that variation is (1) copious—in the sense that natural selection always has a wide-ranging stock of variants to choose from—, (2) undirected, and (3) linear (or continuous). As Beatty writes, 'as long as variation is copious in all directions, then natural selection and not the appearance of variation initiates evolutionary change; there's no waiting' (Beatty 2016: 5). The motivation for this seems clear: natural selection cannot be dependent on the variation provided; it must

---

would muddy the waters further, because it 'carries too much baggage of all kinds to be useful and should only be explicitly used as a historical actor's term, the actor being [George John] Romanes' (Smocovitis 2023: 36).

provide it for itself to keep its primacy. Beatty points out a paradox in this construal: If natural selection in one direction changes the range of available variation, then the variation is not undirected. Yet, variation is supposed to be blind regarding adaptiveness, viz. it is not induced or context-dependent, it does not come about *because* of its adaptive value. It is not supposed to be biased in any way because only natural selection can introduce biases in evolution. This is why the adaptive value can only be decided after natural selection has worked on it.

In short, this view means that beneficial variations cannot be made more probable by processes outside natural selection (Danchin and Pocheville 2017). This assumption shaped the interpretation of research: When artificial selection experiments reached a plateau of relative stasis, it was held that the reason was not that the stock of genetic mutations was depleted; instead,

> when evolution by artificial selection in a particular direction slows to a halt before resuming, this is not because the available variation has been depleted and new variation is required for evolution to re-commence. Rather, it is because it takes time for recessive, otherwise beneficial genes to pair-up and be selectable, together with the time needed for a new combination in the same direction to arise from standing variation. (Beatty 2019: 9)

One might wonder whether this does not introduce a notion of biased variation; but as Beatty emphasises, this apparent contradiction is solved by stating that natural selection does not induce the genetic mutations in single genes or loci. Conversely, it combines mutations (which are deleterious or neutral individually) in a novel way and selects from these combinations: ‘While new mutations are a matter of chance, new gene combinations are often the product of natural selection’ (Beatty 2019: 6). This resolution means that mutations play a miniscule, ‘almost superfluous’ yet indispensable, role in evolution, and that there are always enough mutations for natural selection to combine into novel variants. It is not evident, however, that this makes much of a difference, except semantically. The question of whether (and how) selection maintains or reduces variation is left unanswered. To answer this, we have to know how selection itself is shaped.

François Jacob expresses the idea of natural selection as canalising variation in his influential article ‘Evolution and Tinkering’:

> [N]atural selection does not act merely as a sieve eliminating detrimental mutations and favoring reproductions of beneficial ones, as is often suggested. In the long run, it integrates mutations, and orders them into adaptively coherent patterns adjusted over millions of years, and over millions of generations as a response to environmental challenges. It is natural selection that gives direction to changes, orients chance, and slowly, progressively produces more complex structures, new organs, and new species. Novelties come from previously unseen association of old material. To create is to recombine. (Jacob 1977: 1163)

This means that selection and variation are part of the same process, not distinct stages. It also entails that sexual recombination is subsumed by natural selection, rather than being an independent process (Nei 2013). The question of their relative strength and ability to initiate evolutionary changes crystallises issues we are faced with moving forward.

The idea of creative selection was closely related to the conceptualisation of adaptation described above. Okasha says that ‘adaptationist reasoning is premised on the assumption that developmental constraints do not greatly limit the available variation; for if they did, then well-designed phenotypes could not evolve' (2018: 106). There must be a standing reserve of copious variation on which natural selection can act for adaptive evolution to come about. The differential survival and reproduction of individuals within populations is due to the aptness of phenotypes-in-environments. It is a relational phenomenon. Genotypes are only selected indirectly; via the traits they enable. Yet, it is assumed that phenotypic changes are predicated on genotypic changes (Noble 2011). It is the premise behind the reduction of the former to the latter.

Findings indicating that phenotypes are expressed in and through interaction—both between genes (epistasis) and between genes and environments—are neglected by this view. Instead, the MS adopted an additive approach, which points towards one of its main inventions: viewing natural selection not simply as an agent among others but as a mechanism—the primary mechanism of evolution (Smocovitis 1992; Walsh 2017). This led to what Mayr derisively called 'beanbag genetics' because it considered single genes independently and ignored the effects of gene interaction, which are not necessarily additive (independent from the genetic background) but can also lead to effects that are unpredictable from simple linear models (Fox Keller 2000). It has been shown, for instance, that a gene's effect on fitness depends on the genetic background and epistatic processes—wherein genes produce traits in concert in nonadditive (and thus nonlinear) ways—must be accounted for to understand the generation of phenotypes (Okasha 2018).

I am reminded of the quip that teleology is like the mistress that the biologist cannot do without but hides from the public.[15] Modelling themselves on physics and chemistry, the makers of the MS could not accept anything overtly teleological and framed natural selection as a causal mechanism. Simultaneously, the physical worldview had to be kept in check to preserve some independence, both for biology and for the (genetic) mechanisms they studied. The awkwardness of this position is evident in the causal ideal proposed in their models. They seek reductionism within biology but cannot allow biology itself to be reduced to other sciences. Through this balancing act, natural selection became measurable and concerned with self-contained, discrete entities acting on each other from the outside. Furthermore, since it was postulated—not proven—that the microlevel (alleles) linearly scales up to the macro level (populations), one could 'calculate and determine *evolutionary rates of change*' (Smocovitis 1992: 23, emphasis original) by looking at minor shifts within populations produced by selection. Natural selection exercises its influence mechanically and independently of other processes, with discrete results—suggesting purposive design (Smocovitis 1992: 20). Implicitly, then, purposiveness is maintained in the notion that natural selection act as a rational agent. This is the backdrop of the view that variation is not limited by other processes but produced by natural selection itself.

Let us briefly return to Dobzhansky: He believed, in line with Schmalhausen, that natural selection does not act on isolated, simple, traits. Rather, it acts on complex

[15] The exact origin of this quote is unclear, but it is often ascribed to J.S.B. Haldane (Riskin 2020).

traits exhibited by the whole organism. The creativity of natural selection is intact—*a fortiori*—in this view: '[a] norm of reaction is what natural selection at its most creative evolves: genotypes that have spread through populations because they can produce a range of phenotypes in individuals' (Depew 2011). Dobzhansky thus proposes a subtle version of natural selection, implying that it can promote something akin to what is now called evolvability and developmental plasticity. In this view, selection is not only directional, in the sense that organisms over time are ever better adapted to their environment. If that were the case, it begs the question of how the environment remains stable. Instead, Dobzhansky anticipates facets of niche construction, the view that 'environments are rendered stable in large part by the activities of the organisms themselves' (Depew 2011: 96). We cover this topic in the next chapter.

A Forced Marriage

A theory of evolution must explain the genesis of both stability and variation—the maintenance of biological form and the capacity to accommodate changes. The MS explains both by reference to genes and transcriptional factors, even if they only provide the raw material for natural selection to combine into selectable novelties. Mutations and reorganisation of these mutations enable evolutionary change, while processes like genetic drift, sexual selection, and group selection are acknowledged, but assumed to have negligible effects. Even if the organism can enact changes on a developmental timescale, these changes are not inherited by the next generation. In short, nongenetic processes, even if factored into the equation when explaining adaptive fit, are regarded as 'downstream consequence[s] of genes, and therefore do not represent independent hereditary factors' (Bonduriansky and Day 2018: 137).

Darwin, on his part, was open to multiple causes of variation but struggled to explain where it ultimately stemmed from—finally deciding that it was caused by natural selection alone. His theory could in this respect be considered incomplete as it cannot explain *why* natural selection maintains the fuel on which evolution runs without an explanation of the origin of variation. I will, however, not take for granted that it is correct to say that his theory is incomplete, as it implies that Mendel's theory was needed to complete it. It gives the impression that it was no tension between their views (see Bowler 2009). By contrast, I argue that their marriage was forced.

Without copious blind variation to work on, natural selection would undermine evolution, as it does not produce this stock of novelty by itself. Thus, it would be constrained by the material on which it works—having to wait for fortunate mutations (Lewontin and Levins 2007: 33). But what if it acted on something else than random mutations? Darwin regarded claims about random mutations as the sole cause for variation as an admittance of our ignorance, not a solution (Wagner 2014). This seems to be part of the reason why he changed his mind concerning the origin of variation and was led to the view that natural selection produces the variation it acts on. The dictum *natura non facit salta* (nature does not make saltations) might also explain his reluctance (see Stoltzfus 2017 for a critique). Evolution, in this view, is gradual and concerns the accumulation of minor changes, in contrast to the Mendelians' views, where rapid and discrete, not continuous, macromutations were involved. Since Darwin's dictum did not fit well with Mendel's factors, de Vries' macromutations, or the idea of random genetic mutations driving evolution forward—without predictability and guidance—a compromise was needed.[16]

With Mendel, the seemingly contradictory process of variation and heredity was brought together under a single mechanism of inheritance (Lewontin and Levins 2007: 192).[17] Darwin believed in blending inheritance, where offspring were considered intermediaries of their parent, leading to the depletion of selectable variation. His pangenesis model—with changes to 'gemmules', hereditary particles, as an effect of use and disuse counteracting the effects of this blending (Charlesworth and Charlesworth 2009)—was introduced to explain how this depletion was counteracted. Against his model, the Mendelians proposed that each parent contributes *discrete* material to the progeny. This is the simple model we all learned in school, of dominant and recessive genes, one from each parent. Mendel came to this view by studying peas and finding that they would express either the father's or their mother's traits but not an average of them. Thus, he found that inheritance was not blended as Darwin believed but was inherited in a distinctive manner. Some of this material becomes dominant and thus expressed, whereas some is recessive and latent.[18]

The Mendelians and the Darwinists found common ground by recasting mutations as a constantly growing stock for selection to work on. Hence, the MS came about by neglecting some of Darwin's views in favour of Mendel's. This means that the idea that Mendel was the ideal partner for Darwinism—'merely the filling-in of the missing piece in the jigsaw puzzle that Darwin almost completed' (Bowler 2009: 24)—is false. The compromise modified both Darwin and Mendel's views: it took from the former the notions that natural selection is creative, and that evolution is gradual—and, by utilising the latter, it discarded the evolutionary significance of

[16] Darwin accepted that definite form *could* arise by saltation and that such changes were possibly inheritable in some cases. He simply rejected the idea that this was a widespread phenomenon. Thus, he did not exclude the existence of 'monsters' but he did not find many cases in which they resembled 'normal structures' (Stoltzfus and Cable 2014). He also held that fluctuations—which are tied to effects of the environment on the organism—were inheritable and that inheritance would blend from generation to generation—that a specific trait from one parent would not be inherited in its pure form. Instead, the progeny inherited an average of its parents, not discrete particles.

[17] Heredity concerns the pattern that is inherited, and inheritance concerns the mechanism through which the pattern is inherited (Danchin et al. 2019).

[18] Of course, even if each factor is discretely inherited, many factors contribute to complex traits. They might then appear to be blended but this is only because they are not simple traits decided by one gene. The most common example of the latter is hair colour, where the dominant and recessive genes of the parent decide which alternatives are possible for the offspring. For more complex traits, however, such discreteness is difficult to ascertain. In the case of skin colour, for instance, Francis Galton claimed that this was a case of *phenotypic* blending, but he still held that each parent contributed one hereditary particle (gemmule) and that these particles were not blended (Galton 1889).

environmental factors (fluctuations) and notions of blending inheritance. This mutual modification is often forgotten:

> The classical narrative denies a theoretical clash between Darwinism and Mendelism, and articulates their relationship in terms of the "missing piece" theory, in which Mendelism is the "perfect complement" to Darwin [...], and the decades-long failure of scientists to perceive this perfect fit is an "inexplicable embarrassment" [...]. The popularity of this view is indicated by Bowler's (1988) complaint that "the Darwin industry has followed Fisher's assumption that genetics merely fills in the gaps in Darwin's thinking" (p. 130). (Stoltzfus and Cable 2014: 533)

The 'missing piece theory' assumes that Darwin's theory remained incomplete without Mendelianism, instead of simply arguing that Darwin's theory was wrong regarding the production of variation. The fact that the architects were narrating their own story is evident here: Fisher's claim that some aspects were absent from Darwin's theory is patently incorrect but cemented the view that Darwin and Mendel were perfect partners: 'In reality, Fisher has deleted some of Darwin's deliberate choices. Thus, one must reject the idea of "missing" pieces: at best, this is a theory of *substituted* pieces' (Stoltzfus and Cable 2014: 534, my emphasis).

The compromise proved unstable, as the creativity of natural selection was challenged again during the 60s, when so-called mutationists appeared on the scene. They claimed (1) that discontinuous variation played a large role in evolution, (2) that fluctuations (in contrast to mutations) due to environmental factors were too ephemeral to be inherited and (3) that inheritance was not blended from one generation to the other, as this would lead to the decay of variation from generation to generation and thus undermine the effects of selection by depleting the stock it acted on. By these means, the rate of evolution was mapped onto the rate of mutations. The view can be summed up as: '[M]utation proposes, selection disposes' (Stoltzfus 2017: 4). Selection had no creative role to play.

The challenge surfaced with an article in *Nature* written by Motoo Kimura (1968). According to his theory, called neutralism, most new mutations are detrimental and soon eliminated. Moreover, 'most of the non-detrimental mutations are neutral or nearly so; and a very small proportion of mutations are sufficiently beneficial that their fate is not entirely a matter of random drift' (Beatty 2019: 20). The standing genetic variation in a population, then, is due chiefly to drift, not selection. Kimura's thesis challenged the primacy of selection by attacking the idea that selection always has abundant material to work on (in combination with sexual reproduction) and produces its own variation. Most variations are deleterious or neutral with regard to fitness, and thus not affected by natural selection since it only selects beneficial variants. This means that processes other than selection, like genetic drift, explains much of the spread and fixation of genetic mutations.

Genetic drift denotes random processes that change the allelic frequencies and genetic fixation irrespective of fitness. Through this process, an organism of lower fitness might survive due to chance alone, a fact which threatens the primacy of selection as the sole driver of evolution. According to this view, natural selection acts like a stochastic sieve,[19] preserving useful mutations and getting rid of deleterious ones. It is not creative, not a 'force of nature' (de Vries 1905); instead, creativity is left to mutations, which means that natural selection depends on mutations to produce adaptive changes in populations. Since most mutations are neutral, they cannot be used by natural selection, and thus there is *not* always enough standing variation for natural selection to work on. Masatoshi Nei sums it up: 'Natural selection is an evolutionary process initiated by mutation' (Nei 2013: 196, quoted in Beatty 2019).

Even if the orientation of evolution is steered by natural selection, it cannot be given the sole responsibility for driving evolutionary changes. Mutations do not only provide the raw material for selection but also act as evolutionary agents. The emphasis on mutations as the cause of variation could be called 'neo-Mendelian' instead of 'neo-Darwinian'. Yet, by focussing on the effects of mutations on a population level, instead of the level of individual organisms, the tension was resolved. In short, the mutationists gave up their saltationism to adopt the 'populational thinking' that characterised the MS (Smocovitis 1996). While random mutations occur on an individual level—the level at which natural selection operates—it is the proliferation of alleles within populations which cause evolutionary changes, and this process is gradual and smooth even if the variation which is selected occurs spontaneously (see Turner 2017). Quantitative genetics is about 'smooth changes based on infinitesimal variation' (Stoltzfus and Cable 2014: 535), not abrupt ones based on single 'lucky mutations'.

Conclusion

While it was incorporated into the MS, neutralism reveals how natural selection cannot be the only process inducing variation, even if it might be creative in some cases (Beatty 2019). It reveals how other factors might be on par with selection. It led to a 'widening gap between what formal models allow, and what the prevailing view of the causes of evolution suggests' (Stoltzfus 2012: 1). The neutralist holds that mutations do not just provide raw material, but introduce novelty which is not shaped by natural selection, only stochastically sifted.

This implies that nonadaptive processes may impact the evolution of biological complexity—that not every trait that exists is adaptive by default—, a view that does not fit the view of natural selection as a rational agent, constantly improving its products. Instead, a balance between processes is operative. For example, drift can

[19] Stochasticity is another term for uncertainty, and simply means that differences in evolutionary outcome (survival or death) are due to chance alone, not adaptation. There are many forms of stochasticity: genetic, demographic, environmental etc. In the *Encyclopedia of Ecology* (2008) we find this definition of the latter two forms: 'Demographic stochasticity is found in events within the population that are random and unpredicted and are demonstrated by individual behaviors causing immigration and emigration into or out of the population. Another type of stochasticity is environmental stochasticity—events such as floods, droughts, and other catastrophes that may affect population spatial distribution' (Jørgensen and Fath 2008: 3313).

restrict the variation selection acts on, especially in small populations; but at a certain populational threshold, the allelic variation might be high enough for drift to lose its effect (Cavalli-Sforza et al. 1994). In any case, the attacks on the orthodox view are important to bear in mind when we move to the extension of the MS. How did the proponents of the orthodox view resolve these issues? Were they simply disavowed, or acknowledged but considered marginal?

According to Stoltzfus and Cable (2014), there is a sleight-of-hand involved in the MS view of selection as recombination: Selection is granted credit for new combinations without any empirical observation of how variation is produced. Recombination is cast as natural selection, but it is not evident why this follows from Mendel's theory. It is unclear why recombination is 'a genetic process that produces new genotypes. In this sense, new genotypes produced should be called "mutations"' (Nei 2013: 38). But by 'avoiding the question of how mutations become involved in evolution' (Stoltzfus and Cable 2014: 535), the principle of selection changes subtly: on the one hand, it is changed from acting on variation of different kinds to acting only on mutations; on the other hand, it is regarded as creative while it is unclear how this creativity links up with mutations that supposedly are blind.

Stoltzfus (2017) states that the molecular revolution of the 1960s—which undermined the MS as a master theory—did not lead to reform but a schism, as the architects doubled down on the creativity of selection instead of modifying their theories. Mayr, for instance, claimed that the neutral theory of molecular evolution was not really evolution and thus that there is no non-Darwinian evolution to be spoken of (Mayr 1970; Stoltzfus 2017). According to this view, 'variation-driven trends are impossible because mutation-rates are too small' (Stoltzfus 2017: 2). Again, we see that the idea that Darwinism and Mendelianism were perfect partners should be abandoned:

> The Modern Synthesis was delayed because both partners in the marriage of Darwinism and genetics had to be dragged, kicking and screaming, into the union, which combined an unjustifiably loose interpretation of Darwin with an unjustifiably narrow and speculative view of population genetics. (Stoltzfus and Cable 2014: 536)

It is peculiar to find a dominant research tradition where the most central concepts are so poorly defined. Yet, it also seems to have been advantageous, as it provides the 'theory' with flexibility, making it impossible to falsify. Surely, the orthodox MS view that I depicted above is not intact anymore. No longer is the abundance of variation in 'gene pools' or 'the ability of recombination to produce new variation every generation considered the *sine qua non* of evolution' (Stoltzfus and Cable 2014). Mutations are instead, as seen in textbooks on the matter, considered a central instigator of evolutionary change (Love 2010). In short, the driver of evolution is mutations, natural selection is negative, and all inheritance is vertical. This is the victory of Mendel, not Darwin. We might ask, then, whether an extension of the MS would follow this train of thought and be non-Darwinian as well, or if it would be restoring Darwin's insights. For example, evolutionary developmental biology might pose a greater challenge to Darwin's gradualism and the creativity of natural selection than neutralism.[20] Like Darwin, it places the organism at the centre of evolution.

Undoubtedly, attempts will be made to bring such challenges to the MS into the fold by redefining what Darwinism is (see Hull 1990; Oyama 2000). Where does this leave us? We have seen the ambiguities of the MS, and the compromises made to make it stick together, however loosely. Moreover, we have established that the creative role of selection might be overstated, and that other processes might also be involved in the production of variation. This becomes increasingly important as we move towards the alternative to the MS, as it points in the direction of constraints and self-organisation. We have only begun to grasp what we are dealing with.

Bibliography

Beatty, John. 2016. The Creativity of Natural Selection? Part I: Darwin, Darwinism, and the Mutationists. *Journal of the History of Biology* 49 (4). https://doi.org/10.1007/s10739-016-9456-5.

Beatty, John. 2019. The Creativity of Natural Selection? Part II: The Synthesis and Since. *Journal of the History of Biology* 52 (4). https://doi.org/10.1007/s10739-019-09583-4.

Bonduriansky, Russell, and Troy Day. 2018. *Extended Heredity: A New Understanding of Inheritance and Evolution.* Princeton, NJ: Princeton University Press.

Bowler, Peter J. 1992. *The Eclipse of Darwinism: Anti-Darwinian Evolution Theories in the Decades around 1900.* Baltimore, MD: Johns Hopkins University Press.

Bowler, Peter. J. 1988. *The Non-Darwinian Revolution: Reinterpreting a Historical Myth.* Baltimore, MD: John Hopkins University Press.

Bowler, Peter J. 2009. *Evolution: The History of an Idea.* Berkeley, CA: University of California Press.

Buskell, Andrew, and Adrian Currie. 2017. Forces, Friction and Fractionation: Denis Walsh's Organisms, Agency, and Evolution. *Biology & Philosophy* 32 (6): 1341–1353. https://doi.org/10.1007/s10539-017-9585-z.

Cavalli-Sforza, Luigi Luca, Paolo Menozzi, and Alverto Piazza. 1994. *The History and Geography of Human Genes.* Princeton, NJ: Princeton University Press.

Charlesworth, Brian, and Deborah Charlesworth. 2009. Darwin and Genetics. *Genetics* 183 (3): 757–766. https://doi.org/10.1534/genetics.109.109991.

Danchin, Étienne. 2013. Avatars of Information: Towards an Inclusive Evolutionary Synthesis. *Trends in Ecology and Evolution* 28(6): 351–358. https://doi.org/10.1016/j.tree.2013.02.010.

[20] Depew summarises my point: 'Whenever it is said that natural selection rather than Spencer's environmental pressures or Morgan's mutations or anything else is the novelty- producing, direction-giving, and so "creative" factor in evolution, the implication is that natural selection working over multiple generations is the source, usually in conjunction with other factors, of the adaptations that allow a subpopulation to gain a reproductive edge. This is as true of Mayr's organocentric interpretation of the modern evolutionary synthesis as it is of Dawkins's genocentric version. In denying that natural selection is the creative factor in stickleback evolution, Gilbert and Epel are sending a message that a case like this cannot even be properly described, let alone explained, without detailed ecologically context-dependent knowledge of gene regulation and gene expression and that this knowledge that might run afoul of the basic principles of the modern synthesis' (Depew 2017: 50).

Danchin, Étienne, and Arnaud Pocheville. 2017. Genetic Assimilation and the Paradox of Blind Variation. In *Challenging the Modern Synthesis: Adaptation, Development, and Inheritance*, ed. Philippe Huneman and Denis M. Walsh, 111–136. Oxford, UK: Oxford University Press.

Danchin, Etienne, Arnaud Pocheville, Olivier Rey, Benoit Pujol, and Simon Blanchet. 2019. Epigenetically Facilitated Mutational Assimilation: Epigenetics as a Hub within the Inclusive Evolutionary Synthesis. *Biological Reviews* 94 (1): 259–282. https://doi.org/10.1111/brv.12453.

Depew, David J. 2011. Adaptation as Process: The Future of Darwinism and the Legacy of Theodosius Dobzhansky. *Studies in History and Philosophy of Science Part C: Studies in History and Philosophy of Biological and Biomedical Sciences* 42 (1). https://doi.org/10.1016/j.shpsc.2010.11.006.

Depew, D. J. (2017). Natural Selection, Adaptation, and the Recovery of Development. In *Challenging the Modern Synthesis: Adaptation, Development, and Inheritance*, ed. P. Huneman and D. M. Walsh, 37–67. Oxford, UK: Oxford University Press. https://doi.org/10.1093/oso/9780199377176.003.0001.

Depew, David J., and Bruce H. Weber. 1995. *Darwinism Evolving: Systems Dynamics and the Genealogy of Natural Selection.* Cambridge, MA: MIT Press.

Depew, David J., and Bruce H. Weber. 2011. The Fate of Darwinism: Evolution After the Modern Synthesis. *Biological Theory* 6 (1): 89–102. https://doi.org/10.1007/s13752-011-0007-1.

de Vries, Hugo. 1905. *Species and Varieties, their Origin by Mutation.* Chicago, IL: The Open Court Publishing Co.

Dobzhansky, Theodosius. 1951. *Genetics and the Origin of Species.* New York, NY: Columbia University Press.

Dobzhansky, Theodosius. 1973. Nothing in Biology Makes Sense Except in the Light of Evolution. *The American Biology Teacher* 35 (3). https://doi.org/10.2307/4444260.

Fábregas-Tejeda, Alejandro, and Francisco Vergara-Silva. 2018. Hierarchy Theory of Evolution and the Extended Evolutionary Synthesis: Some Epistemic Bridges, Some Conceptual Rifts. *Evolutionary Biology* 45 (2). https://doi.org/10.1007/s11692-017-9438-3.

Futuyma, Douglas J. 2017. Evolutionary Biology Today and the Call for an Extended Synthesis. *Interface Focus* 7 (5). https://doi.org/10.1098/rsfs.2016.0145.

Gawne, Richard, Kenneth Z. McKenna, and H. Frederik Nijhout. 2018. Unmodern Synthesis: Developmental Hierarchies and the Origin of Phenotypes. *BioEssays* 40 (1). https://doi.org/10.1002/bies.201600265.

Gayon, Jean. 1997. The "Paramount Power of Selection": From Darwin to Kauffman. In *Structures and Norms in Science, Synthese Library*, ed. M. L. D. Chiara, K. Doets, D. Mundici, and J. Van Benthem, vol. 260, 265–282. Dordrecht: Springer. https://doi.org/10.1007/978-94-017-0538-7_16.

Gilbert, Scott F. 1988. Cellular Politics: Ernest Everett Just, Richard B. Goldschmidt, and The Attempt to Reconcile Embryology and Genetics. In *American Development of Biology*, ed. J. Rainger, R. Benson, and K. R. Maienschein, 311–346. Philadelphia, PA: University Of Pennsylvania Press.

Gilbert, Scott F. 2000. Genes Classical and Genes Developmental: The Different Use of Genes in Evolutionary Syntheses. In *The Concept of the Gene in Development and Evolution*, ed. Peter J. Beurton, Raphael Falk, and Hans-Jörg Rheinberger, 178–192. Cambridge, UK: Cambridge University Press. https://doi.org/10.1017/cbo9780511527296.010.

Gilbert, Scott F. 2007. Fate Maps, Gene Expression Maps, and the Evidentiary Structure of Evolutionary Developmental Biology. In *From Embryology to Evo-Devo*, ed. Jane Maienschein and Manfred Laubischler, 358–374. Cambridge, MA: MIT Press. https://doi.org/10.7551/mitpress/3128.003.0013.

Gilbert, Scott F., and David Epel. 2015. *Ecological Developmental Biology: The Environmental Regulation of Development, Health, and Evolution.* 2nd ed. Sunderland, MA: Sinauer Associates, Inc. Publishers.

Gould, Stephen Jay. 1980. G. G. Simpson, Paleontology, and the Modern Synthesis. In *The Evolutionary Synthesis; Perspectives on the Unification of Biology*, ed. Ernst Mayr and William B. Provine, 153–172. Cambridge, MA: Harvard University Press.

Gould, Stephen Jay. 1983. The Hardening of the Modern Synthesis. In *Dimensions of Darwinism. Themes and Counterthemes in Twentieth Century Evolutionary Theory*, 71–93. Cambridge, UK: Cambridge University Press.

Gould, Stephen Jay. 1991. *Ever Since Darwin: Reflections in Natural History*. Harmondsworth: Penguin.

Gould, Stephen Jay. 2002. *The Structure of Evolutionary Theory*. Cambridge, MA: Harvard University Press. https://doi.org/10.2307/j.ctvjsf433.

Gould, Stephen Jay, and Richard Lewontin. 1979. The Spandrels of San Marco and the Panglossian Paradigm: A Critique of the Adaptationist Programme. *Proceedings of the Royal Society of London. Series B, Biological Sciences* 205: 581.

Haldane, J. B. S. 1953. Foreword. In *Evolution. Symposia of the Society of Experimental Biology*, ix–xix. Cambridge, UK: Cambridge University Press.

Hosken, David J., and Clarissa M. House. 2011. Sexual Selection. *Current Biology* 21 (2): R62–R65. https://doi.org/10.1016/j.cub.2010.11.053.

Hull, David L. 1990. *Science as a Process: An Evolutionary Account of the Social and Conceptual Development of Science*. Chicago, IL: University of Chicago Press. https://www.degruyter.com/isbn/9780226360492.

Huneman, Philippe, and Denis M. Walsh, eds. 2017a. *Challenging the Modern Synthesis: Adaptation, Development, and Inheritance*. New York, NY: Oxford University Press. https://doi.org/10.1093/oso/9780199377176.001.0001.

Huneman, Philippe, and Denis M. Walsh. 2017b. Introduction: Challenging the Modern Synthesis. In *Challenging the Modern Synthesis: Adaptation, Development, and Inheritance*, ed. Philippe Huneman and Denis M. Walsh, 1–33. New York, NY: Oxford University Press.

Huxley, Julian. 2010. *Evolution: The Modern Synthesis*. Cambridge, MA: MIT Press.

Jablonka, Eva, and Marion Lamb. 2020. *Inheritance Systems and the Extended Synthesis. Elements in the Philosophy of Biology*. Cambridge University Press. https://doi.org/10.1017/9781108685412.

Jacob, François. 1977. Evolution and Tinkering. *Science* 196 (4295): 1161–1166. https://doi.org/10.1126/science.860134.

Jaeger, Johannes. 2024. Artificial Intelligence Is Algorithmic Mimicry: Why Artificial “Agents” Are Not (and Won’t Be) Proper Agents. *Neurons, Behavior, Data Analysis, and Theory* 1: 21. https://doi.org/10.48550/arXiv.2307.07515.

Jaeger, Johannes, David Irons, and Nick Monk. 2012. The Inheritance of Process: A Dynamical Systems Approach. *Journal of Experimental Zoology Part B: Molecular and Developmental Evolution* 318 (8). https://doi.org/10.1002/jez.b.22468.

Jørgensen, Sven Erik, and Brian D. Fath. 2008. *Encyclopedia of Ecology, Five-Volume Set*. Amsterdam: Elsevier.

Juarrero, Alicia. 1999. *Dynamics in Action: Intentional Behavior as a Complex System*. Cambridge, MA: MIT Press.

Keller, Evelyn Fox. 2010. *The Mirage of a Space Between Nature and Nurture*. Durham, NC: Duke University Press.

Keller, Evelyn Fox. 2000. *The Century of the Gene*. Cambridge, MA: Harvard University Press.

Kimura, Motoo. 1968. Evolutionary Rate at the Molecular Level. *Nature* 217: 624–626. https://doi.org/10.1038/217624a0.

Lakatos, Imre. 1978. *The Methodology of Scientific Research Programmes*, ed. John Worrall and Gregory Currie. Cambridge, UK: Cambridge University Press. https://doi.org/10.1017/CBO9780511621123.

Lala, Kevin N., Tobias Uller, Nathalie Feiner, Marcus W. Feldman, Scott F. Gilbert. 2024. *Evolution Evolving: The Developmental Origins of Adaptation and Biodiversity*. Princeton, NJ: Princeton University Press.

Laland, Kevin, Tobias Uller, Marc Feldman, Kim Sterelny, Gerd B. Müller, et al. 2014. Does Evolutionary Theory Need a Rethink? *Nature* 514 (7521): 161–164. https://doi.org/10.1038/514161a.

Levins, Richard, and Richard Lewontin. 1985. *The Dialectical Biologist*. Cambridge, MA: Harvard University Press.

Lewontin, Richard, and Richard Levins. 2007. *Biology Under the Influence: Dialectical Essays on Ecology, Agriculture, and Health*. New York, NY: Monthly Review Press.

Longo, Giuseppe. 2020. Information at the Threshold of Interpretation: Science as Human Construction of Sense. https://doi.org/10.1007/978-3-030-25001-0_5.

Longo, Giuseppe, and Matteo Mossio. 2020. Geocentrism vs Genocentrism: Theories Without Metaphors, Metaphors Without Theories. *Interdisciplinary Science Reviews* 45 (3). https://doi.org/10.1080/03080188.2020.1798588.

Love, Alan C. 2010. Rethinking the Structure of Evolutionary Theory for an Extended Synthesis. In *Evolution, The Extended Synthesis*, 403–441. Cambridge, MA: MIT Press. https://doi.org/10.7551/mitpress/9780262513678.003.0016.

Love, Alan C. 2017. Evo-Devo and the Structure(s) of Evolutionary Theory. In *Challenging the Modern Synthesis: Adaptation, Development, and Inheritance*, ed. Philippe Huneman and Denis M. Walsh, 159–187. New York, NY: Oxford University Press.

Malafouris, Lambros. 2015. Metaplasticity and the Primacy of Material Engagement. *Time and Mind* 8 (4): 351–371. https://doi.org/10.1080/1751696X.2015.1111564.

Marques, Victor, and Carlos Brito. 2014. The Rise and Fall of the Machine Metaphor: Organizational Similarities and Differences Between Machines and Living Beings. *Verifiche* 43 (1–4): 77–111.

Martínez, Maximiliano, and Maurizio Esposito. 2014. Multilevel Causation and the Extended Synthesis. *Biological Theory* 9 (2). https://doi.org/10.1007/s13752-014-0161-3.

Mayr, Ernst. 1970. *Populations, Species, and Evolution*. Cambridge, MA: Harvard University Press.

Moss, Lenny. 2003. *What Genes Can't Do, What Genes Can't Do*. Cambridge, MA: MIT Press. https://doi.org/10.7551/mitpress/7205.001.0001.

Moss, Lenny. 2012. Is the Philosophy of Mechanism Philosophy Enough? *Studies in History and Philosophy of Science Part C: Studies in History and Philosophy of Biological and Biomedical Sciences* 43 (1): 164–172. https://doi.org/10.1016/j.shpsc.2011.05.015.

Müller, Gerd B. 2017. Why an Extended Evolutionary Synthesis Is Necessary. *Interface Focus* 7 (5): 20170015. https://doi.org/10.1098/rsfs.2017.0015.

Nei, Masatoshi. 2013. *Mutation-Driven Evolution*. Oxford, UK: Oxford University Press. http://site.ebrary.com/id/10695420.

Nicholson, Daniel J. 2012. The Concept of Mechanism in Biology. *Studies in History and Philosophy of Science Part C: Studies in History and Philosophy of Biological and Biomedical Sciences* 43 (1): 152–163. https://doi.org/10.1016/j.shpsc.2011.05.014.

Noble, Denis. 2006. *The Music of Life: Biology Beyond Genes*. Oxford, UK: Oxford University Press.

Noble, Denis. 2011. Neo-Darwinism, the Modern Synthesis and selfish genes: Are they of use in physiology? *Journal of Physiology*. https://doi.org/10.1113/jphysiol.2010.201384.

Noble, Denis, and Raymond Noble. 2021. Rehabilitatin of Karl Popper's Ideas on Evolutionary Biology and the Nature of Biological Science. In *Karl Popper's Science and Philosophy*, 193–209. Cham: Springer.

O'Malley, Maureen A., and John Dupré. 2005. Fundamental Issues in Systems Biology. *BioEssays*. https://doi.org/10.1002/bies.20323.

Odling-Smee, F. John, Kevin N. Laland, and Marcus W. Feldman. 2003. *Niche Construction: The Neglected Process in Evolution*. Princeton, NJ: Princeton University Press.

Okasha, Samir. 2006. *Evolution and the Levels of Selection*, XLI. Oxford, UK: Oxford University Press.

Okasha, Samir. 2018. Agents and Goals in Evolution, Agents and Goals in Evolution (Oxford, UK: Oxford University Press) https://doi.org/10.1093/oso/9780198815082.001.0001.

Oyama, Susan. 2000. *The Ontogeny of Information: Developmental Systems and Evolution*. Durham, NC: Duke University Press.

Pigliucci, Massimo, Courtney J. Murren, and Carl D. Schlichting. 2006. Phenotypic Plasticity and Evolution by Genetic Assimilation. *Journal of Experimental Biology* 209:2362–2367. https://doi.org/10.1242/jeb.02070.

Riskin, Jessica. 2020. Biology's Mistress, a Brief History. *Interdisciplinary Science Reviews* 45 (3): 268–298. https://doi.org/10.1080/03080188.2020.1794388.

Sepkoski, David. 2019. The Unfinished Synthesis?: Paleontology and Evolutionary Biology in the 20th Century. *Journal of the History of Biology* 52 (4): 687–703. https://doi.org/10.1007/s10739-018-9537-8.

Smocovitis, V. B. 1992. Unifying Biology: The Evolutionary Synthesis and Evolutionary Biology. *Journal of the History of Biology* 25 (1): 1–65. http://www.jstor.org/stable/4331201.

Smocovitis, Vassiliki Betty. 1996. *Unifying Biology: The Evolutionary Synthesis and Evolutionary Biology*. Princeton, NJ: Princeton University Press.

Smocovitis, V. Betty. 2023. Every Evolutionist Their Own Historian: The Importance of History, Context, and the Extended Evolutionary Synthesis. In *Evolutionary Biology: Contemporary and Historical Reflections Upon Core Theory*, ed. Thomas E. Dickins and Benjamin J. A. Dickins, 25–54. Cham: Springer. https://doi.org/10.1007/978-3-031-22028-9_2.

Stoltzfus, Arlin (2012). Constructive neutral evolution: Exploring evolutionary theory's curious disconnect. *Biology Direct*. https://doi.org/10.1186/1745-6150-7-35.

Stoltzfus, Arlin. 2017. Why We Don't Want Another "Synthesis". *Biology Direct* 12 (1): 23. https://doi.org/10.1186/s13062-017-0194-1.

Stoltzfus, Arlin, and Kele Cable. 2014. Mendelian-Mutationism: The Forgotten Evolutionary Synthesis. *Journal of the History of Biology* 47 (4): 501–546. https://doi.org/10.1007/s10739-014-9383-2.

Stotz, Karola. 2014. Extended Evolutionary Psychology: The Importance of Transgenerational Developmental Plasticity. *Frontiers in Psychology* 5. https://www.frontiersin.org/article/10.3389/fpsyg.2014.00908.

Sultan, Sonia E. 2015. *Organism and Environment: Ecological Development, Niche Construction, and Adaptation*. Oxford University Press.

Turner, J. Scott. 2017. *Purpose & Desire: What Makes Something 'Alive' and Why Modern Darwinism Has Failed to Explain It*. New York, NY: HarperOne.

Wagner, Andreas. 2014. *Arrival of the Fittest: Solving Evolution's Greatest Puzzle*. London, UK: Oneworld Publications.

Walsh, Denis M. 2015. *Organisms, Agency, and Evolution*. Cambridge, UK: Cambridge University Press. https://doi.org/10.1017/CBO9781316402719.

Walsh, Denis M. 2017. "Chance Caught on the Wing": Metaphysical Commitment or Methodological Artifact? In *Challenging the Modern Synthesis: Adaptation, Development, and Inheritance*, ed. P. Huneman and D. M. Walsh, 239–260. Oxford, UK: Oxford University Press. https://doi.org/10.1093/oso/9780199377176.003.0008.

Whitehead, Alfred North. 2015. *The Concept of Nature*. Cambridge, UK: Cambridge University Press. https://doi.org/10.1017/CBO9781316286654.

Winther, Rasmus G. 2000. Darwin on Variation and Heredity. *Journal of the History of Biology*. https://doi.org/10.1023/A:1004834008068.

Woodger, J. H. 2010. *Biological Principles: A Critical Study*. London and New York: Routledge. https://doi.org/10.2307/2016523.

Zambrana, Rocío. 2015. *Hegel's Theory of Intelligibility*. Chicago, IL: University of Chicago Press.

Chapter 3
A Foray into Organicism

Abstract This chapter critically examines the limitations of the Modern Synthesis (MS) in evolutionary biology, which reduces evolution to genetic variation and natural selection while neglecting organism-environment dynamics. It introduces current reiteration of organicism, the organisational approach, as an alternative framework, emphasising that living systems are organised wholes shaped by historicity, agency, and emergent properties. The analysis highlights how ecological dialectics and niche construction theory demonstrate organisms' active role in modifying their selective pressures, challenging the MS's passive view of adaptation. The discussion contrasts weak and strong emergence, arguing that biological complexity requires recognizing top-down causation. While dynamical systems theory models nonlinear evolutionary processes, its inability to predict critical transitions underscores the need for more flexible frameworks. The chapter concludes by assessing proposals like the Extended Evolutionary Synthesis (EES), which integrates developmental processes, epigenetic inheritance, and multi-level selection. Ultimately, it calls for replacing the MS's reductionism with a dialectical approach centred on organism-environment reciprocity—one that better accounts for the creativity, historicity, and agency inherent in evolution.

Introduction

> [T]he progress of biology […] has probably been checked by the uncritical assumption of half-truths. If science is not to degenerate into a medley of *ad hoc* hypotheses, it must become philosophical and must enter upon a thorough criticism of its own foundations. (Whitehead 1967: 16–17)

Whitehead's words above are unforgiving but warranted. Half-truths abound and should be excised by philosophical critique. In this chapter, I continue this work by introducing trends in current biology. But let us first reiterate the main facets of the MS. Within this tradition, natural selection, variation, adaptation, and inheritance are considered independent processes. Among these, natural selection is assumed

R. Sandnes Haukedal, *Agency and Organisation: The Dialectics of Nature and Life*, Library of Ethics and Applied Philosophy 44,
https://doi.org/10.1007/978-981-95-1924-8_3

to be primary. Since it produces the variation that it acts on, it is not only necessary but sufficient for evolutionary change. Furthermore, natural selection is genetic selection enacted by a (mostly) stable and non-modifiable environment outside the organism. These aspects explain why development is deemed non-evolutionary, i.e. as the simple unfolding of the blueprint written down in the genes, and why adaptation is considered a passive process. The MS thereby undermined any form of explanatory teleology, even if there is a covert form of purposiveness in (1) the gene's-eye view of evolution, (2) the primacy of natural selection, and (3) the irreducibility of biology as an autonomous science.

The MS gained dominance by 'simultaneously refuting and co-opting new forms of saltationism, Lamarckism, and developmentalism' (Depew 2017a: 63). Furthermore, many of the concepts it employs are metaphors, not theories, as we will see more clearly as we proceed. We should bear this in mind as we delve deeper into the MS by contrasting it to alternative views, like organicism, a view which—according to Daniel Nicholson and Richard Gawne—has largely been suppressed by the common narrative of biological history (2015), where the opposition was between vitalism and mechanism. With organicism as a theoretical bridge, we move from a discussion of the MS towards its alternative, the EES. To accomplish this, I discuss how the views of the MS's heretics prefigure the ongoing changes in evolutionary thinking.

I structure the chapter thusly: I begin with organicism and argue that dialectical materialism and ecology share similarities, especially in their emphasis on the entangled interplay between the organism and the environment. I then introduce a current *repetition* of organicism, the organisational approach, which will occupy us throughout this work. This perspective is formulated as a critique of modern organicism, which, in their view, does not provide a critical corrective to the MS since it has largely adopted the 'pervasive language of molecular biology, a language that forces causative power to molecules, and in particular, to genes' (Soto et al. 2016: 78).

While critical of the diluted organicism that the MS absorbed, the organisational approach draws its theoretical principles from organicism's core ideas. The notion of *closure of constraints* is central in this concern, as are the three basic principles of (1) variation, (2) organisation, and (3) the default state of proliferation with variation and motility (Soto et al. 2016). After discussing these principles, especially the first two, we move to the question of how we should understand causation and emergence in biology. I then get more specific about the alternatives to the MS, first discussing niche construction and epigenetics; then, after a short interval on the difference between developmental and selective niche, I discuss an approach that shares some of the concerns of the organisational approach, namely dynamical systems theory. Following this exegesis, I aim to show why this approach fails to grasp the historicity of living systems. Subsequently, I return to questions of what the alternative to the MS might look like, how they are already implicit in the tradition of evolutionary biology, how radical the different alternatives are, and what success would mean for its proponents—before trying to summarise some of the main issues we have discussed thus far and chart a way forward.

Ecological Dialectics

> Biology could not seriously explore an organicism in response to its own crises until its relation to physics was changed. (Haraway 2004: 24)

There is an argument to be made that ecology and dialectics are not only compatible but inextricable. I cannot do justice to this argument here, but some of the most prominent ecologists of the twentieth century were inspired by dialectical materialism, the works of Friedrich Engels in particular (Foster 2020; Sheehan 2018, 2022).[1] According to Parsons (1977), dialectics is a generalisation of the principles of ecology beyond living systems, while ecology is the application of dialectical principles to nature. This suggests why there is considerable conceptual overlap. I will touch more upon historical interconnections in the next chapter. (My presentation will, however, be lopsided since my aim is not primarily to survey the historical link between these traditions but to examine the philosophical implications of their convergence.)

The basic principle of dialectical materialism is that dialectics applies to nature itself, not just to the interaction between humans and their natural and social environments. The contradiction that the organism maintains itself only by engaging with its surroundings is an example of this. This outlook challenges the stark opposition between human and natural history. Applying such a view, Levins and Lewontin observe that there are two theories of evolution, one microscopic and one macroscopic, and claim that this is analogous to physics. There cannot be contact between these theories until the 'dichotomy between organism and environment is broken down' (Lewontin and Levins 2007: 63). They call for a dialectical approach to evolution:

> The divorce between the relative fitness of genotypes and the fitness of populations arises from the fiction that new varieties are selected in a fixed environment, so that the only issue is whether, given that environment, they will produce fewer or more offspring. But in reality, a new variety means a new environment, a new set of relations among organisms and with inorganic nature. On the other hand, each mutational change cannot result in a totally new relation between organism and environment, or else no cumulative evolutionary change could ever take place. (Lewontin and Levins 2007: 63)

This quote takes us back to the notion that the environment is inherently unstable. It also prepares the ground for the discussion of organicism by highlighting the MS's decoupling of the organism and the environment. The close connection between the organism and its environment, and how a change in the organism is effectively a change in its environment as well, indicates a relational perspective that I expand on later.

At face value, modern biology rid itself of organicism. But we have seen that a kind of organicism is implicit in the very fact biology is an autonomous science. What happened, says El-Hani and Emmeche (2000), was that mainstream organicism

---

[1] For example, Arthur Tansley, the botanist that coined the term 'ecosystem' was influenced by Marxism through his tutor Ray Lankester, as documented by Foster (2020).

tacitly accepted the centrality of the organism, yet ignored its actual functioning and explanatory role.[2] It was, therefore, an ambiguous organicism without the organism, or with the organism as a mere object—as a meeting point for two other processes: The internal production of mutations, and the external culling done by natural selection. This led to a situation where evolutionary theory was

> both reductionistic and emergentist: reductionist in the constitutive sense of looking at the plant or animal (or any other living being) simply as a special material aggregation of molecules historically organized by means of evolution by natural selection, an entity that can, in principle if not in practice, be understood in terms of the operation of pure efficient causality, internally (producing phenotypes under the rules of genotypes) as well as externally (producing an evolving lineage by variational changes); and emergentist in the sense of acknowledging the complexity and specific properties of organisms (such as self-reproduction), appearing only at the level of the cell and the body as a system of cells. (El-Hani and Emmeche 2000: 340)

The perspective outlined below is less ambivalent than the MS since it does not share the reductionist premise where parts are self-contained while simultaneously maintaining that they cannot be further reduced. It demands, however, that we discard the view that explanations of purpose and emergence are heuristics, and consider them to be about reality itself.

Organicism brings together many of the issues we have discussed, concerning 'the interdependence of the parts and their relation to the whole' (Montévil et al. 2016: fn12, 39). It emphasises both the role of the organism and organisational constraints acting on the parts of a system and is about how parts, conceived as processes, both enable and are sustained by the whole they give rise to (and vice versa). Moreover, it concerns self-organisation, emergent properties (resulting from interactions) and nonlinear causation. Soto and Sonnenschein (2018) enumerate four characteristics of organicism: (1) organisation, (2) historicity, (3) normative agency, and (4) specificity. I do not explain the first now, as it is covered extensively below. Historicity simply points to the fact that living beings are produced through lineages of predecessors, and thus cannot be viewed apart from their reproduction. Normative agency is intrinsic in the purposive activity of organisms and the ascription of functions. Finally, specificity concerns how life is not determined by the general laws of physics, even if they must obey them. Different principles obtain at the biological level, and these cannot be reduced to genetics since the genetic code contained in DNA lacks the causal specificity to linearly code for complex phenotypes. It lacks the *expressive power* needed to specify (code for) all the effects it causes. This does not cancel its causal relevance but acknowledges that it is dependent on factors that it did not determine.

Another way of saying this is by reference to Juarrero (1999), which points out how initiating an event is not equivalent to maintaining it. The DNA can initiate

---

[2] Moss (2012) writes: 'However tacit it may become in practice, the point of departure of any functional, let alone mechanism-based, analysis is the holistic assumption of a unified entity that acts flexibly and contingently to sustain its own existence. Implicit in the very meaning of biological mechanism [...], the presupposition of a self-sustaining entity (only in relation to which some activity can count as a biological mechanism) is not a contingent feature of a mechanism-based account but is rather an a priori feature of any such possible investigation' (166).

the formation of proteins, but it does not by itself maintain (or regulate) the process leading from such proteins to distal and complex phenotypes. In other words, it acts as a constraint, not a determining factor. Therefore, 'additional principles may be necessary to understand organisms as living beings' (Sonnenschein and Soto 2018: 498). In short, it is the view that the overall organisation and parts of a living structures are 'the result of the reciprocal interplay among all its components' (Moreno and Mossio 2015). This means that '[i]n biology, *histories* and *contexts* (sometime strongly) canalize and constraint random evolutions' (Longo and Montévil 2013: 72, emphases original). For example, cells do not have a function outside their context:

> To heirs of nineteenth-century holism, autonomy was understood in terms of "totipotency," the possession by the cell of the potential of the whole. The autonomy of the cell understood this way is then the precondition for either normal or aberrant growth and a prior guarantee of neither. What determines which way it will go, normal or aberrant, is not its internal features but the subsequent history of its interactions. (Moss 2003: 129)

Ana Soto and Carlos Sonnenschein (2004) cite this paragraph from Moss in an article outlining an organicism explanation of cancer, which exemplifies the consequences of adopting the organicist view. Instead of considering cancer a consequence of mutations causing genes to proliferate detrimentally, they regard cell proliferation to be the default state of the organism. Cancer, then, has more to do with higher-level tissue organisation than with lower-level mutations causing cell proliferation. It is a function of organisation going awry, not deleterious mutations by themselves. This means that studying the lowest possible level, without considering the higher-level context—which the lower levels both *give rise to and are moulded by*—will not reveal what causes cancer (Soto and Sonnenschein 2004). Another example of how context influence cell function are stem cells, which have the pluripotency of turning into any cell within the body in response to the demands of the whole organism (Gilbert 2000).

Both examples indicate how phenotypes emerge from a complex network—are *facilitated* by a developmental system (Lala et al. 2024). They also reveal what is missed by the logic of decomposition and separability at work in the MS. In the latter view, organisms are separated from environments, evolution from development, evolutionary adaptation from physiological adaptability, and so on. Separating factors undergird the reductionistic and deterministic trajectory of the MS where 'allele frequency change caused by natural selection is the only credible process underlying the evolution of adaptive organismal traits' (Charlesworth et al. 2017: 10). This means that evolution as a mechanistic process, about neatly distinguished entities acting on each other from the outside. The cost of this outlook is that adaptability, development, and the action of organisms are reduced to 'epiphenomen[a] of gene regulation' (Gilbert and Faber 1996: 139).

The MS seems to have been 'conflating a successful methodology with a valid ontology', as Joseph Needham (1929) puts it. Processes may appear reducible without remainder when experimenting, but it does not follow that the lowest possible level of explanation is the locus with ultimate causal force (nor that the lowest level is simple, as discussed in the next chapter). The idea that form has no effect on the entities in this arrangement is, however, just a postulation. It is not shown empirically, even if it is the *sine qua non* for decomposition as a viable strategy. It is tied to the view that 'composition doesn't add anything to the capacities of the parts, or at least that the capacities of the parts are relatively unaffected by their contexts' (Walsh 2015: 38). This is what we are up against, and the challenge is to conceptualise how composition affects the parts in a manner that does not simply reiterate the common conceptions of causality of mechanistic science but reimagines it.

Emergence, Strong and Weak

> What distinguishes a butterfly from a lion, a hen from a fly, or a worm from a whale is much less a difference in chemical constituents than in the organization and the distribution of these constituents. (Jacob 1977: 1165)

Despite its long history, and the importance it played in the formation of evolutionary theory, the organicist tradition has become the forgotten third in the debate between August Weismann (mechanism) and Hans Driesch (vitalism). Oscar Hertwig, for example, tried to merge epigenesis with preformation by highlighting how many of the characteristics of an organism were produced by epigenetic interaction between the egg and the environment.[3] Within genetics, Weismann's views won out but, according to Gilbert, 'Hertwig's "organicism" (an epigenetic materialism) was adopted by embryologists as a reasonable explanation of development'. It involved 'the extension of epigenesis from interactions between cells of the embryo to the interactions between developing organisms and their respective environments' (Gilbert 2002: 204).

This hints at a broad and a narrow version of epigenetics: The latter pertains to 'the study of the mechanisms that determine which genome sequences will be expressed in the cell', whereas the former concerns 'the causal mechanisms by which genotypes give rise to phenotypes' (Griffiths 2017: 4). The broad definition is agnostic about what causes the change. It is not limited to molecular changes in transcription but 'anything that affects development and its evolution' (Reid 2009: 210). It accounts for the complex process that leads to functional phenotypes. But the neglect of embryology within the MS meant discarding the importance of epigenetic interaction in both senses—not only *within* the organism but *between* the organism and its environment. (I return to this topic at the end of the chapter.)

Against the continuity of levels postulated by the MS, a more hierarchical and complex model of evolution—like organicism—holds that different mechanisms,

---

[3] He exemplified this through ant colonies, where the determinants of each caste are not contained wholly within the egg (as Weismann thought) but that each 'caste is merely a nutritional polyphenism. Here, each larva has the potential to be a member of any caste, and what it becomes is determined by the diet the larva is fed' (Gilbert 2002: 205). Polyphenism means a qualitatively distinct and discrete phenotype of the same genome, switching abruptly at a certain threshold, in contrast to polymorph (or adaptive plasticity), which is gradual and quantitative.

laws, and modes of evolution obtain at different levels, and that selection operates differently at them (see Sepkoski 2019). This indicates the link between organicism and emergence, as the latter is the view that an entity can have qualities that are not found in the parts. Furthermore, there are two basic versions of emergence: strong and weak. The latter is epistemological, in that it says that we must treat different levels as theoretically distinct to understand them, but still holds that emergent levels are ultimately reducible to a basic level and that no changes can occur except if there are changes at this level. It is thus an ambiguous view, where the emergent levels lack causal autonomy. Strong emergence, on the other hand, says that emergent levels are real, not just heuristic. In other words, they have causal autonomy from the processes they depend on. Later, I argue that this view is more useful for systems biology.

Organicism was the unspoken metaphysics of biology that provided it with autonomy. Yet, it was not committed to *ontological* organicism, which involves strong emergence of new qualities and levels. Instead, it held that the organisms were just highly organised instances of the same elementary particles found in physics (Emmeche 2004). Going beyond the epistemological form of organicism means that biology is not only a science of the emergent level above chemistry and physics. It is not enough to say that biology is about systems that organise chemical and physical processes while leaving these processes intact. There are also emergent levels within biology that are irreducible to one another *and* may retroact on the processes they emerge from. (See Chap. 5.)

Gilbert and Sarkar state that 'the properties of any level depend both on the properties of the part "beneath" them and the properties of the whole into which they are assembled' (2000: 2). The prevalent strategy of the MS, analytical decomposition, is founded on the opposite view—and employs individualistic assumptions concerning biological and molecular interaction (O'Malley and Dupré 2005). Thus, one of the impediments to a more holistic (or relational) view of evolution concerns the units of evolution. The move towards 'holism' entails softening the distinctions between variation, selection, and adaptation, demanding that we examine the intertwinement of such processes (O'Malley and Dupré 2005). As we shall see, we cannot take for granted that these units are individual genes (or genomes). Our genomes are not isolated. Instead, all organisms are *multigenomic* (Gilbert et al. 2012). If we look at the genome(s) of humans, we consist mainly of microbes. This fact seems to warrant 'radical reappraisals between biological entities and the organisation of life itself' (O'Malley and Dupré 2005: 1274).

Waddington's notion, *epigenotypes*, encapsulates how genes both act and are acted upon: 'In the epigenotype, the gene is not an autonomous entity; it is part of a network of interacting components' (Gilbert 2000: 731). We must move from talking about 'genes for x' to 'genes involved in x' and acknowledge the buffering effects of networks and regulatory genes. Emergent complex causal properties of systems must be accounted for, as the components of a system exhibit novel behaviour that they did not outside it. The organisation is not an epiphenomenon, an aggregate of its parts, as 'single components cannot be understood in terms of their intrinsic properties but must be seen as simultaneously determined by features of the systems of which they are part' (O'Malley and Dupré 2005: 1274). From this perspective, it is meaningless to ascribe functions to anything without accounting for their context, i.e. the organisation they both give rise to and is affected by. Parts and wholes are mutually enabling. This points to the notion of developmental constraints, how systems properties constrain the behaviour of individual components and thus are not only negative. If, for instance, two expression patterns are mutually exclusive, they 'repress each other's expression forming a double negative (and thus positive) regulatory feedback loop' (Jaeger et al. 2012: 594). The function of one component cannot be isolated from its constitutive interaction with other components. The whole system self-organises in a manner which makes the continuation of the initiating process dependent on its products (cf. Juarrero 1999).

Organising Constraints

> The billiard ball model of causality has lured biologists towards linear, unidirectional, unilevel models of biological systems that are not actually built this way. Linearity is broken by branching pathways and unidirectionality is broken by feedback. Models of single-level interactions are broken by biological hierarchy. (Bizzarri et al. 2019: no pagination)

The genome does nothing by itself. Without proofreading, it would be too unstable to evolve, and an 'error catastrophe' undermining the whole organisation would quickly ensue. DNA is *only functional within a larger whole*—a logic that we will revisit many times. Or, as Griesemer (2006) says, genes are not master molecules but 'prisoners of development'. Furthermore, treating the selection of random mutations as the main source of variation is not an explanation. Jesper Hoffmeyer summarises my point: 'As opposed to the organism, selection is a purely external force while mutation is an internal force, engendering variation. And yet mutations are considered to be random phenomena and hence independent of both the organism and its functions' (Hoffmeyer 1997: 56).[4]

Matteo Mossio et al. (2016) propose organisation as a theoretical principle and place the organism in the centre of evolution. They write that

> [f]rom an organicist perspective, organisms are the main object of biological science because they are the systems that underlie biological phenomena and—crucially—they cannot be reduced to more fundamental biological entities (such as the genes or other inert components of the organism). (Mossio et al. 2016: 25)

Again, irreducibility and emergence are central, as organised systems integrate the interplay of bottom-up and top-down processes (see Noble 2006). Organisms are

[4] This assumption is increasingly untenable as new methods within molecular biology have made it possible to study *de novo* mutations, viz. mutations not inherited by either parent but novel, at high resolution. One study has revealed that mutations for malaria are not randomly distributed among populations but occur more frequently in populations that are more exposed to the threat than others (Melamed et al. 2022). This suggests that mutations are not completely random with respect to function.

characterised by thermodynamical openness while simultaneously displaying *organisational closure*: 'they realize closure, i.e. a mutual dependence between a set of constituents which maintain each other through their interactions and which could not exist in isolation' (Mossio et al. 2016: 27). This duality enables organisms to exchange energy with the outside world while maintaining their identity. It is not enough that some of the processes of the system are stable, however; the overall organisation must sustain itself. The authors endorse Kauffman's view of self-organisation in which the work done by the system is constrained by what is called a 'work-constraints cycle':

> When a (W-C) [work-constraint] cycle is realized, constraints that apply to the system are produced and maintained by the system itself. Hence, the system needs to use the work generated by the constraints in order to generate those very constraints, by establishing a mutual relationship—a cycle—between constraints and work. (Mossio et al. 2016: 28)

Organisation is when 'some of its constituents acting as constraints realize a regime of mutual dependence between them' (Mossio et al. 2016: 28). Montévil and Mossio (2015) call this *closure of constraints*, which is how a biological system stabilises and maintains itself. Organisational closure is a principle of stability accounting for the fidelity of organismic development (Mossio et al. 2016). As we shall see, this idea—of constraints restricting the degrees of freedom of processes—challenges common notions of causality and thus offers us a new way to think about the interplay of processes.

But let us take a step back: The basic definition of constraints is that they pertain to boundary conditions which selectively harness matter to perform functions (Pattee 1972). They often take the form of spatially distributed physical factors acting as organising principles (Bizzarri et al. 2019). Some of these conditions, such as gravity, are relatively independent of what they constrain, while others concern components which inhere in the organism, such as a cell membrane, and thus are more easily modulated. Furthermore, an organism as a whole might act as a constraint on other organisms (see next chapter).[5] Constraints channel the flux of energy and matter while remaining locally unaffected by the processes they constrain, 'at the time-scale in which it takes place' (Bich et al. 2020: 5).[6] They 'contribute to determine the behavior of the system (be it physical, chemical or biological), by reducing the degrees of freedom of the processes and dynamics on which they act' (Mossio et al. 2016: 28). Importantly, constraints do not only inhibit the processes; they canalise them to make possible constellations which would not be possible without them (Brigandt 2020).[7] We are therefore dealing with an alternative kind of causality,

[5] Constraints can be divided into two classes: '(a) holonomic (independent of the system's dynamical states, as being established by the space-time geometry of the field) and, (b) non-holonomic (modified during those biological processes to which they contribute in shaping)' (Bizzarri et al. 2020: 8). I speak only about the latter of these classes now, as most constraints are non-holonomic in living systems.

[6] Moreno and Mossio (2015) use the example of a riverbank. Although it is affected by the river it constrains, at shorter intervals, it is correct to characterise the riverbank as a constraint on the flow of the river.

[7] Such an understanding of constraints contrast with the critique that Lala et al. (2024) raise, but their critique is lopsided. They take for granted that constraints only denote restrictions, instead of

a regime of *enablement* different from the efficient causality of externally related processes.

Constraints are not static or given, but processes that change over time. While they are unperturbed by the processes they constrain (at the specific time interval of such processes), they change at other time scales, and are thus transient or historically contingent (Montévil 2020; Bich 2025). In living systems reaching closure, constraints depend on the dynamics on which they act.[8] This circulation enables them to *self-determine*, as their existence depends on their own activity and not an entity outside them. From this view—and this fits with Sultan's notion of a *genetic repertoire*—DNA is a constraint, a template which enables certain phenotypes instead of others. Like constraints in general, DNA is also a process subject to changes by its interplay with other processes. It is, in short, reused, or exapted, in multiple ways in different contexts (Longo 2019), in contrasts to the notion that it is a blueprint for development.

Besides organisational closure, the organisational approach highlights the differentiation of parts. This distinguishes biological organisations from dissipative systems (such as flames) where all the components contribute to a single macroscopic pattern. Differentiation is not the same as being composed of different material structures (this is also the case for flames) if all components in this system share function.[9] By contrast:

> Organizational differentiation implies not only that different material components are recruited and constrained to contribute to self-maintenance but, in addition, also that the system itself *generates* distinct structures contributing in a different way to self-maintenance. (Mossio et al. 2009: 826)

This means that functions can only be ascribed by considering the whole it contributes to—only insofar as the organisation maintains itself through differentiated structures that contribute to the overall organisation in distinct ways. Hence, by combining closure and differentiation the organisational approach offers a way of naturalising functions, of grounding their ascriptions (Cooper 2020). This indicates why teleology is not simply an analogy but based on the circular regime of the object itself, a point I make more forcefully in subsequent chapters. By contrast, a part of an artefact exists for the sake of the whole but does not depend on it for its further existence. Through organisation, livings organisms achieve a specific mode of self-determination and normativity, which can also be extended to ecosystems (El-Hani et al. 2024).

---

understanding the notion more 'dialectically' and in line with negativity, which is both positive and negative. But they do not engage with the literature on constraints, such as Pattee (1972) and the authors I cite above, and therefore miss how the notion is rather similar to 'bias', which is their preferred term.

[8] This is the difference between closure of constraints and closure of processes: For the latter, the constraints are external in the sense that they do not depend on the dynamics they constrain (El-Hani and Nunes-Neto 2020). Closure of processes is what is called operational closure in theories of autopoiesis (more in next chapter).

[9] Hegel uses the example of fire when speaking about why chemical processes are finite, unable to maintain themselves (see Stein 2024).

On Causality and Constitution

Longo, Montévil, and Kauffman note the obvious mistake of assigning causal agency to genes as if a mutation might by itself produce a specific phenotype:

> A classical mistake is to say: this mutation causes an idiot child [...], thus ... the gene affected by the mutation is the gene of intelligence, or: here is the gene that causes/determines the intelligence. In logical terms, it consists in deducing from "notA implies notB", that "A implies B": an amazing mistake. All that we know is a causal correlation of differences. (Longo et al. 2012: 14)

If you vary one factor and get a novel result, you may consider this factor the cause of the change. But such an ascription is inconclusive and assumes that causality is discrete and linear. At most, what is disclosed is that this factor is involved. The above authors argue that insofar as there are causes in biology, they are *differential.* A differential cause is not efficient, it does not acts directly on another object in a billiard ball manner; instead, a differential cause alters the space of possibilities for a specific organism-environment interaction. Longo and Montévil (2013) write, 'the differential causes modify the always reconstructed coherence structure of an organism, a niche, an ecosystem', which in turn modifies enablement, such that 'a niche may be no longer suitable for an organism, an organism to the niche' (12). Either natural selection culls away the modified organism or changes to the niche select against or enable certain organisms. In the latter scenario, where natural selection is not responsible for the change, it is not the physical modifications of the environment (leading to altered selection pressures) that matter, but changes in enablement relations. Likewise, causality does not reside in a contingent event—like a mutation—by itself, but in the differential relations that it may or may not affect. Only by modifying the space of possibilities of a system do events exert causal effects.[10] The information it embodies relies on a 'system of patterns' that 'assemble it into an organism' (Farnsworth 2022). If we think in terms of enablement, it is these patterns, differential relations, that are causal. We should keep the question of the causal status of relations in mind as we proceed.

A common argument against the systems perspective of irreducible emergent levels is that any systemic capacity must be founded on capacities that parts must also have, i.e. that the higher levels have no explanatory autonomy since the capacity must already be present in the part (Kim 2006). Explanatory autonomy, it is said, is premised on the system's causal autonomy—an autonomy claimed to be paradoxical, since it must be both dependent on its parts and autonomous from them. But this argument takes for granted that 'any autonomous emergent explanation must be a causal-mechanistic one' (Walsh 2012: 177), and that the former is reducible to the latter. Only if emergent properties are themselves efficient causes does the argument

[10] This is not to say that mutations cannot effect causal changes, only that they do not act in isolation. It also depends on the location and timing of the mutation, and they whole regulatory system in which the mutation occurs. As a general rule, concrete knowledge of the context, timing, position etc. in which it emerges is needed to gauge its effect. As Lala et al. (2024) say: 'the genome is not a code to be deciphered but a score to be interpreted' (78).

stick, but as we shall see, the story is not that simple—as the goals of organismic activities can be achieved by multiple mechanisms, which are constitutively co-evolving, and do not follow the strictures of efficient causality.

Causal analyses, according to the Newtonian model, demands clearly delineated particles or things acting on each other from the outside. But if we show that entities are just sliced out of more complex processes in which these distinctions do not hold, this model is weakened. Also, it is not the case that the system *causes* the underlying mechanisms to work in a specific way on a micro level; rather, the system regulates their behaviour (Walsh 2012). It is constraining their action, not strictly speaking causing them—even if constraints also mould causal relations over time (Pigliucci 2008). The idea that 'the development of a part is dependent on the development of the whole' (Hertwig 1892: 480) points to the difference between and causal and a constitutive relation: while the whole does not initiate the activity on the lower levels—since these levels have their own rules—, these rules are generated through the interplay of levels. Thus, there exists *constitutive* interlevel relations between them.

This undermines the argument that 'reflexive downward causation' is ultimately dependent on the activity of the subjacent processes. Downward causation 'does not require that the system introduces a bias into the parts' already existing causal repertoire' (Walsh 2012: 179), nor that their activity is ultimately caused by the emergent system. Instead, subsystems give rise to but are simultaneously constrained by the systems they enable. They are involved in systems of intertwined enablement relations. (I explore emergence more in Chap. 5.)

If constraints are processes and DNA is a constraint, it follows that we should not view genes as synchronic things, but as diachronic processes (Gilbert 2000). The same goes for constitutive relations: if we take for granted that they are synchronic in the sense that they form a static background on which diachronic and causal processes can act, we lose sight of the *causality of constitution* (Kirchhoff 2015). Non-linear, reciprocal, forms of causation such as co-constitution are hereby ignored (Lewens 2019; Buskell 2019). This is untenable if we adopt the notion of causation as changes in enablement relations, in which the whole organisation and its constitution exert causal power on its components. Even if evolution were to happen only through the selection of genes, this alone would tells us nothing about the process through which genes appear on the evolutionary stage. Since life existed before genes, they must themselves be the outcome of evolution—selected for their qualities as storage units (see Deacon 2012). It is thus an example of a *result becoming a cause*. We are dealing with processes; fixed results are abstractions from deeper dynamics (Dupré and Nicholson 2018).

This also seems to be Walsh's (2015) point when he says that we cannot know in advance whether something will have evolutionary significance.[11] It can only be

[11] He writes: 'As there is no distinction in kind between evolutionary characters and nonevolutionary characters—as there is no difference in kind between evolutionary changes and nonevolutionary changes—we should expect that there is no criterion of evolution. Typically, one cannot say of a

decided retrospectively. Furthermore, selection pressures are also processes—transient results becoming causes. Natural selection is not an independent force acting from the outside, immune to evolution itself. It is therefore arbitrary to begin the explanation with citing selection in the past, treating variation and selection as independent forces (Lala et al. 2024). We must instead account for how developmental interaction 'imposes limits on the directional action of natural selection' (Alberch 1982: 313).

Developing Niches

> Evolution occurs by changing development. (Gilbert and Bard 2014: 140)

As we have seen, if the fit between the organism and the environment comes down to a coincidental fit between genes and their environment, anything the organism does in its lifetime is cancelled out on an evolutionary timescale. Dobzhansky's process-view of adaptation endows the organism with more wiggle room but *ultimately* the capacity to actively adapt (adaptability) is written in the genes and displays latent possibilities already there—provided by previous cycles of natural selection. Waddington stuck to the MS view that mutations accumulate over time and constitute a hidden stock of possibilities. Perturbations release the 'cryptic variability' inscribed in the genes. Against the main tendency of the MS, however, he gave the organism a directive role in evolution. The leeway left by genes—that do not map onto phenotypes in a one-to-one manner—accounts for the robustness and plasticity of organisms. Such responses can later be stabilised by *genetic assimilation*, making the trait more faithfully reproduced. But this does not mean that something is evolutionary only when it has become genetically fixated; it only implies that it is inherited discretely and stably, which does not exclude other forms of inheritance that also display high degrees of stability (Fabris 2018).

The idea of 'nonrandom genetic mutations might not pose a particular challenge to the orthodox modern synthesis' (Huneman and Walsh 2017: 17) if the centrality of genetic factors is maintained. These authors argue that only biased phenotypic variation could pose a challenge, but only insofar as phenotypic plasticity is not reduced to an adaptation (genetically fixed and discrete) favoured by natural selection. We might have to ask, then, whether natural selection is an independent process or just the emergent result of other processes, which also means that it does not have a specific locus—the environment—but is the product of multiple interacting processes. How is natural selection itself produced?

Niche construction theory (NCT) underscores how natural selection is affected by other processes and how variation is therefore less random with regards to fitness than the MS holds. By viewing selection as an independent mechanism introducing

change in a population occurring at a time, whether or not it constitutes an evolutionary change. In general, evolutionary events can only be judged to be so in retrospect' (Walsh 2015: 240).

adaptive bias into evolution, the latter cast natural selection as a cause and not a consequence—whereas development is seen as the consequence of selection that cannot counteract it (see Walsh 2015). But this is misleading, insofar as evolution and development, selection and variation, are less separate than the MS assumes. Furthermore, natural selection might not be a cause for change—acting on variants produced by random mutations—but a predominantly stabilising force (Lecointre 2018; Dupré and Nicholson 2018; Alberch 1982). In this view, natural selection simply preserves 'pre-oriented options' (Ochoa 2024).

Proponents of niche construction claim that it is not subordinated to natural selection but an additional source of directionality in evolution. By itself, however, this does not mean that selection is not creative. It might exhibit, *sensu* Dobzhansky above, natural selection at its most creative, and is therefore still be within the realm of neo-Darwinism (Weber 2011). Nonetheless, it represents a move beyond the adaptationist programme, since it does not assume that the constructive activity is *an adaptation*—in the sense of a fixed trait—but a process:

> [A]ny mechanism of phenotypic plasticity in conjunction with reliably present signals from the environment can generate the same niche-constructing activity generation by generation, with evolutionary consequences, without that activity itself being an adaptation, or precisely specified by genes [...]. Mechanisms other than mutation can underlie evolutionary novelties [...]. (Laland et al. 2008: 552)

NCT is simple in its basic structure: It concerns how organisms actively alter their environment to further their existence. The environment provides the organism with some aspects which are modifiable and others that are not (some constraints which are more plastic than others, we might say). Some of these are reworked—intentionally or unintentionally—to lessen or redirect selection pressures. When it comes to those that are less modifiable (at least on the scale of one individual's life), the organism can alter *the effects* of these circumstances, by migration, dispersal, or habitat selection (see Odling-Smee et al. 2003). This allows them to *experience* another environment, as I will expand upon later.

The first example, in which the organism modifies its environment, is called *perturbation*, the latter are examples of *relocation*. Furthermore, these categories are divided into inceptive and counteractive construction, which is when the organism initiates a change or reacts to a change in its niche (Odling-Smee et al. 2003; Laland et al. 2009). They denote active and reactive processes. There are thus four basic kinds of niche construction in the original model: the active and reactive versions of perturbation and relocation. The activity of the organism causes changes to the environment, modifies selection pressures, and thus ultimately explains genetic changes in populations. Other phenotypes could be selected as an effect of changes in the environment reaching a threshold at which they cascade to gain evolutionary significance.

NCT should not be confused with the extended phenotype model proposed by Dawkins (1982). He considers niche construction a consequence of other evolutionary processes but not as a cause, e.g. as explanatory (Laland et al. 2008). In Dawkins' view, the environment is *not* changed by the organism but by environmental states only, even though evolutionary change depends on *both* the state of the organism and of the environment. The importance of niche construction is hereby diminished, explained through natural selection in the past (Uller and Helanterä 2019). This discrepancy was pointed out by Lewontin (1983) and called *externalist* by Peter Godfrey-Smith (1996)—since the external environment is the only 'explanatory reference device'. The organism's internal properties are caused by the external environment. As soon as they have been internally fixed, they work like a programme, awaiting specific external cues which trigger it.

By contrast, niche construction proper considers the causes of development not as determinants but *affordances*—opportunities for action—for the organism to reinforce or overturn (Laland et al. 2008). This dovetails with Sultan's notion of the genotype as a repertoire for organisms to exploit. It also means that development is not only permissive, i.e. allowing genetic factors to unroll, but plays a constructive role by causally *specifying* the course of evolution (Griffiths and Stotz 2013). Hence, NCT says that the niche of an organism is not an independent variable but tied up with the functioning of the organism itself. This is a crucial principle for a dialectical approach to evolution. Not only does an organism physically modulate its surroundings, but it can also take measures that change the causal impact of the niche. We may call this its 'effective niche', which is irreducible to the physical ambience of the environment (Sultan 2015).[12] An organism can relocate to a different niche, but it can also create pockets within its current niche which are more beneficial to its maintenance. What is more, phenotypic changes which do not alter the environment may change the way the organism *experiences* the niche, and thus affects 'its selective impact on the organism' (Sultan 2015: 10). This suggests an extension of NCT which we return to in Chap. 7.

There is an important distinction here: whereas NCT concerns the *selective* niche, developmental niche construction is about the *developmental* niche. As Griffiths and Stotz point out: when the water flea *Daphnia* signals to its offspring to grow defences against predators, this means that

> the parent Daphnia is structuring the developmental environment of its offspring. But this is no more a case of selective niche construction than is the inheritance of an advantageous mutation! The Daphnia embryo alters itself to fit the selective environment rather than altering the selective environment. (Griffiths and Stotz 2018: 237)

Developmental niche construction explains how parents reliably influence the phenotype of their offspring to promote healthy development, which is another example of a non-random process—and a specifying cause of normal development. This is not about modifying how selection influences the reproductive success of the current generation, however, but concerns exploiting the ecological information—which emerges from the organism's interaction with the niche—to secure stable development. Here, we again see how developmental factors instruct the course of evolution, and are not without expressive power.

---

[12] This is similar to G. Evelyn Hutchinson's (1957) distinction between fundamental and realised niche, where the latter is a subset of the former.

The differences between these forms of niche construction should not be overstated (see Uller and Helanterä 2019). Developmental and selective niche construction share parameters, even if they are conceptually dissimilar (Griffiths and Stotz 2018). Whereas the latter is focused on altered selection pressures—the parameters that decide selective fitness—, the former is restrained by the conservation of the life cycle that undergirds normal development. It also pertains to inheritance, something the original formulation of NCT did not. Otherwise, niche construction could be limited to the *current* population, not its progeny.

The larger point is that developmental niche construction blurs the distinction between construction and adaptation since the example above shows signals from the parent 'that induce transgenerational adaptive phenotypic plasticity' (Griffiths and Stotz 2018: 237). (The significance of undermining this distinction will be brought out later.) Also, combining selective and developmental niche construction troubles the distinction between development and evolution. If niche construction has effects going beyond one generation—revealing non-genetic inheritance not produced by natural selection, we see why it poses a stark challenge to the MS. Not only does it heighten the probability of survival, but it also induces heritable phenotypic changes that are transmitted non-genetically (Uller and Helanterä 2019).

Dynamical Systems and Their Limitations

> On the one hand, dynamical systems theory is a powerful approach to study the behaviour of complex regulatory systems. On the other hand, traditional dynamical systems theory is fundamentally limited when dealing with the co-constituting dynamic structures of living systems. (Jaeger 2019)

Most biological systems—indeed most systems that exist in the universe—are complex nonlinear systems, where a given input does not yield a proportional output. They are non-additive; the parts of the system cannot be added up but display emergent properties. In other words, the parts change in and through their behaviour, and cannot be isolated from the system and studied by themselves. Dynamical systems theory (DST) is often used to visualise this, but it has limitations when it comes to living systems, as I will show.

DST analyses attractors, which are changing patterns of behaviours which are interdependent. Attractors form as state space, consisting of basins and hills that make out a space of possibilities. Such spaces are *emergent collectives* (see Juarrero 2010).[13] They are formed by attractor landscapes around the different attractors and

[13] The most common is the point attractor, a single point in the space of possibilities. We can think of it as a simple trait, like eye colour. Besides the simplest attractor, a specific point or state, some attractors display limit cycles or loops, oscillating either in a regular interval (of stable) or spiralling away if unstable. This is a periodic attractor. Finally, you have Lorenz's infamous strange attractors, from chaos theory, in which the trajectories of the attractor might diverge greatly based on minor differences in initial conditions. This is the butterfly effect (see Kauffman 1993).

repellors—deep, shallow, broad, or narrow. An attractor is a state or point towards which the system converges but never reaches (an asymptote) (Kauffman 1993). They vary in their degree of complexity—an index of interconnectedness—making some systems stable and others unstable. Basins represent all the initial conditions which converge on the attractor, the area where its attractive force works. Their depth and width indicate the variance (width) and the stability (depth). At some point, a system can become rigid, trapped in certain attractor basins. At other points, the landscape evens out, and multiple basins become equally deep or attractive, i.e. equally *plausible.* In these periods, rapid nonlinear changes may occur, even if the perturbations are minor.

A *phase portrait* predicts the direction in which the system might be perturbed at a certain point in time. It defines the *dynamical repertoire* of the system, a *space of possibilities* (Jaeger and Monk 2014). The stability of the system depends on the number of possible attractors it exhibits, their width, and the depth of their basins.[14] Hence, there is a trade-off between stability and capacity for change. The structural stability of a system is defined by 'the persistence of a specific attractor and its associated basin over a large range of changing parameter values.' It '[u]nderlies the robustness and canalization of biological regulatory processes' (Jaeger and Monk 2014: 2279). A stable point is represented as a valley and an unstable point as a mountain top (Kauffman 1993). Combined to form a topology, it visualises how an evolutionary system must find the right balance between stability and change in order both to maintain itself and respond to changes in its conditions of existence.

The overarching features of DST are defining a state space, examining the importance of regulatory processes, and explaining how and why systems exhibit robustness in the face of perturbations (Jaeger 2019). We saw this emphasis on regulatory constraints in organicism as well. The causal factor of most importance at a certain time is a function of its constitutive interplay with other factors. It is neither self-caused nor static, but dynamic and regulated. For example, niche construction will impose itself on other processes and have more influence in some cases; in others, it will not, depending on the overall state of the system. The interactions between the attractors are responsible for the *differential causation* of the event.

This undercuts a characteristic feature of the MS, namely its belief that the processes of evolution are relatively independent and thus possible to study in isolation. This view is combatted by *entangled causation* (Vecchi et al. 2019), where assigning causal power to one factor over another is deceptive since DST 'in its multi-causal, multilayered, historical approach sees biology and environment as continually meshed and inseparable' (Thelen and Smith 2006: 305; also Lehrman 1953).[15] In this view, there is no privileged level of causality. As there is not enough causal specificity

[14] Width has to do with specificity: the broader the attractor basin, the more complex the phenotype. If the basins are deep, it means that it takes more effort to change from one attractor to the other and that a change to another attractor would be more drastic (Thelen and Smith 1996).

[15] It does not mean that any cause is as specific as another. Vecchi et al. (2019) opine that since DNA could be more stable and causally specific than other factors there is not absolute parity between causes. It is, however, doubtful that this can be conclusively ascertained due to complex interactions between organisational levels.

in the initial conditions to specify the trajectory of a biological system in advance, the information is not given but produced through the activity of the organism.

As we move to the fundamental limitation of DST, my purpose is not to criticise it specifically, but to point to a more general problem; namely, that a historical system cannot be predefined. It is not only that you cannot calculate the probabilities, but you cannot even know the possibilities of the state space. The unpredictability of the state space itself is one of the main motifs of the organisational approach, as it brings out the critical historicity of living systems, and how their possibilities are not positively given in advance (Longo and Montévil 2013). Johannes Jaeger writes that in such cases

> the very existence of the system and its configuration space become mutually dependent. This type of co-dependence goes far beyond mere mutual causation between two separately existing processes [...]. Neither the organism nor its configuration space exist without each other. They are dependently co-originated or co-constituted in a dialectical relationship through the autonomous activities of the living system. (Jaeger 2019: 345)

Drawing on the dialectical perspective of Lewontin and Levins, Jaeger argues that biological systems are self-determining—meaning that 'the rules of change are immanent to the system' (Jaeger 2019: 345). It is not that probabilistic analyses are irrelevant, but we should be aware of their inherent limitations. Such analyses require that the possibilities of the system are known in advance, as that is the only way to measure their relative weight. In living systems, however, the probabilities are actively maintained by the organisation in question. They are not derived from theoretical principles. As Montévil (2018) writes, 'When relevant, probabilities and possibility spaces in biology are defined by constraints and should thus be interpreted as the result of an active stabilization' (379). So, the validity of the probabilistic framework cannot be presupposed but depend on the activity of organisms—and is thus not applicable to situations where the constraints are overturned, and evolutionary changes occur. Again, a diachronic perspective is necessary.

Attractors constrain the possibilities of a system, by making certain states less likely to occur. They represent stable states and give rise to landscapes of such states. Moreover, attractors at any level might be constrained by attractors at other levels. But an attractor landscape could be overturned by something which could not be predicted from the current landscape. This is called a *critical transition* (see Longo et al. 2012). These transitions occur when a bifurcation point is crossed and establishes new structures.

According to Longo et al. (2012), biological systems undergo such transitions continually. They write:

> [A]s for the pertinent observables, the phenotypes, we propose to understand evolutionary trajectories as cascades of critical transitions, thus of symmetry changes. In this perspective, one cannot pre-give, nor formally pre-define, the phase space for the biological dynamics, in contrast to what has been done for the profound mathematical frame for physics. (11)

We may understand, then, why the proponents of the organisational approach propose variation as a theoretical principle of biology. They argue that biological organisations are characterised by variability—undergoing constant symmetry changes—, unlike physical systems which are mostly inert. As such, biological systems display randomness, historicity and contextuality. They occupy a space between chaos and order, or between flexibility and rigidity. If the regime were too ordered, it would be too rigid to react to changes; if it is too disordered, it would not allow any reliable response (Roli et al. 2018). Poised at *the edge of chaos*, we cannot know in advance which states might be occupied by the system based on the current phase space. This introduces a radical contingency into the heart of biology.

Expansion, Extension, or Rejection?

> We have proposed to invert the epistemic strategy of physics. Physics understands changes by invariance: the equation and their invariants describe changes of states but do not change themselves. By contrast, in biology, we argue that variations come first, and that (historicized) invariants come second. (Montévil 2021: 12)

I have yet to formally introduce the core facets of the EES, but we have touched upon some of them already. As noted, the calls for extensions coincided with the formulation of the MS itself. And as we shall see, the scope and radicality of the alternatives are not settled. To put it bluntly, they have mostly involved adding factors such as multilevel selection, epigenetic inheritance, niche construction, and evolvability to the MS. Common to most of them is putting the organism and its purposive actions at the centre of evolution once more. There are multiple ways of conceptualising this centrality, of course, and I have focussed on the model I find most promising. With the re-centring of the organism comes the de-centring of other factors, such as genes. The role of natural selection is more contentious, as many would still consider it the dominant feature in evolution but add that it does not only act on the level of genes and that it is not creative. I now get clearer about the challenges to the MS before concluding.

Depew and Weber (2013) depict three scenarios for the replacement of the MS: expansion by way of recognising that natural selection acts on multiple levels; extension, by including neglected processes; and rejection and replacement with a new framework. The question often boils down to what should be considered causes and results in evolution. The EES brings out the incomplete ontology of the MS, by incorporating more evolutionary processes into its gamut. Even if there is no consensus about the alternative(s) to the MS, we do find commonalities. Depew sums up the topics we have already discussed when he writes:

> To extend the synthesis, it seems, requires throwing something out and freeing what remains from the influence of the dogma that evolution reduces to natural selection considered as a mechanistic process of optimization ranging over random mutations in self-replicating molecules. This conception of natural selection does scant justice to variation's many sources, natural selection's many modes, the roles of nonselectionist processes in evolution, and the telic nature of agency-enhancing adaptation. It does even less justice to human behavioral evolution. (2017: 59)

Could it be, then, that the very notion of natural selection should be discarded? According to Walsh (2015) many seeming challenges to the MS do not strike at its core. But the fact that many of the ideas put forward by the nascent alternatives to the MS may well be consistent with essential facets of the MS does not mean that they are part of it (cf. Pigliucci and Müller 2010). As we have seen, one strategy is admitting the existence of a phenomenon, while restricting its impact, as was done with drift and multilevel selection. In other cases, one acknowledges processes like epigenetic inheritance—but adds the caveat that they are too labile to count as ultimate causes. Against this, we could argue that lability is not necessarily a weakness since it makes rapid changes in response to environmental fluctuations possible. Furthermore, even if they are short-lived, their effects might not be. They may 'strongly affect phenotypic variation and fitness and therefore infect the course of natural selection' (Danchin et al. 2019: 10).

An example of something going beyond the MS without challenging its core is recasting development as an active process, as Dobzhansky did. Here, development involves organismic activity but is regarded as proximate—cut off from selection and inheritance. Even if it afforded more explanatory power, it does not count as evolution proper. In line with Depew (2017b), we might need to challenge the notion 'that trans-generational natural selection is the cause of the adaptations that organisms considered as members of adapted populations possess' (29) since this is a central conceptual boundary it cannot cross.

Extensions are made more challenging by the fact that the MS has already undergone several revisions, mapped above. Take Mendel's discovery of how discrete traits were not blended in beans, and how it influenced our current view of selection:

> The idea of "natural selection" invoked by scientists today has been reconfigured to fit a world of discrete inheritance [...]. In such a world, it is natural to imagine that change might begin with some rare trait that confers a small advantage, e.g., 2%, so that individuals with the trait tend to leave more offspring that, in turn, inherit the 2% advantage and leave more offspring. Over time, the trait increases in frequency until it prevails. To a contemporary reader, this is "natural selection". (Stoltzfus and Cable 2014: 510)

This view contradicts Darwin's, who believed that blending occurs from generation to generation, e.g. that there is no discrete inheritance of traits but a loss of identity. Consequently, describing contemporary mainstream biology as fully aligned with Darwinism 'shows disregard for scholarly rigor' (Stoltzfus 2017: 6). Are we opposing Darwinism and Mendelianism?

This points to a more general problem concerning the debate about the MS; namely, a lack of historical awareness or biased descriptions. Again, treating the MS a coherent research tradition might be setting up a strawman; and there is, according to Smocovitis (2023), something ahistorical about the whole debate, making the stakes involved overstated, as current debates resemble those of the 80s. These circumstances make calls to extend the MS increasingly difficult to pin down.

Noble (2021) opts for a rejection of the MS, claiming that the MS 'has been undermined, not supported, by molecular biology' (21). Tim Lewens (2019) asks whether the success of the theories that run against the MS and the degree to which they have already been integrated into the existing framework means that calls for extensions are superfluous. It could be argued that they are victims of their success, co-opted by a research programme that they meant to undermine. Are these examples abundant, and do they constitute a proper challenge to the MS? How much of a change in how biology is theorised and practised is needed for the EES to become reality, and what would constitute success for the proponents of the EES?

Lewens distinguishes between what he calls radicals and accretionists. On the one side, the accretionists accept the core features of the MS, searching for additional causes of evolution but not disputing its overall functioning. This would entail extension in the weak sense of enlarging the scope of natural selection and including more processes in evolutionary explanations. If this is what the EES amounts to, it would have largely succeeded already, and no fundamental reforms are needed.

By contrast, the radical camp does not simply add mechanisms and processes to the MS view, while leaving the basic functions intact. They seek fundamental revisions of the MS. In Lewens view, Laland et al. (2015) are representative of this view since to them the EES 'is not just an extension of the MS but a distinctively different framework for understanding evolution' (Laland et al. 2015: 3, quoted in Lewens 2019: 708). The radical proponents of the EES state that nothing less than a new synthesis is needed—that the MS is not plastic enough to take the phenomena and processes they study into account. They opt for rejection and replacement.

These calls for extension seem to mean different things: first, it concerns the acknowledgement of perspectives and disciplines neglected when the MS was formulated—like embryology and palaeontology; and, secondly, the integration of these and other disciplines that have been regarded as separate and independent even when they were recognised. The former only seemingly challenges the MS framework. It is an additive approach, stating that the extension is just a matter of including more factors, not undermining the main tenets of the MS. Some hold, for example, that genetic, epigenetic, cultural ecological inheritance mechanisms and parental effects are not opposed but complementary. All these processes favour adaptation on different timescales, making organisms able to accommodate environmental changes (Danchin et al. 2019). But since genetic factors are still privileged and seen as largely unaffected by the other processes, it does little to take us beyond the MS.

The radical position attempts to 'de-fragment' the MS and question the causal decomposability of processes. This might mean, among other things, that not only the creativity but also the causal primacy of natural selection is at stake. In any case, both sides call for the integration of processes often treated as separate and acknowledge that evolution is more complex than the MS made it out to be. While there are examples that might indicate how the MS is already being undermined from within by research going against the grain, insofar as such findings have yet to be systematically integrated with discoveries in associated disciplines, they do not comprise an extended synthesis, even if they might constitute necessary steps towards it. Some argue that these developments are more on par with Darwin's original theory and curtail Mendel's influence (see Walsh 2015).

An ambitious victory for the radicals is improbable, says Lewens, as there are many different strands of the critique of the MS that needs to be agreed upon. Nevertheless, the mere fact that researchers are coming together under the EES heading, and that their work is acknowledged by mainstream biology, is already a success, in his view. Insofar as they outline a research agenda that differs from the MS, and work as a collective, there are reasons to celebrate, even if the fight is not (and may never be) over. Lewens concludes:

> Even when this happens, the very malleability of the tools we use for evolutionary enquiry offers accretionists multiple ways to accommodate—and even to co-opt—insights and approaches generated by those working under the banner of the EES. These are victories for the EES as such, for they are victories that are the distinctive product of a coming-together of researchers working on the themes stressed by EES enthusiasts. But they are victories that can be celebrated by all evolutionists. (Lewens 2019: 719)

Accordingly, we should not expect some final blow to the MS. Lewens also notes that experimental constraints are part of the story. Scientists pose questions they can answer. A lack of methodological tools may explain why some processes were neglected and have only recently come to the fore. For example, 'technological advances in the past decade have revealed the incredible degree of plasticity in gene expression in response to diverse environmental conditons' (Laland et al. 2014: 163; see also Sultan 2015). These advances have pushed 'fringe' theories into the mainstream, accepting them at the high table. The question of how many changes must be made to the MS to accommodate new insights must be kept in mind—even if it might be too early to decide this (Sultan 2015: 141).

How malleable are the tools provided by the MS? Is it, for instance, possible to argue for functional context-dependence without fundamentally recasting the role of genes?

Another take is represented by Gawne et al. (2018), who suggest that the debate is not about the MS at all. Rather, it is

> primarily about the extent to which genetic mechanisms suffice to explain phenotypes. In simple terms, researchers intent on extending the synthesis are of the belief that factors other than DNA sequences play an important role in the production of phenotypes. [...] The goal of the movement is to obtain an integrative understanding of developmental and evolutionary change that acknowledges the importance of mechanisms and processes, which according to its proponents, were not appreciated until recently. (Gawne et al. 2018: 2)

Also, the proponents of the MS do not really defend the MS; instead, they focus on showing that the claims of the advocates of the EES are false:

> In doing so, they have often shown support for ideas that were developed during the early 1900s, but it remains unclear whether these efforts constitute a defense of some systematically organized body of thought we might meaningfully refer to as the field's "modern synthesis." (Gawne et al. 2018: 2)

This raises the question of whether the MS was ever a coherent research tradition, an issue that I return to in the conclusion. Simultaneously, the extended synthesis has yet to be 'consistently or coherently defined', but if it is correct to say we may move toward the EES simply by acknowledging the full implications of the MS, perhaps we understand why it is so difficult to formulate a proper challenge to the MS. It is caused by the difficulty of defining the MS. Perhaps it is correct to say that what the debate simply about the relative importance of genetic factors.

Gawne, McKenna, and Nijhout emphasise the curious fact that both sides of the debate 'agree that genes are the ultimate proprietors of phenotypes' (2018: 2). While the proponents of the MS state this explicitly, it is implicit 'in the claim that phenotypes are the result of processes such as genotype x environment interaction and epigenetic modification of gene activity' (Gawne et al. 2018: 2). They suggest that overcoming the MS means breaking with this gene-centrism.

While it is a start, I wonder whether this is enough. If what they say is the case, the alternatives to the MS seem to aim at amending some of the more controversial aspects of the MS while leaving the core intact. If it is mostly about the relative weight of genetic factors, it seems more aimed at mutationism that the MS as such. It leaves the primacy of natural selection unscathed, even if it might admit that it acts on more than the genetic level. According to Ochoa (2024), the creativity of natural selection is even maintained, even if more sources of creativity are added. Moreover, although they place less emphasis on genetic factors, 'the advocates of the EES still support the gene-centered perspective when the concept of "inheritance" is mentioned' (Ochoa 2024: 432).[16]

In any case, Stoltzfus (2017) warns that we might end up replacing the grand theory of the MS with another grand theory. In his view, '[t]he era of master theories based on ruling principles and grand schemes is long past' (6). We should thus be wary if the proponents of the EES propose it as another general theory of evolution. So, how can the EES constitute a productive alternative to the MS? Should we opt for another synthesis, and if so, what do we mean by synthesis? We return to these questions in the concluding chapter.

Conclusion

> Ideas do not have to be correct in order to be good; it is only necessary that, if they fail, they do so in an interesting way. (Rosen 1991: 248)

Part of a radical challenge to the MS means adopting the idea sketched earlier, of epigenesis both as something intra-organismic *and* as happening between and across organismic boundaries. Here, epigenesis is not merely about molecular events then but also always about the interaction between an organism and its environment

[16] '[R]eformists argue that they are organism-centered rather than gene-centered, but apparently, they do not deny that genes are the main system of heredity. Rather, they merely include other entities that are "heritable": "Heredity extends beyond genes"' (Ochoa 2024: 424). In Ochoa's view, the notion of replication, in which genes are the units of stability, must be replaced by a modular way of thinking, in which developmental constraints ensure a trait (e.g., a limb, an eye) develops consistently across generations, even if the underlying genes change (see also DiFrisco et al. 2020).

(Gilbert 2002). Gilbert observes how epigenetic 'interactions include those between DNA and proteins, between neighboring cells, among tissues within the body, and between the body and its environment' (Gilbert 2002). The latter of these examples represent a broader view of epigenesis in which epigenesis 'constrains the patterns of variation in phylogeny' (Oster and Alberch 1982: 444).

Regarding development, not selection, as the main creative factor in evolution, would constitute a radical break with the MS. Not only does it strike at the gene-centrism of the MS, but it also challenges the primacy of natural selection as *the* instigator of change by emphasising the role of development and ecology in the production of novelty.[17] Gilbert relates this to Hertwig, who argues for 'the extension of epigenesis from interactions between cells of the embryo to the interactions between developing organisms and their respective environments' (Gilbert 2002: 204). A shift of focus, from fully formed organisms—interacting externally with a given environment—to symbiosis, the co-evolution of species, is suggested by this move. We turn to this topic in the next chapter.

An example of a change in thinking is discarding the idea that every phenotype is adapted to its environment. The reason is that the word 'adaptation' means that something is formed by natural selection to (*ad*) a specific function, and we now have countless examples of traits that were selected for a specific purpose but later co-opted for something else. We also have instances of non-adaptive or maladaptive traits being selected indirectly, as part of a larger developmental system (Lala et al. 2024).

Pigliucci and Kaplan (2000) enumerate different aspects of this shift away from adaptation and selection towards constraints: (1) the trait might be a consequence of genetic drift, in which neither selection nor adaptation is operative; (2) indirect selection: a trait comes about as an indirect effect of being associated with another trait (leaving it undecided whether this trait was a target of natural selection); (3) selection without adaptation: a trait that increases in frequency without being adaptive to the population, for instance if resources are sparse and a mutation that increases fecundity comes about (Lewontin's example); (4) adaptation without selection (plasticity-driven adaptation), when behavioural flexibility produces the trait (even if the plasticity might itself have been selected for, the trait is not the direct outcome of selection); (5) adaptation and selection but multiple adaptive peaks make it impossible to distinguish between the different forms of a trait based on adaptiveness alone; and (6) exaptation: adaptation and selection, but the adaptation in question is a secondary use (re-functionalisation) of a trait that was previously selected for other reasons.

---

[17] This seems to align with Ochoa's (2024) notion of *unmodern* synthesis, and his critique of the EES as not really breaking with the MS since it maintains that natural selection is a creative force, and that it also maintains the gene-centrism of the MS, as well as the extrapolation from microevolution to macroevolution. While there is something to it, I think his critique only applies to some of the proponent of the EES, as this book testifies. He furthermore argues that the latter dimension, that of extrapolation, might be the most intractable, as it seems to require that we adopt some kind of saltationism. While I do not discuss this matter directly, I think we find resources—also for overcoming the latter hurdle—in the authors I deal with, especially Lala et al. (2024).

Based on this inconclusive list, let us return to adaptationism before ending. To be sure, some of the above alternatives to adaptationism overlap. The differences between exaptation, plasticity-driven adaptation and indirect selection seem minuscule. The first includes both adaptation and selection but concerns derived effects of this interplay; plasticity-driven adaptation, on the other hand, implies that the trait could be a secondary effect of selection, but in this case not for a specific trait but overall plasticity (or evolvability); indirect selection entails that one trait is produced as the by-product of another or by being part of a larger system, which is the direct target of selection. I do not bother much with the exact differences (and possible theoretical confusion) between these processes presently, as the main point of Gould and Lewontin's critique of adaptationism still stands: We must account for other mechanisms than gradual natural selection producing adaptations if we want a more comprehensive theory of evolution. Moreover, we must explicate not only 'how a mechanism operates here and there, but [...] the ontological production of that mechanism' (Drake et al. 2007: 164). In short, we must explain the historical constitution of mechanisms, and not simply by reference to further mechanisms.

The idea formulated by spandrels is that constraints can effectuate what is produced by evolution, encapsulated by Terrence Deacon (2012) as 'evolution by progressive constraints'. These might of course be caused by natural selection but are produced by processes which are not directly selected for, and side-effects that are not themselves adaptive—i.e. do not provide more fitness in the environment in which the organism finds itself. This entails that selection is indirectly constructive, through the constraints it imposes on other processes. It also suggests that selection is itself constrained, both by the variation provided by processes outside its remit and by the niche constructive activities of organisms. Jacob spoke about 'evolution as tinkering', in which 'natural selection only works on the materials available and within the constraints present at a particular time and place' (Pigliucci and Kaplan 2000: 68). The *bricolage* nature of recombination means that not only does natural selection impose constraints; its creativity is also constrained by the sources of variation.[18] Moreover, the co-production of constraints makes it impossible to decide where constraints *ultimately* come from.[19]

Variation seems to be less random than presumed by the MS: not only can stress induce changes in the direction and rapidity of genetic mutations, but we must include other forms of non-genic variation as well, which cannot be assumed to be random by default (McClintock 1984). Hence, the adaptive fit is not produced by selection alone (even if it might channel it), and environmental selection pressures are plastic.

---

[18] Stoltzfus (2012) says that the notion that natural selection is constrained only makes sense for the Darwinian conception of selection as a creative factor, not for the notion of selection as (Mendelian) frequency-shifting force or as a stochastic sieve. As for the eliminative notion, he says 'invoking the absence of a variation as a constraint or limit is nonsensical, like saying that the absence of mass is a constraint on the force of gravity' (3).

[19] Gawne et al. (2018) imagine the sceptic's retort to their emphasis on hierarchies in development: '"Sure, genes themselves don't produce phenotypes but the regulatory systems you just described do, and they're under genetic control"' (3). This retort begs the question by presupposing that development is conservative.

While natural selection might be creative, is not omnipotent. The variation it acts on might be partly produced by it, but it is also constrained by other processes, which 'impose limits on the direction and action of natural selections' (Alberch 1982: 313). consequently, natural selection cannot produce whichever organism it wants, as emphasised by Gould and Lewontin (see Depew and Weber 1995).

Guillaume Lecointre (2018) argues that natural selection is primarily a source of stabilisation. In his view, natural selection only a source of change in periods where the environment changes significantly. Overlooking this role of natural selection led biologists to search for other sources of biological stability. This principle 'was found in a metaphor imported from cybernetics to biology, *that of the genetic program*, and its corollary, *genetic information*' (Lecointre 2018: 143, emphases original). But this is not a solution since it does not explain where the programme comes from or how it is actualised.

Another source of information is the environment: Gilbert and Epel (2015) claim that environments are not only selective; they are also *informative* (also see Gilbert and Epel 2009). They provide ecological information, information for affordances—which is not some substance objectively present in the environment, but the product of interaction between organism and environment (Heras-Escribano and de Jesus 2018; Pharoah 2020). It is 'the result of a combination between the physical forces of the environment and the active, exploratory character of organisms' (Heras-Escribano and de Jesus 2018: 254). We already touched upon this topic when we spoke of the cues provided by the developmental niche. Similarly, as Sultan points out:

> This notion of a neutral environment, as well as the research strategy that notion dictates, loses its utility once it is recognized that environmental conditions also provide critical, precise developmental information that is implemented via specific effects on gene expression. Accordingly, environmental and genetic factors play similar roles as determinants of development [...]: both environmental and genetic factors inform—that is, lend form to—phenotypic outcomes. In other words, there need be no qualitative distinction between the type of information provided to the developmental system by genetic elements and that provided by environmental factors [...]. (Sultan 2015: 18)

The MS has largely disregarded this symmetry between forms of information and ended up interpreting environmental input as noise—irrelevant to the study of organismic development, and simply there to trigger programmed reactions. It has obfuscated the direct, inductive, role that the environment plays in evolution, a role that it plays irrespective of the agents of selection (Reid 2009). As we shall see, this means that organisms do not have to sit around and wait for natural selection to act to say that evolution has occurred. Adaptability—a source of novelty that provides the organism with the capacity to respond to changes—is conferred independently of natural selection. It might be the lessening of selection that produces such capacities. In any case, natural selection stabilises but does not drive such evolutionary changes.

Several alleys are open to the critics of the MS at this juncture: They could discard the two-stage model (of variation preceding selection) and admit that natural selection is open to influences such as the niche construction work done by organisms; they could argue that there is evolution without natural selection, for example of self-organised or autopoietic systems, keeping with the two-stage model but undermining the importance of selection; they could accept that variation is blind with regards to adaptive value, keep the basic model and just extend variation to include other modes than genetic mutations (and thus render it less blind and more plastic); they could highlight how plastic responses during development can later be stabilised by natural selection, and thus initiate evolutionary changes, etc.

The last option introduces another two-stage model: here, the *adaptable* responses of the organism happen without natural selection and are stable enough to persist over generations, before non-plastic variation may later take over to stabilise the trait (genetic assimilation) (Depew and Weber 2011). It is still, however, possible to claim that this capacity was itself an adaptation furthered by natural selection in the past; but as I have tried to show, this is not only arbitrary but quickly leads to an infinite regress. A notion of reciprocity or circular causation should instead be introduced.

The question is whether any of these options go far enough to break with the MS; for example, is it enough to include more modes of inheritance or is a complete redefinition of the concept of inheritance needed? Merlin (2017) says that if one opts for the former, one still accepts the discreteness of the different modes, whereas the latter means that inheritance also is relational or distributed. Unfortunately, I must cut this discussion short as we will only have the means to answer these questions after I have developed the organisational position further. To embark on this task, I turn to Donna Haraway's dissertation on twentieth-century organicism and her critique of autopoiesis. This will take us further into the history of organicism and its modern iteration in the organisational approach, and reveal how closure of constraints is not about complete closure but is instead a tendency towards it.

Bibliography

Alberch, Pere. 1982. Developmental Constraints in Evolutionary Processes. In *Evolution and Development: Report of the Dahlem Workshop on Evolution and Development Berlin 1981, May 10–15*, ed. John Tyler Bonner. Heidelberg: Springer. https://doi.org/10.1007/978-3-642-45532-2_15.

Bich, Leonardo. 2025. *Biological Organization* (Elements in the Philosophy of Biology). Cambridge, UK: Cambridge University Press. https://doi.org/10.1017/9781009393959.

Bich, Leonardo, Matteo Mossio, and Ana M. Soto. 2020. Glycemia Regulation: From Feedback Loops to Organizational Closure. *Frontiers in Physiology* 11. https://doi.org/10.3389/fphys.2020.00069.

Bizzarri, Mariano, Douglas E. Brash, James Briscoe, Verônica A. Grieneisen, Claudio D. Stern, et al. 2019. A Call for a Better Understanding of Causation in Cell Biology. *Nature Reviews Molecular Cell Biology*. https://doi.org/10.1038/s41580-019-0127-1.

Bizzarri, Mariano, Oleg Naimark, José Nieto-Villar, Valeria Fedeli, and Alessandro Giuliani. 2020. Complexity in Biological Organization: Deconstruction (and Subsequent Restating) of Key Concepts. *Entropy* 22 (8). https://doi.org/10.3390/E22080885.

Brigandt, Ingo. 2020. Historical and Philosophical Perspectives on the Study of Developmental Bias. *Evolution and Development* 22 (1–2). https://doi.org/10.1111/ede.12302.

Buskell, Andrew. 2019. Reciprocal Causation and the Extended Evolutionary Synthesis. *Biological Theory* 14 (4). https://doi.org/10.1007/s13752-019-00325-7.

Charlesworth, Deborah, Nicholas H. Barton, and Brian Charlesworth. 2017. The Sources of Adaptive Variation. *Proceedings of the Royal Society B: Biological Sciences* 284 (1855). https://doi.org/10.1098/rspb.2016.2864.

Cooper, Andrew. 2020. Do Functions Explain? Hegel and the Organizational View. *Hegel Bulletin* 41 (3): 1–18. https://doi.org/10.1017/hgl.2020.14.

Danchin, Etienne, Arnaud Pocheville, Olivier Rey, Benoit Pujol, and Simon Blanchet. 2019. Epigenetically Facilitated Mutational Assimilation: Epigenetics as a Hub within the Inclusive Evolutionary Synthesis. *Biological Reviews* 94 (1): 259–282. https://doi.org/10.1111/brv.12453.

Dawkins, Richard. 1982. *The Extended Phenotype: The Long Reach of the Gene*, Popular Science from Oxford. Oxford, UK: Oxford University Press.

Deacon, Terrence. 2012. *Incomplete Nature: How Mind Emerged from Matter*. New York, NY: W.W. Norton & Co.

Depew, David J. 2017a. Darwinism in the Twentieth Century: Productive Encounters with Saltation, Acquired Characteristics, and Development. In *The Darwinian Tradition in Context: Research Programs in Evolutionary Biology*, ed. Richard G. Delisle, 61–88. Cham: Springer.

Depew, David J. 2017b. Natural Selection, Adaptation, and the Recovery of Development. In *Challenging the Modern Synthesis: Adaptation, Development, and Inheritance*, ed. Philippe Huneman and Denis M. Walsh, 37–67. Oxford, UK: Oxford University Press. https://doi.org/10.1093/oso/9780199377176.003.0001.

Depew, D. J., & Weber, B. H. 1995. *Darwinism Evolving: Systems Dynamics and the Genealogy of Natural Selection*. Cambridge, MA: MIT Press.

Depew, D. J., & Weber, B. H. 2011. The Fate of Darwinism: Evolution After the Modern Synthesis. *Biological Theory*, 6 (1): 89–102. https://doi.org/10.1007/s13752-011-0007-1.

Depew, David J., and Bruce H. Weber. 2013. Challenging Darwinism: Expanding, Extending, or Replacing the Modern Evolutionary Synthesis. In *The Cambridge Encyclopedia of Darwin*. Cambridge, UK: Cambridge University Press.

DiFrisco, James, Alan C. Love, and Günter P. Wagner. 2020. Character Identity Mechanisms: A Conceptual Model for Comparative-Mechanistic Biology. *Biology & Philosophy* 35 (4): 44. https://doi.org/10.1007/s10539-020-09762-2.

Drake, James A., Michael Fuller, Craig R. Zimmerman, and Javier G. P. Gamarra. 2007. Emergence in Ecological Systems. In *From Energetics to Ecosystems: The Dynamics and Structure of Ecological Systems*, ed. Neil Rooney, K. S. McCann, and D. L. G. Noakes, 157–183. Dordrecht: Springer. https://doi.org/10.1007/978-1-4020-5337-5_8.

Dupré, John, and Daniel J. Nicholson. 2018. A Manifesto for a Processual Philosophy of Biology. In *Everything Flows: Towards a Processual Philosophy of Biology*, 3–45. Oxford, UK: Oxford University Press. https://doi.org/10.1093/oso/9780198779636.003.0001.

El-Hani, Charbel Niño, and Claus Emmeche. 2000. On Some Theoretical Grounds for an Organism-Centered Biology: Property Emergence, Supervenience, and Downward Causation. *Theory in Biosciences* 119: 234–275. https://doi.org/10.1007/s12064-000-0018-0.

El-Hani, Charbel Niño, and Nei Nunes-Neto. 2020. Life on Earth Is Not a Passenger, but a Driver: Explaining the Transition from a Physicochemical to a Life-Constrained World from an Organizational Perspective. In *History, Philosophy and Theory of the Life Sciences*, XXVI. https://doi.org/10.1007/978-3-030-39589-6_5.

El-Hani, Charbel N., Felipe Rebelo Gomes de Lima, and Nei de Freitas Nunes-Neto. 2024. From the Organizational Theory of Ecological Functions to a New Notion of Sustainability. In *Organization in Biology*, ed. Matteo Mossio, 285–328. Cham: Springer. https://doi.org/10.1007/978-3-031-38968-9_13.

Emmeche, Claus. 2004. Organicism and Qualitative Aspects of Self-Organization. *Revue Internationale De Philosophie* 58 (228): 205–217.

Fabris, Flavia. 2018. Waddington's Processual Epigenetics and the Debate over Cryptic Variability. In *Everything Flows: Towards a Processual Philosophy of Biology*, ed. John Dupré and Daniel J. Nicholson, 246–263. Oxford, UK: Oxford University Press. https://doi.org/10.1093/oso/978 0198779636.003.0012.

Farnsworth, Keith D. 2022. How an Information Perspective Helps Overcome the Challenge of Biology to Physics. *Biosystems* 217: 104683. https://doi.org/10.1016/j.biosystems.2022.104683.

Foster, John Bellamy. 2020. *The Return of Nature: Socialism and Ecology*. New York, NY: Monthly Review Press.

Gawne, Richard, Kenneth Z. McKenna, and H. Frederik Nijhout. 2018. Unmodern Synthesis: Developmental Hierarchies and the Origin of Phenotypes. *BioEssays* 40 (1). https://doi.org/10.1002/bies.201600265.

Gilbert, Scott F. 2000. Genes Classical and Genes Developmental: The Different Use of Genes in Evolutionary Syntheses. In *The Concept of the Gene in Development and Evolution*, ed. Peter J. Beurton, Raphael Falk, and Hans-Jörg Rheinberger, 178–192. Cambridge, UK: Cambridge University Press. https://doi.org/10.1017/cbo9780511527296.010.

Gilbert, Scott F. 2002. The Genome in Its Ecological Context: Philosophical Perspectives on Interspecies Epigenesis. *Annals of the New York Academy of Sciences* CMLXXXI:202–218. https://doi.org/10.1111/j.1749-6632.2002.tb04919.x.

Gilbert, S. F., and J. Bard. 2014. Formalizing Theories of Development: A Fugue on the Orderliness of Change. In *Towards a Theory of Development*, ed. A. Minelli and T. Pradeu, 129–43. Oxford, UK: Oxford University Press. https://doi.org/10.1093/acprof:oso/9780199671427.003.0008.

Gilbert, S. F., and S. Sarkar. 2000. Embracing Complexity: Organicism for the 21st Century. *Developmental Dynamics* 219 (1): 1–9.

Gilbert, Scott F., and Marion Faber. 1996. Looking at Embryos: The Visual and Conceptual Aesthetics of Emerging Form. In *The Elusive Synthesis: Aesthetics and Science*, ed. Alfred I. Tauber, 125–151. Dordrecht: Springer. https://doi.org/10.1007/978-94-009-1786-6_6.

Gilbert, Scott F., Jan Sapp, and Alfred I. Tauber. 2012. A Symbiotic View of Life: We Have Never Been Individuals. *The Quarterly Review of Biology* 87 (4): 325–341. https://doi.org/10.1086/668166.

Gilbert, S. F., & Epel, D. 2015. *Ecological Developmental Biology: The Environmental Regulation of Development, Health, and Evolution* (2nd ed.). Sunderland, MA: Sinauer Associates, Inc. Publishers.

Godfrey-Smith, Peter. 1996. *Complexity and the Function of Mind in Nature*. Cambridge, UK: Cambridge University Press.

Griesemer, James. 2006. Genetics from an Evolutionary Process Perspective. In *Genes in Development*, ed. E. M. Neumann and C. Rehmann-Sutter, 199–237. Durham, NC: Duke University Press.

Griffiths, P. E. 2017. Genetic, Epigenetic and Exogenetic Information in Development and Evolution. *Interface Focus* 7 (5): 20160152. https://doi.org/10.1098/rsfs.2016.0152.

Griffiths, P. E., and K. Stotz. 2013. *Genetics and Philosophy: An Introduction* (Cambridge Introductions to Philosophy and Biology). Cambridge, UK: Cambridge University Press.

Griffiths, P., and Stotz, K. 2018. Developmental Systems Theory as a Process Theory. In *Everything Flows: Towards a Processual Philosophy of Biology*, ed. J. Dupré and D. J. Nicholson, 225–45. Oxford, UK. https://doi.org/10.1093/oso/9780198779636.003.0011.

Haraway, Donna J. 2004. *Crystals, Fabrics, and Fields: Metaphors That Shape Embryos*. Berkeley, CA: North Atlantic Books.

Heras-Escribano, Manuel, and Paulo de Jesus. 2018. Biosemiotics, the Extended Synthesis, and Ecological Information: Making Sense of the Organism-Environment Relation at the Cognitive Level. *Biosemiotics* 11 (2): 245–262.

Hertwig, Oscar. 1892. Urmund Und Spina Bifida: Eine Vergleichend Morphologische, Teratologische Studie an Missgebildeten Froscheiern. *Archiv Für Mikroskopische Anatomie* 39 (1). https://doi.org/10.1007/BF02961535.

Hoffmeyer, Jesper. 1997. *Signs of Meaning in the Universe*. Bloomington, IN: Indiana University Press.

Hutchinson, G. Evelyn. 1957. Concluding Remarks. *Cold Spring Harbor Symposia on Quantitative Biology* 22: 415–427.

Huneman, P., and D. M. Walsh. 2017. Introduction: Challenging the Modern Synthesis. In *Challenging the Modern Synthesis: Adaptation, Development, and Inheritance*, ed. P. Huneman and D. M. Walsh, 1–33. New York, NY: Oxford University Press.

Jacob, F. 1977. Evolution and Tinkering. *Science* 196 (4295): 1161–1166. https://doi.org/10.1126/science.860134.

Jaeger, Johannes. 2019. Dynamic Structures in Evo-Devo: From Morphogenetic Fields to Evolving Organisms. In *Perspectives on Evolutionary and Developmental Biology*, ed. Giuseppe Fusco. Padova: Padova University Press. https://doi.org/10.31219/osf.io/mprc8.

Jaeger, Johannes, and Nick Monk. 2014. Bioattractors: Dynamical Systems Theory and the Evolution of Regulatory Processes. *Journal of Physiology*. https://doi.org/10.1113/jphysiol.2014.272385.

Jaeger, Johannes, David Irons, and Nick Monk. 2012. The Inheritance of Process: A Dynamical Systems Approach. *Journal of Experimental Zoology Part B: Molecular and Developmental Evolution* 318 (8). https://doi.org/10.1002/jez.b.22468.

Juarrero, A. 1999. *Dynamics in Action: Intentional Behavior as a Complex System*. Cambridge, MA: MIT Press.

Juarrero, Alicia. 2010. Complex Dynamical Systems Theory. *Cognitive Edge* 8.

Kauffman, Stuart. 1993. *The Origins of Order: Self-Organization and Selection in Evolution*. Oxford, UK: Oxford University Press.

Kim, Jaegwon. 2006. Emergence: Core Ideas and Issues. *Synthese* 151 (3). https://doi.org/10.1007/s11229-006-9025-0.

Kirchhoff, Michael D. 2015. Extended Cognition & the Causal-Constitutive Fallacy: In Search for a Diachronic and Dynamical Conception of Constitution. *Philosophy and Phenomenological Research* 90 (2): 320–360. https://doi.org/10.1111/phpr.12039.

Lala, K. N., T. Uller, N. Feiner, M. Feldman, D. Andrews, et al. 2024. *Evolution Evolving: The Developmental Origins of Adaptation and Biodiversity*. Princeton, NJ: Princeton University Press.

Laland, Kevin N., John Odling-Smee, and Scott F. Gilbert. 2008. EvoDevo and Niche Construction: Building Bridges. *Journal of Experimental Zoology Part B: Molecular and Developmental Evolution* 310 (7). https://doi.org/10.1002/jez.b.21232.

Laland, Kevin N., John Odling-Smee, Marcus W. Feldman, and Jeremy Kendal. 2009. Conceptual Barriers to Progress within Evolutionary Biology. *Foundations of Science* 14 (3): 195–216.

Laland, Kevin, Tobias Uller, Marc Feldman, Kim Sterelny, Gerd B. Müller, Armin Moczek, Eva Jablonka et al. 2014. Does Evolutionary Theory Need a Rethink? *Nature* 514 (7521): 161–164. https://doi.org/10.1038/514161a.

Laland, Kevin N., Tobias Uller, Marcus W. Feldman, Kim Sterelny, Gerd B. Müller, et al. 2015. The Extended Evolutionary Synthesis: Its Structure, Assumptions and Predictions. *Proceedings of the Royal Society B: Biological Sciences*. https://doi.org/10.1098/rspb.2015.1019.

Lecointre, Guillaume. 2018. The Boxes and Their Content: What to Do with Invariants in Biology? In *Life Sciences*, Information Sciences. https://doi.org/10.1002/9781119452713.ch14.

Lehrman, Daniel S. 1953. A Critique of Konrad Lorenz's Theory of Instinctive Behavior. *The Quarterly Review of Biology* 28 (4): 337–363. https://doi.org/10.1086/399858.

Lewens, Tim. 2019. The Extended Evolutionary Synthesis: What Is the Debate About, and What Might Success for the Extenders Look Like? *Biological Journal of the Linnean Society*. https://doi.org/10.1093/biolinnean/blz064.

Lewontin, R. 1983. The Corpse in the Elevator. *The New York Review of Books*. https://www.nybooks.com/articles/1983/01/20/the-corpse-in-the-elevator/.

Lewontin, Richard, and Richard Levins. 2007. *Biology Under the Influence: Dialectical Essays on Ecology, Agriculture, and Health*. New York, NY: Monthly Review Press.

Longo, Giuseppe. 2019. Interfaces of Incompleteness. In *Systemics of Incompleteness and Quasi-Systems*, ed. Gianfranco Minati, Mario R. Abram, and Eliano Pessa, 3–55. New York, NY: Springer. https://doi.org/10.1007/978-3-030-15277-2_1.

Longo, Giuseppe, and Maël Montévil. 2013. Extended Criticality, Phase Spaces and Enablement in Biology. *Chaos, Solitons and Fractals* 55. https://doi.org/10.1016/j.chaos.2013.03.008.

Longo, Giuseppe, Maël Montévil, and Stuart Kauffman. 2012. No Entailing Laws, but Enablement in the Evolution of the Biosphere. In *ACM Proceedings of GECCO*, 1379–1392.

McClintock, Barbara. 1984. The Significance of Responses of the Genome to Challenge. *Science* 226 (4676). https://doi.org/10.1126/science.15739260.

Melamed, Daniel, Yuval Nov, Assaf Malik Malik, Michael B. Yakass, Evgeni Bolotin, et al. 2022. De Novo Mutation Rates at the Single-Mutation Resolution in a Human HBB Gene Region Associated with Adaptation and Genetic Disease. *Genome Research* 32 (3): 488–498. https://doi.org/10.1101/gr.276103.121.

Merlin, Francesca. 2017. Limited Extended Inheritance. In *Challenging the Modern Synthesis: Adaptation, Development, and Inheritance*, ed. Philippe Huneman and Denis M. Walsh, 263–277. Oxford, UK: Oxford University Press.

Montévil, Maël. 2018. A Few Pending Challenges from the Perspective of a Theory of Organisms. *Constructivist Foundations* 13: 377–379.

Montévil, Maël. 2020. Historicity at the Heart of Biology. *Theory in Biosciences.* https://doi.org/10.1007/s12064-020-00320-8.

Montévil, Maël. 2021. Entropies and the Anthropocene Crisis. *AI and Society.* https://doi.org/10.1007/s00146-021-01221-0.

Montévil, M., and M. Mossio. 2015. Biological Organisation as Closure of Constraints. *Journal of Theoretical Biology* 372; 179–191. https://doi.org/10.1016/j.jtbi.2015.02.029.

Montévil, Maël, Matteo Mossio, Arnaud Pocheville, and Giuseppe Longo. 2016. Theoretical Principles for Biology: Variation. *Progress in Biophysics and Molecular Biology.* https://doi.org/10.1016/j.pbiomolbio.2016.08.005.

Moreno, Alvaro, and Matteo Mossio. 2015. *Biological Autonomy* (History, Philosophy and Theory of the Life Sciences, xii). Dordrecht: Springer. https://doi.org/10.1007/978-94-017-9837-2.

Moss, L. 2003. *What Genes Can't Do. What Genes Can't Do.* Cambridge, MA: MIT Press. https://doi.org/10.7551/mitpress/7205.001.0001.

Moss, L. 2012. Is the philosophy of mechanism philosophy enough? *Studies in History and Philosophy of Science Part C: Studies in History and Philosophy of Biological and Biomedical Sciences* 43 (1): 164–172. https://doi.org/10.1016/j.shpsc.2011.05.015.

Mossio, Matteo, Cristian Saborido, and Alvaro Moreno. 2009. An Organizational Account of Biological Functions. *The British Journal for the Philosophy of Science* 60 (4): 813–841. http://www.jstor.org/stable/25592036.

Mossio, Matteo, Maël Montévil, and Giuseppe Longo. 2016. Theoretical Principles for Biology: Organization. *Progress in Biophysics and Molecular Biology* 122 (1): 24–35. https://doi.org/10.1016/j.pbiomolbio.2016.07.005.

Needham, Joseph. 1929. *The Sceptical Biologist.* New York, NY: W. W. Norton & Company.

Noble, D. 2006. *The Music of Life: Biology Beyond Genes.* Oxford, UK: Oxford University Press.

Noble, D. 2021. The Illusions of the Modern Synthesis. *Biosemiotics* 14 (1): 5–24. https://doi.org/10.1007/s12304-021-09405-3.

Nicholson, Daniel J., and Richard Gawne. 2015. Neither Logical Empiricism nor Vitalism, but Organicism: What the Philosophy of Biology Was. *History and Philosophy of the Life Sciences* 37 (4): 345–381. https://doi.org/10.1007/s40656-015-0085-7.

Ochoa, Carlos. 2024. Deconstructing the Extended Evolutionary Synthesis: Do We Need a New Theory of Evolution? In *Unity and Disunity in Evolutionary Biology: Deconstructing Darwinism*, ed. Richard G. Delisle, Maurizio Esposito, and David Ceccarelli, 403–439. Cham: Springer. https://doi.org/10.1007/978-3-031-42629-2_14.

Odling-Smee, F. John, Kevin N. Laland, and Marcus W. Feldman. 2003. *Niche Construction: The Neglected Process in Evolution.* Princeton, NJ: Princeton University Press.

Oster, G., and P. Alberch. 1982. Evolution and Bifurcation of Developmental Programs. *Evolution* 36 (3): 444–459. https://doi.org/10.1111/j.1558-5646.1982.tb05066.x.

O'Malley, M. A., and J. Dupré. 2005. Fundamental Issues in Systems Biology. *BioEssays*. https://doi.org/10.1002/bies.20323.

Parsons, Howard L. 1977. *Marx and Engels on Ecology*. Wastport, CT: Greenwood Press.

Pattee, Howard H. 1972. Laws and Constraints, Symbols and Languages. In *Biological Process in Living Systems: Toward a Theoretical Biology*, ed. C. H. Waddington, 248–258. London, UK: Routledge.

Pharoah, Mark. 2020. Causation and Information: Where Is Biological Meaning to Be Found? *Biosemiotics* 13 (3): 309–326. https://doi.org/10.1007/s12304-020-09397-6.

Pigliucci, Massimo. 2008. Is Evolvability Evolvable? *Nature Reviews Genetics* 9: 75–82. https://doi.org/10.1038/nrg2278.

Pigliucci, Massimo, and Jonathan Kaplan. 2000. The Fall and Rise of Dr Pangloss: Adaptationism and the Spandrels Paper 20 Years Later. *Trends in Ecology & Evolution* 15 (2): 66–70. https://doi.org/10.1016/S0169-5347(99)01762-0.

Pigliucci, M., and G. Müller, eds. 2010. *Evolution, the Extended Synthesis*. Cambridge, MA: MIT Press.

Reid, R. G. B. 2009. *Biological Emergences: Evolution by Natural Experiment*, Vienna Series in Theoretical Biology. Cambridge, MA: MIT Press.

Roli, Andrea, Marco Villani, Alessandro Filisetti, and Roberto Serra. 2018. Dynamical Criticality: Overview and Open Questions. *Journal of Systems Science and Complexity* 31 (3): 647–663. https://doi.org/10.1007/s11424-017-6117-5.

Rosen, Robert. 1991. *Life Itself: A Comprehensive Inquiry Into the Nature, Origin, and Fabrication of Life*. New York, NY: Columbia University Press.

Sepkoski, D. 2019. The Unfinished Synthesis? Paleontology and Evolutionary Biology in the 20th Century. *Journal of the History of Biology* 52 (4): 687–703. https://doi.org/10.1007/s10739-018-9537-8.

Sheehan, Helena. 2018. *Marxism and the Philosophy of Science: A Critical History* (Radical Thinkers). London and New York: Verso Books.

Sheehan, Helena. 2022. Marxism, Science, and Science Studies from Marx and Engels to COVID-19 and COP26. *Monthly Review*.

Smocovitis, V. B. 2023. Every Evolutionist Their Own Historian: The Importance of History, Context, and the Extended Evolutionary Synthesis. In *Evolutionary Biology: Contemporary and Historical Reflections Upon Core Theory*, ed. T. E. Dickins and B. J. A. Dickins, 25–54. Cham: Springer International Publishing. https://doi.org/10.1007/978-3-031-22028-9_2.

Sonnenschein, Carlos, and Ana M. Soto. 2018. An Integrative Approach toward Biology, Organisms, and Cancer. In *Methods in Molecular Biology*, MDCCII. https://doi.org/10.1007/978-1-4939-7456-6_2.

Soto, Ana M., and Carlos Sonnenschein. 2004. The Somatic Mutation Theory of Cancer: Growing Problems with the Paradigm? *BioEssays* 26 (10): 1097–1107. https://doi.org/10.1002/bies.20087.

Soto, A. M., Longo, G., Miquel, P. A., Montévil, M., Mossio, M., Perret, N., Pocheville, A. Sonnenschein, C. 2016. Toward a Theory of Organisms: Three Founding Principles in Search of a Useful Integration. *Progress in Biophysics and Molecular Biology* 122: 77–82. https://doi.org/10.1016/j.pbiomolbio.2016.07.006.

Stein, Sebastian. 2024. Nature and Its Limits: Hegel's Idealist Critiques of Physicalism, Naturalism, and Essentialism. In *Hegel's Philosophy of Nature: A Critical Guide* (Cambridge Critical Guides) ed. Marina F. Bykova, 36–57. Cambridge, UK: Cambridge University Press. https://doi.org/10.1017/9781009023160.004.

Stoltzfus, A. 2012. Constructive Neutral Evolution: Exploring Evolutionary Theory's Curious Disconnect. *Biology Direct*. https://doi.org/10.1186/1745-6150-7-35.

Stoltzfus, Arlin. 2017. Why We Don't Want Another "Synthesis". *Biology Direct* 12 (1): 23. https://doi.org/10.1186/s13062-017-0194-1.

Stoltzfus, Arlin, and Kele Cable. 2014. Mendelian-Mutationism: The Forgotten Evolutionary Synthesis. *Journal of the History of Biology* 47 (4): 501–546. https://doi.org/10.1007/s10739-014-9383-2.

Sultan, S. E. 2015. *Organism and Environment: Ecological Development, Niche Construction, and Adaptation.* Oxford University Press.

Thelen, Esther, and Linda B. Smith. 1996. *A Dynamic Systems Approach to the Development of Cognition and Action.* Cambridge, MA: MIT Press.

Thelen, Esther, and Linda Smith. 2006. Dynamic Systems Theories. In *Handbook of Child Psychology: Theoretical Models of Human Development*, ed. William Damon and Richard M. Lerner, 258–312. Hoboken, NJ: John Wiley & Sons. https://doi.org/10.1002/9780470147658.chpsy0106.

Uller, Tobias, and Heikki Helanterä. 2019. Niche Construction and Conceptual Change in Evolutionary Biology. *British Journal for the Philosophy of Science* 70 (2). https://doi.org/10.1093/bjps/axx050.

Vecchi, D., Miquel, P.-A., & Hernández, I. 2019. From Biological Determination to Entangled Causation. *Acta Biotheoretica*, 67 (1): 19–46. https://doi.org/10.1007/s10441-018-9339-6.

Walsh, Denis M. 2012. Mechanism and Purpose: A Case for Natural Teleology. *Studies in History and Philosophy of Science Part C: Studies in History and Philosophy of Biological and Biomedical Sciences* 43 (1): 173–181. https://doi.org/10.1016/j.shpsc.2011.05.016.

Walsh, D. M. 2015. *Organisms, Agency, and Evolution.* Cambridge, UK: Cambridge University Press. https://doi.org/10.1017/CBO9781316402719.

Weber, Bruce H. 2011. Extending and Expanding the Darwinian Synthesis: The Role of Complex Systems Dynamics. *Studies in History and Philosophy of Science Part C: Studies in History and Philosophy of Biological and Biomedical Sciences* 42 (1): 75–81.

Whitehead, Alfred North. 1967. *Science and the Modern World.* New York, NY: The Free Press.

Chapter 4
Poesis and Closure

Abstract This chapter explores the philosophical and biological concepts of poiesis and closure, focusing on Donna Haraway's critique of autopoiesis and her advocacy for *sympoiesis* as a more comprehensive framework for understanding living systems. Haraway's work challenges the reductionist and mechanistic paradigms of traditional evolutionary theory, emphasizing instead the symbiotic and relational nature of organisms. Drawing on endosymbiotic theory and organicism, the chapter argues that biological systems are best understood as dynamic, multispecies consortia (*holobionts*) rather than isolated individuals. The discussion contrasts autopoiesis, which prioritises internal self-maintenance, with sympoiesis, which highlights the constitutive role of interspecies interactions in development and evolution. The organisational approach, particularly the notion of closure of constraints, is presented as a way to reconcile these perspectives, offering a dialectical model where organisms and environments co-constitute each other. Ultimately, the chapter advocates for a processual and ecological view of life, underscoring the limitations of systems theory and the need for a more integrative, diachronic biology.

Introduction

> In a sense, organisms have ceased to exist as objects of knowledge, giving way to biotic components, i.e., special kinds of information-processing devices. (Haraway 1991: 164)

As I now turn to the more explicitly philosophical side of my dissertation, some caveats are in place: I will not try to reconstruct the theories of Haraway or the other philosophers I discuss comprehensively. I take from them what I need to make my case and simultaneously aim to make sense of some of their core concept through biology. In this chapter, I elaborate on some of the key topics discussed previously, while tying them to more philosophical debates concerning autopoiesis and symbiosis. The epistemological, ethical, and political dimensions of Haraway's work are left untouched, as I focus on her initial and most recent work. When discussing these, I neglect debates she raises about ideology, genetics, the Anthropocene and so on. While this might leave the reader unsatisfied, I am guided by the overall

R. Sandnes Haukedal, *Agency and Organisation: The Dialectics of Nature and Life*,
Library of Ethics and Applied Philosophy 44,
https://doi.org/10.1007/978-981-95-1924-8_4

aim of this thesis: bringing philosophy in dialogue with contemporary evolutionary science—not providing a systematic reading of the philosophers I engage with.

Some of the questions of evolutionary theory raised above, about variation and organisation, could be answered by reference to Lynn Margulis' endosymbiotic theory, which is about the symbiotic relationship between a host cell and a smaller cell (endosymbiont), where the latter lives inside the former. Through this process, organelles such as mitochondria and chloroplasts are produced. Such symbioses threaten the borders between individuals by revealing that we are consortia of different genomes (Gilbert et al. 2012). This perspective is the backdrop of much I will say, although I do not discuss it at length. Instead, I present Haraway's PhD concerning metaphors applied within biology, as it shows how her thinking has been infused by ecology from the start. Since it is also one of the most comprehensive works on the history of organicism published (Emmeche 2004), it serves as a historical introduction to this tradition.[1] Hence, this chapter expands on organicism and the importance of organisation in biology, both seen through the lens of Haraway and used to illuminate her philosophy.

The related topics of self-organisation of self-production (autopoiesis) will be examined, but I mostly focus on the latter. I use it as a speculum for getting at Haraway's understanding of living organisms—and juxtapose this view with contemporary theories of organisation introduced already. I assert that the EES might benefit from adopting another notion of organisation than is provided by autopoietic theory. The main task is thus to show the limitations of this framework by utilising notions of sympoiesis and organisational closure and grasp how they can be combined. I assess the notion of closure by juxtaposing sympoiesis and autopoiesis—and propose that these terms point to a contradiction within systems theory. This contradiction, I claim, partly explains Haraway's growing uneasiness with autopoiesis and the cybernetic theory it is based on.

I will begin with a brief description of endosymbiosis and its biological significance and tie this to the broad conception of epigenesis and interspecies relations, and some of the discussions we have previously had. I move to Haraway's doctoral thesis on metaphors within biology and the history of organicism. I recap some of the main assumptions of organicism and compare it to mechanism and vitalism, to see how they relate to the central metaphors discussed by Haraway. I also go more deeply into the authors discussed in the dissertation and why they emphasised the importance of viewing form as an ongoing process, not a static thing. I will, however, focus mostly on their general outlook, not their concrete research. In passing, I touch upon the significance of Whitehead to their views, even if I cannot systematically analyse this influence. I then spend some time on Harrison's views and how he warned against reifying wholeness. His notion that organisation is not an answer but a question is a critique of reductionist models.

[1] A notable exception is Erik L. Peterson's *The Life Organic: The Theoretical Biology Club and the Roots of Epigenetics* (2017), published after Emmeche made this claim.

This anti-reductionistic bias is also found in Needham's thinking, as seen in his crucial concept: *levels of organisation*, which is also central to the recent organisational approach. His views combat the notion that there are no levels of organisation within biology and thus challenges the view that biological systems can be reduced to the actions of their parts. After discussing the core elements of organicism, I move to more current questions, about the status of systems theory, and what makes Haraway suspicious of this approach. This leads to a discussion of autopoiesis, which I contrast with sympoiesis, to judge what can be gained from adopting the latter term. I then return to closure of constraints, to see how it may inform our discussion and how it may shed light on Haraway's critique of autopoiesis. After this, I go back to sympoiesis to show how the organisational view hints at a multispecies or symbiotic view, where closure is reached by cooperation between organisms.

In the next part, I discuss the meaning of 'structural coupling' within autopoiesis and ask how this notion fits with the organisational notion that closure is never completed but is a tendency which may involve multiple partners who make each other reciprocally. Based on this, I argue that we should move from external structural couplings to a more dialectical conception of the interplay between organisms and environment. In this view, constitutive interactions between organisms and species are non-optional since organisms maintain and survive only by being extended into their environment and cooperating with others. This troubles the mechanistic understanding of their interaction.

Symbiosis and Organisation

> The reductionist approach that would always explain the whole in terms of the parts leads to an infinite nihilistic regress in which eventually everything is held to be explainable in terms of essentially nothing. (Sperry 1982: fn6, 261)

Trained as a biologist, Haraway's works have had a direct impact on evolutionary theory. Scott F. Gilbert and David Epel, for example, have taken up her perspective in textbooks such as *Ecological Developmental Biology* (2015). As we shall see, the ecological outlook leads Haraway to criticise the notion of autopoiesis, asserting that sympoiesis (making-with) is a more adequate model.[2] It builds on endosymbiosis, which accounts for major transitions in evolution by reference to symbiotic relationships between bacteria. The transition from unicellularity to multicellularity is explained by one microbe partially ingesting the another; while the smaller of these new partners became the mitochondria, the larger one 'provided the DNA that would become encapsulated by the nucleus' (Gilbert 2019: 3). The eukaryotic cell, which provided the basis for cell evolution, was a result of endosymbiosis between different

[2] Gilbert et al. (2010) recommended that one calls interspecies epigenesis—the co-development of symbiotic partners—*symbiopoiesis*, but I will stick with the former cognate as it is the one Haraway applies.

bacterial kingdoms. This was a fundamental event in the evolution of multicellular organisms.

Endosymbiosis challenges some of the core assumptions of the MS, especially that only the gradual process of selection of genetic mutation drives speciation. It disputes the idea that macroscopic evolution is just microscopic evolution over a longer period, i.e. that evolutionary changes occur simply due to the accumulation and selection of mutations (O'Malley 2017). While the MS might have allowed such metabolic events as proximate causes of evolution, as something which might have made certain events possible (qua releasing cause) but lacks the heritable stability to count as an ultimate cause of evolution. Above, we saw this in relation to development as a proximate cause. But this led to problems, as the proponents of the MS 'do not causally or statistically explain why such [metabolic] events had the evolutionary outcomes they did' (O'Malley 2015: 10275). They simply assume it to be caused by genetic selection. Margulis, on the other hand, claims that metabolic events where one microbe partially swallows another provides an ultimate explanation of *why* evolution had proceeded as it did since symbiosis had provided natural selection with organelles that made new forms of life possible.

We have seen that epigenesis in the broad sense is an ecological concept—that it does not only involve the intra-organic interaction between cells, but also the interaction of organism and environment. It also has an interspecies dimension, revealing how symbiotic relations underlie the formation and functioning of all organisms (see Gilbert and Epel 2015; Gilbert et al. 2010; McFall-Ngai 2014). Interspecies relationships give rise to higher-order regimes of causality, through which the overall life cycle is maintained. As such, they exemplify the broad interpretation of epigenesis. The emergent holobiont (the whole organism plus its symbionts) encapsulates the close integration between individuals—composite organisms formed and maintained by their relations with other species. It is a result of permanent or transient associations between its constituent bionts (Margulis 1993), a complex system formed by the organism and multiple other species it is made from. Thus, it exemplifies the co-evolution of species, how 'gene expression in one species is regulated by products from another species, and the species have co-evolved to maintain this developmental relationship. The two (or more) organisms work together to develop each other' (Gilbert 2002: 211). To repeat, this is not only a proximate cause but a fundamental event in the history of life that 'both enabled and constrained genetic novelty' (O'Malley 2015: 10275).

From this perspective, we see why Weismann's division of germ and somatic cells, where the former is unaffected by the development of the organism's somatic cells and thus inherited in a pristine condition, cannot be maintained. Individuality does not mean genetic homogeneity. Only very few species are genetically homogenous (Pradeu 2012). It counters the notion that germ cells cannot be affected by somatic cells: 'many organisms are heterogeneous in the sense that, contrary to Weismann's main idea, their somatic cells can undergo mutations and give birth to germ cells' (260). Even the few organisms that are homogeneous and thus abide by Weissman's barriers, 'comprise entities of different origins that can influence their evolution' (Pradeu 2012: 260). Therefore, they are also heterogeneous. In this view, there are no truly homogenous organisms, as that would imply that they were completely isolated from their surroundings, making their ability to maintain their life cycle mysterious. It also underscores how interspecies relations are not only the products of evolution but also causes, creative factors. Such interactions enable an organisation which cannot be reduced to its parts, and which 'exerts downward agency over the entities which compose it' (Gilbert 2019: 3).

Organicism constitutes 'a third way between the experimentally invalid reductionism of form to gene and the unscientific vitalism that gave consciousness to inorganic matter. It provided a scientific way of describing those processes whereby directionality could be demonstrated' (Gilbert, in Haraway 2004: xii). Both organicism and epigenetics provide alternatives to the MS explanation of the drivers of evolution. From these perspectives, natural selection acts on viable life cycles, a complex 'trait' involving the complete organism. Also, organisms from one species depend on symbiotic relations with others, not only to construct their niche but to *maintain* themselves long enough to reproduce. Not only that, it is not just one life cycle that is selected, but the organism as an integration of multiple life cycles, maintaining themselves through their interplay (Gilbert 2019).

Organic Metaphors

In the preface to the volume based on Haraway's dissertation, *Crystals, Fabrics, and Fields*, Gilbert says that it provides 'insights into how Donna came to her present views' (Haraway 2004: xi). He maintains that an idea that Haraway formulates explicitly later—that 'the relation is the smallest possible pattern of analysis'—is nascent in this volume. He goes as far as stating that while her philosophy is not reducible to epigenesis, it cannot be understood without it (Haraway 2004: xiii). Hence, her dissertation is an interpretive key for her intellectual trajectory. I follow this cue to get at the situatedness of livings systems at the centre of her thinking about how organisms are produced and maintained.

A quick summary of Haraway's dissertation: By interrogating the works Ross G. Harrison, Joseph Needham, and Paul Weiss—and to formulate the organicist alternative to vitalism and mechanism—Haraway focuses on the metaphors employed by these approaches. She holds that any useful theoretical perspective demands a central metaphor, and argues that by knowingly employing metaphors, the limitation of a paradigm is exposed since anomalies that challenge it appears. Also, she maintains that metaphors are central to research, that they decide what scientists regard as relevant or not, and thus also shaped their expectations. It enabled them to see connections between phenomena that 'might otherwise have appeared unrelated' (Haraway 2004: 107).

Haraway distinguishes four pathways of enquiry: (1) the primacy of form, (2) 'roles of symmetry, polarity, and pattern concepts in the old and new metaphoric systems' (Haraway 2004: 17), (3) the field-particle duality, and (4) how organicism in biology is connected to structuralism in philosophy as 'the focus on the fundamental concept of organizing relationships' (Haraway 2004: 17). In what follows, I focus mainly on the first three, but I cannot give them equal emphasis.

Haraway's study of metaphors in science considers how the machine metaphor shaped the relationship between scientists and their objects; not only how the scientists relate to what they study, but also how research is imbued with aesthetic commitments. Metaphors are not just 'props to the imagination', says Haraway. Rather, 'they have been intrinsic to explanations of basic properties of life' (Haraway 2004: 41), guiding the actions of scientists. Mobilising Thomas Kuhn's paradigm concept to ascertain the metaphors which are operative in scientific research, she highlights his insistence on the 'communal and exemplar nature of paradigms' (Haraway 2004: 3). This interpretation of science means that '[t]here is no absolute court of appeal; there are only alternate world views with fertile basic metaphors' (Haraway 2004: 7). Scientific paradigms shape what can be made intelligible.

The shift from 'the machine to the organic system' involves a critique of positivism, which does not take metaphors in science seriously. Haraway mentions the gene as an example of how scientific metaphors are predictive. But they also carry within themselves the seeds of the paradigm's destruction: As they push the logic of the paradigm to its limits, they generate anomalies.

The atomistic view of the world hinges on the aesthetics of the machine, which can be decomposed into smaller and smaller parts, and the crystal—the 'smaller, simpler version of the organism in a nearly literal sense' (Haraway 2004: 11). But there were other aesthetic commitments present during the formulations of the MS, related to organisation and patterns, and in opposition to self-contained units. Haraway opines: 'It is impossible to miss the pointed imagery derived from the machine paradigm. The ultimate simple unit did not get involved in entangling "internal organizing relations"' (Haraway 2004: 29). Each cell counts as an embryo unto itself.

Haraway claims that a fully-fledged organicism could only be formulated after new developments in physics, defying strict determinism, and when biology changed its relation to it. By reference to von Bertalanffy, she distinguishes between two conceptualisations of mechanism: the first one is the one we encountered in the formulation of the MS, tied to the familiar ideal of reducing biological phenomena to physical or chemical laws. She mentions Whitehead's observation, that the mechanistic worldview was made untenable by relativity theory and quantum mechanics, with 'vastly different conceptions of organization, causality, and determinism and thus of fundamental explanation' (Haraway 2004: 26). Simple reductionism was challenged, and physics could not offer the stable foundation for the unification of science.

The second understanding considers the organism as a machine, abiding by the laws of mechanics. It means, says Haraway, that all parts were preformed, and that organisation did not introduce any changes into the parts:

> Thus, sum means aggregate without additional principles or regularities specified or necessary to explain the animal-machine's operation. All "organizing relations" are external in J.H. Woodger's sense. An internal relation implies that the "parts" themselves are different depending on whether they are in or out of a particular context. (Haraway 2004: 28)

According to this view, an organisation is an aggregate of the intrinsic capacities of the parts, which have pregiven functions apart from the whole. Vitalism, the idea that there is a vital force that cannot be reduced to physical or chemical processes, never challenged this view directly. It merely added that there is some force which transgresses material elements and relations—which is outside the gambit of science. Therefore, the organicists 'saw vitalism as part of the mechanistic paradigm rather than opposed to it because both were limited by the same images and metaphors' (Haraway 2004: 38). While both vitalism and organicism ran against reductionism, only the latter challenged the mechanistic views of regulation and determination directly. But they shared an emphasis on how 'the form of the whole' influences its part and is embodied in the purposive behaviour of organisms. In contrast to organicism, however, vitalism is a compromise which 'allows a free run to mechanism throughout the whole of inanimate nature and holds that the mechanism is partially mitigated within living bodies' (Whitehead 1967: 79). It cemented the dualistic view of nature instead of challenging it.

The organismic perspective, on the other hand, forecloses the view that biology is just a postscript to physics. Indeed, the primacy of the physical view of nature is at stake. So why was this view not (openly) adopted by the architects of the MS? They wanted autonomy but still adopted the physical, mosaic view of development with external relations between the parts only. One explanation might stem from the status of physics, and the 'physics envy' displayed by many biologists. For their theory to count as scientific at all they might have seen no other choice (see Depew and Weber 1995). Also, emulating physics made their job easier, as they could ignore aspects not abiding by such models. As we have seen, this ultimately left them in a contradictory position, forcing them to rely on a notion of purpose, however implicit.

Fields and Particles

> At no point in the scheme does a single cause determine a single effect; systems of oriented processes differentiate into subsystems coordinated by their mutual functioning. (Haraway 2004: 90)

As we go more deeply into Haraway's dissertation, I will not underline differences between the scientists but highlight their shared commitment of the organicists, as representatives of a developmental and non-vitalist perspective on biology. In short, Weiss concretised what Needham and Harrison had only 'metaphorised'. He introduced the field metaphor, which he drew from physics, into biology, to articulate a biology of interaction: 'His experimental work on cell shape and movement and interaction has been constantly informed by an appreciation of the organism as a system-whole' (Haraway 2004: 16). Weiss made organisation the object of study instead of using it as an explanatory device.

While vitalism postulated an immaterial force, organicism was a materialist and realist position, even if it shared vitalism's concerns with wholeness and autonomy.

According to organicism, what makes biological phenomena irreducible was emergent levels of organisation, not a vital force.[3] A strict mechanistic view does not allow for such downward regulation, nor for identifying functions within a whole. Only organicism offered a new paradigm for understanding organisation and regulation by an emphasis on *form as process*:

> From an organismic perspective, the central and unavoidable focus of biology is form. Every other consideration of the biological sciences leads up to the task of at last stating the laws of organic form. Form is more than shape, more than static position of components in a whole. For biology the problem of form implies a study of genesis. (Haraway 2004: 39)

According to this view, form is a moving target: 'The structure-function and part-whole relations, understood in a constant dialectic interplay, constitute the cornerstone of the developmental edifice' (Haraway 2004: 44). It is not enough to say that function produces structure, as the structure is also 'the basis of function' by constraining it in constitutive interplay. Insofar as one deals with organisation at a different level from physics and chemistry, the debate concerning reducibility is irrelevant.

This was also Needham's view. Through his studies into muscle fibres, it became evident that the function-structure dichotomy was one of interdependence and reciprocity. This leads us back to questions of causal specificity—accentuating how geneticists conjectured that one could extrapolate complex traits from genes in the same manner as simple ones—'if one were only clever enough', as Haraway says. Specificity is stressed by Whitehead as well, who deems the mechanistic approach too abstract to grasp the concreteness of organisms. In his *Science and the Modern World*, he laid out the foundations of a process view of nature and warned against the unwarranted reification of processes into things. He considered the world as a unified whole in which processes had ontological primacy. Whatever is seen as stable things are just frozen processes, abstracted from the concrete system in which they are formed. In his view, biology must free itself from the shackles of physics by becoming philosophical and 'enter upon a thorough criticism of its own foundations' (Whitehead 1967: 25).

Haraway states: 'Considerations of symmetry, polarity, and gradient are part of a general treatment of pattern. The single most important organizing principle developed in this [the twentieth] century to treat pattern was that of *field*' (Haraway 2004: 54). Shunned before, it now became indispensable, but it was critical to ground its material basis in biochemistry to avoid the anti-scientific tendency of vitalism. The field metaphor acted as a synthesiser:

> Field notions [...] were meant to bridge the gap between the polar opposites. To do so necessitated rethinking several historical dichotomies such as structure and function, crystal and organism, part and whole. The organicist biology developed by Harrison, Needham, and Weiss found a way beyond the antithesis; organism and organization required a union of field and particle. (Haraway 2004: 56)

---

[3] Haraway notes that the first self-identified organicist might have been J.S. Haldane (the father of J.B.S. Haldane), but that Needham considered him a neo-vitalist because he 'left word such as organization so vague' (2004: 37).

The field metaphor made contradictions such as simultaneous closedness and openness intelligible as part of larger processes of becoming. We find it in earlier ideas of organisational closure, like Jean Piaget's theory, where the whole is not imposed on the parts but emerges from the relations between them. Even if the organisation has primacy, and the emergent whole constrains the actions of its parts, its power is not total: 'The laws of substructures are conserved in the more complex whole. The elements of a structure are themselves organized in their own right. Thus, the search for the unorganized starting point is meaningless.' (Haraway 2004: 63), and an infinite regress is avoided. The parts are both *structured* and *structuring*; they self-organise, 'which in turn implies conservation and a certain closure of the structure to the exterior' (Haraway 2004: 62). This principle of modification of the part by the whole—tied to organisational closure—is increasingly important as we move along. It also hints at why relationism is a more apt term than holism.

Fields or 'formative tendencies' individuate particles by constraining their actions and affording them with functions. But the field metaphor was never intended to be conclusive in Harrison's view; rather, 'it was meant to stimulate further research, not to provide verbal explanation where full understanding was lacking' (Haraway 2004: 59). Waddington's developmental topologies also depict biological fields as more complex than physical fields. Haraway writes: 'An embryological field is more complex than electromagnetic or gravitational fields in two ways: time can never be left out of the picture and the material substratum is immensely more complicated' (Haraway 2004: 60). This underscores the inadequacy of directly transposing physical models onto biological systems.

Advancing the notion that there are no simple natural systems in the mechanist sense, Harrison also warns against fetishising wholeness. Analysis is as important as ever but must be recognised as a cumbersome process. Haraway summarises his views:

> Organization and wholeness for Harrison were not answers to biological questions; they were the biological questions *par excellence*. The crux of his analytical approach lay in his insistence that analysis was performed in order to allow refocusing on the whole organism, on its problems of integration. His own use of a given method was always to ask a question of the organism, never to exploit a tool of analysis for its own sake. (Haraway 2004: 83–84, emphasis original)

This illustrates the difference between the *reductive* and *integrative* phases of research (cf. Griffiths and Stotz 2013). The first involves identifying parts, the second seeks to show how they fit together as embedded in a larger whole. It should be added, however, that these phases could also be part of a mechanistic approach, insofar as it can be a matter of aggregation. But if we follow the organicist approach, these phases are not distinct since no part is simple or unorganised. Although crystals are not as complex as organisms, they still represent an intermediate organisational level. The fabric metaphor represents a further level, where the organised elements, cells, are integrated to behave as wholes, as tissues. Weiss studies tissue organisation as a process 'welding units into higher order patterns' (Haraway 2004: 158).

So, while breaking a system into parts is necessary, it must be with the caveat that its parts are organised and influenced by the whole in which they are part. 'Units themselves, Harrison implied, were of greater or lesser complexity. He never talked in terms of total reductionism' (Haraway 2004: 84). Organisation and interaction run all the way down.[4] There exists neither an stable, ultimate level of causation nor reason to maintain the dichotomy between autonomous parts and determination by the whole. As Harrison puts it: 'It is impossible to develop science wholly from the top down or from the bottom up. The investigator enters where he can *gain a foothold* by whatever means may be available' (Harrison 1949: 4, quoted in Haraway 2004: 95, my emphasis[5]).

Similarly, Saetzler et al. (2011) argue that if systems biology wants to be an alternative to the reductionism of the MS, it cannot ignore the interplay of upward and downward causation. We find their view echoed in Denis Noble's version of systems biology, and the *middle-out* approach, merging bottom-up and top-down strategies (see Noble 2006), an issue we return to in the next chapter. Whether ordering processes are called organisations or fields is not the issue; more important is the fact that these terms 'were not explanations in themselves' (Haraway 2004: 164), but steps on the way towards explanations. We are not excused from causal analysis but the whole must be kept in view since each part is endowed with specific capacities through its immersion in the whole to which they also give rise.

The Ecological Theatre of the Particle

> Remarking that Marx wanted to dedicate part of Capital to Darwin, Needham found in the dialectic a way out of mechanism and vitalism, a way to approach history in embryology. (Haraway 2004: fn28, 136)

In Needham's thinking we find a dynamic way of understanding different levels of organisation and their interplay. First, he shared Harrison's view on analysis: Functions were realised by complexes, not simple mechanisms, as there were organising principles even in simple biological systems. These 'laws of integration' were instantiated by dynamical structures, not 'static scaffolding'. He abandoned the machine metaphor in favour of the field because he came upon totalities that were not modular in the sense of functioning in relative independence from the larger whole. Hence, they were not decomposable without remainder.

Secondly, by proposing the notion of integrative levels, Needham challenges Mayr's view that there is no emergence *within* biology, only *between* biology and the

---

[4] While crystals exhibit organisation, their complexity cannot be compared to organisms. Crystals constitute an intermediary step. In a letter to Needham, J. H. Woodger emphasised that the difference should not be considered one of degree since they exhibit qualitatively different types of organisations (Haraway 1997: fn5, 110–11).

[5] 'The complete Silliman Lecture notes are unpublished. I am indebted to Sally Wilens, long-time research associate of Harrison, and to the archive section of Sterling Memorial Library for xerox copies of the lectures' (Haraway 2004: fn21, 93).

physical sciences. Integrative levels are new levels of organisation emerging from lower ones. He writes:

> This deadlock [between mechanism and vitalism] [...] was overcome when it was realized that every level of organization has its own regularities and principles not reducible to those appropriate to lower levels of organization, nor applicable to higher levels, but at the same time in no way inscrutable or immune from scientific analysis and comprehension. (Needham 1943: 18)

One level does not simply determine another, even if it might partake in its construction. The notion that there are different levels of organisation with laws and regularities that differ from others was, as we have seen, crucial to the development of biology as an autonomous science. But in Needham's view, there are also emergent levels within biology, and interactions between these.

We saw above that relationism was a way of breaking with vitalism since '[p]roperties of parts were determined by the sorts of relationships maintained with neighboring parts within the whole' (Haraway 2004: 113). Internal relationships are enough; no vital principle is needed because organisation is immanent in the particles themselves. As Whitehead says: 'For each relationship enters into the essence of the event; so that, apart from that relationship, the event would not be itself. This is what is meant by the very notion of internal relations' (Whitehead 1967: 123). If the relation is part of the definition of the essence of the event, substance is itself a process of individuation, not given at the outset. Whitehead extends this to the relation between part and whole. Understanding this interplay as internal and constitutive is a central component of a dialectical view of nature.

As the unit and proper level of analysis for biology, the organism became a point of departure leading to the acknowledgement of 'organising relations at all levels, higher and lower, coarse and fine, of the living structure' (Needham 1942: 656, quoted in Haraway 2004: 139). The organism was an important node in the network but not privileged. More important was the fact that all levels display organisation. This 'hierarchical continuity of biological order' (Haraway 2004: 140) took precedence over discontinuity. Simultaneously, each level had a different form of organisation, which means that the continuity was not seamless or linear. We see, then, how the dialectical approach to biology requires topological thinking in the form of landscapes of possibilities produced by the interplay of causes. It also emphasises the idea that development introduces biases, canalising the potential it is provided by evolution. As Needham writes: 'Development, then, consists of a progressive restriction of potencies by determination of the parts to pursue fixed fates' (Needham 1936: 58, quoted in Haraway 2004: 141). This aligns with what we have said about determination versus enablement:

> To say that the operation of a higher level is *based on* the proper functioning of the components of a lower is not to say the operation of the higher can be *reduced to* the lower. Rather, the higher complex is given as a unit that requires its own laws. (Haraway 2004: 149, emphasis original)[6]

---

[6] Haraway remarks that Weiss was a precursor to Michael Polanyi's distinction, embellished in the title of his article 'Life's Irreducible Structure' (1968), where 'there may be more variability in each

According to Weiss, parts generate an ‘ordered matrix’, that constrains their functioning. But the parts contribute reciprocally to this matrix. Such self-organisation is a process of individuation, the constitution of a field of action which is specific to each organism. This constitution involves not only interactions of internal units, says Weiss (1968), but also between these units and their environment. From ‘the fact that the interacting units themselves have a distinctly nonrandom, patterned constitution’ (Weiss 1968: 61, quoted by Haraway 2004: 169) emerges *nonrandom* systems of higher order.

Since the variation displayed at the level of phenotypes is the outcome of highly organised developmental processes, it is not random (Stewart 2019). For instance, tissue organisation does not emerge from simple, unorganised, molecules but from macromolecules which are already organised at the lower level by being embedded in a system of constraints. They exhibit autonomy from physical laws, even if they depend on and are constrained by physical and chemical processes. Their macro-crystalline structure is a basic property, not derived from something simpler. Haraway identifies a chicken-and-egg problem here. It is, however, eschewed by the fact that ‘all the components had at some point a necessary connection with an already organized system’ (Haraway 2004: 171). Yet, a suspicion might arise, leading one to ask whether the particle which gives rise to the organisation is not primary, as it seems to be the building block without which no organisation would emerge at all. But if organisation goes all the way down, and each particle is a self-organised process then the problem of what comes first is moot. Weiss proposed the term ‘molecular ecology’ to underscore this situation. Hence, organicism is not just about relating the parts to the whole. This is, according to Haraway, a commonplace that all philosophical approaches to biology admit: ‘The crux of the issue is the nature of the units: Is the search for the ultimate uncomposed unit justified?’ (Haraway 1997: 94). From what we have learned thus far, it is not.

In summary, the organicists place the organism at the centre of evolution by opposing reductionism and arguing for organising principles that cannot be reduced to physical or chemical levels. But they go further, as they also propose that biological systems consist of different interactive levels. Endorsing Whitehead’s process, they opt for a diachronic approach to biological systems and propose that we adopt the field metaphor not as a solution but as a challenge. Finally, they argue for a relational and dialectical approach to biological systems, which does not give the whole primacy over the parts but acknowledges their mutual influence.

---

component of a whole than exists in the behavior of the system itself; in this sense the whole is more than the sum of its parts’ (Haraway 2004: 149).

What Is Wrong with Systems Theory?

> If nature were atomized and inherently chaotic, only creative mind could see and carve into it and from it those patterns of higher order to which we concede consistency and beauty. But nature is not atomized. Its patterning is inherent and primary, and the order underlying beauty is demonstrably there; what is more, [the] human mind can perceive it only because it is itself part and parcel of that order. (Weiss 1960, quoted by Haraway 2004: 147)

The quotation above summarises the view adopted by Weiss, Needham, and Harrison. It also hints at our fundamental involvement with nature—how it is intelligible for us because we do not stand outside it (a notion that I discuss more in Chap. 7). As should be clear, the perspective above was overridden by the MS. This partly explains why it lacked a robust bulwark against reductionism, having to fend of threats to the autonomy of biology without a conclusive victory, only unstable compromises. We see this in how the MS architects could allow organicist explanations, but would insist that 'the truth is elsewhere' (Haraway 2004: 199)—viz. that it does not reflect the ontological structure of the world even if helps us make sense of it. Again, it is only heuristics.

After outlining the organicism discussed in Haraway's dissertation, I now move to her more recent works. While there is continuity, there is also tension. For example, in her dissertation, Haraway views organicism as a systems theory of organised levels and hierarchies. Later, however, she parts ways with systems theory, seemingly considering it too static to account for the emergent topologies, 'more venturesome [...] than organic systems theory can offer' (Haraway 2004: xix). Why this shift? To propose an answer, I turn to autopoiesis, a theory that is useful as a counterweight to externalism but ultimately goes too far in the other direction.

Systems theory proposes that the world consists of different kinds of systems—open, closed, isolated, operating at different scales. It usually adheres to the principle that a system is irreducible to the sum of its parts, and thus a kind of holism. We already discussed one kind of systems theory above, namely dynamical systems theory. Autopoiesis is another version. It states that living cells maintain themselves by recursively achieving what the authors named *operational closure* (Varela et al. 1974). It has two main features: self-maintenance through circular organisation and structural coupling, which is an ongoing exchange with the environment which affects the structure of the system. From these features emerges life which 'cannot be reduced to the properties of the components' (Žukauskaitė 2020: 145).

By self-assembly, the components produce a physical boundary, a membrane, by which they preserve their identity. Hence the term autopoiesis, self-making, which has to do with how cells internally produce the components needed to sustain their physical boundary, bestowing autonomy upon biological systems. The system is thus self-referential, as every part depends on the organisation of the other parts (Hooker 2011). Autopoiesis shares facets with organicism, as seen in how a biological system emerges from a continuous flux of interrelated processes (Raimondi 2021). And like the organisational approach, it emphasises how the organisation of a biological system remains invariant through constant change and regeneration of its parts. There are, however, some differences...

For our purposes, two criticisms of this view should be noted initially: Firstly, the interaction of the autopoietic system with its environment is not considered constitutive but decided from within the system itself. There is thus a causal asymmetry involved, where the autopoietic system permanently has the upper hand. While this move combats determinism—by providing the organism with autonomy from efficient causality—it has the negative effect of casting the constitution of the organism as independent from its environment (Moreno and Mossio 2015). Secondly, the theory is formulated in abstract terms, ignoring the concrete material and energetic instantiation of the system—which is therefore considered 'purely contingent to its realisation' (Moreno and Mossio 2015: xxviii). Both these criticisms dovetail with Haraway's concerns.

On the face of it, autopoiesis may seem an obvious perpetuation of the organicist view, as it highlights the minimal organisation needed for producing a membrane which individuates cells. Yet, Haraway is hesitant:

> I'm sympathetic to certain kinds of cybernetic efforts to think through autopoiesis. [...] I am nonetheless deeply resistant to systems theories of all kinds, including so-called third-order cybernetics and the autopoiesis and structural coupling approaches. I'm not really happy there, but I remember that there is much more than Norbert Wiener in cybernetics. (Gane and Haraway 2006: 139)

Here, we see a reluctance toward systems theory, and autopoiesis seems to explain why. What does Haraway repudiate here? When she comments on her dissertation in the same interview, she emphasises that systems are not exhaustive, nor are they optional. From this, it seems the problem with systems theory and cybernetics has to do with a non-committal ontological stance, pretending to be outside the system it regards—not 'part and parcel of that order', as Weiss says. In other words, systems theory entails a *logic of domination* instead of participation. It hides the fact that we are situated within the systems which affect us, and that we perform 'worlding operations' whether we are aware of it or not:

> Our systems are probabilistic information entities. It is not that this is the only thing that we or anyone else is. It is not an exhaustive description but it is a non-optional constitution of objects, of knowledge in operation. [...] It is a statement that we had better get it—this is a worlding operation. Never the only worlding operation going on, but one that *we had better inhabit as more than a victim.* (Gane and Haraway 2006: 139, my emphasis)

Systems are not (only) theoretical constructs affecting us from the outside but *worldings*. We are agents in and of their becoming, not innocent bystanders. Worldings concern embodied processes through which a world is produced. It is a blending of material and semiotic dimensions (matter and meaning)—an ongoing production of a context through which things appear meaningful to organisms. Since 'natures, cultures, subjects and objects do not pre-exist their intertwined worldings' (Haraway 2016: 13), it goes beyond self-production to involve the production of other organisms and species. Thus, it is profoundly symbiotic—an ecological concept having to do with the material effects of interaction.

This might not seem very different from the idea—found in autopoietic theory—that knowing and doing are inseparable processes, and that an observer of the world also brings this world into existence (Maturana and Varela 1980). In Weber and Varela's (2002) words, 'an autonomous being who does not suffer passive world encounter but fashions a world of meaning from within' (115). I nonetheless argue that there is something lacking in this view: While it acknowledges the unfinished state of any worlding, it does not account for relationalities and its multispecies dimension since it suggests that the world is made meaningful from *within* as if cut off from the outside world.

From Auto- to Sym-

> [I]f developmental symbioses represent the rule and not merely the exceptional case, then the entire notion of "autopoiesis" must be abandoned. We are not adults entering into symbiotic relationships with other adults or microbes. Rather, the processes that made us adults are already the interactions between us and our microbes. (Gilbert 2002: 213)

Sympoiesis was proposed by Beth Dempster (2000). She writes: 'The essential differences between the two types of poietic systems relate to the presence and lack of self-defined boundaries and their different degrees of organizational closure' (Dempster 2000: 2). Instead of discarding autopoiesis, however, she opts for plurality. She touches upon restrictions in autopoietic theorising and how they should be complemented. Autopoietic theories do not theorise ecological relationships all the way down, and therefore cannot explain how a biological system maintains itself (Di Paolo 2009). It resorts to metaphors and intuitions, and thus fails to account for the richness between the organism and world, overlooking 'the organism-making aspects of the world at the expense of the worldmaking aspects of the organism' (Di Paolo 2009: 13). In short, it states that the organism is self-producing without accounting for how this production is maintained.

Maturana and Varela (1980) hold that individual unities cannot evolve, since they keep their identity only by maintaining their organisation; evolution only pertains to systems evincing a 'succession of identities generated through sequential reproduction' (104). The notion that identity is maintained in an ongoing fashion through constitutive interaction with the outside is not entertained. Saying there cannot be any evolution within an organisation, only between them, implies a strict division between ontogenesis and evolution, challenged by studies on the formative interaction that both processes rely on.[7] Sympoietic systems are evolutionary, whereas autopoietic ones are only developmental. An autopoietic system cannot be evolutionary since, in this view, self-contained organisms cannot undergo evolution; only

[7] Varela has since acknowledged the shortcomings of his original formulation and now opts for an enactive approach to cognition and life. Historical reciprocity has thus replaced the emphasis on the internal definition of autopoietic systems (see Di Paolo 2018).

lineages of different organisms do. Sympoiesis, on the other hand, concerns the distributed maintenance of a viable life cycle.

Dempster explains that while autopoietic systems concern growth from less to more developed stages, 'sympoietic systems are continually, though not necessarily consistently, changing' (Dempster 1998: 36). Their trajectory is unpredictable and can be subject to 'dramatic and surprising change'. Furthermore, autopoietic systems are centrally controlled while sympoietic systems evince distributed control. Another facet of sympoietic systems is that they are *homeorhetic*, not homeostatic. This means that

> their balance is maintained by dynamic tension: the mechanism for holding the position is self-organization. Interactions among system components, coupled with various 'external' influences, generate an evolutionary dynamic balance, which provides a degree of stability and predictability in these systems. (Dempster 2000: 12–13)

Or, as Buiatti and Longo (2013) say, it means that 'during their cycles, they keep changing' (148). They need to maintain their dynamic connection within levels of organisation, a notion reminding us of Needham's perspective. In line with Stuart Kauffman (1993), whose perspective we explore more in the next chapter, the balance that a system needs to maintain itself cannot be stated in advance but is a function of the system's self-organisation. Homeostasis, by contrast, is not necessarily produced by the system itself; it can be also imposed externally. Strictly speaking, autopoiesis refers to abstract synchronic, homeostatic models (see Maturana and Varela 1980; Juarrero 1999).

While not rejecting autopoiesis wholesale, Haraway prefers sympoiesis or *making-with*. Autopoiesis does not concern itself with self-organisation, says Haraway (2008)[8]; instead, it assumes organisation to already be in place. Sympoiesis, on the other hand, explains how biological organisation is generated and maintained collectively. By situating autopoiesis in an ecological context, it accounts for the dynamic self-organisation of a system. In a sense, there is no autopoiesis without sympoiesis, without the distributed agency which emerges from the interplay of processes. Furthermore, while autopoiesis accounts for how symbiosis is required for normal *functioning*, it fails to comprehend how symbiosis is also 'necessary for normal animal development' (Gilbert 2019: 15). Stated differently, it does not account for the coordination between 'relational interactions and internal compensations' (Di Paolo 2009: 17), it simply assumes it since autopoiesis would not be feasible without it.[9] By contrast, sympoiesis follows the logic of symbiosis, of how

[8] This might be a misreading, as the authors of autopoietic theory never said that autopoiesis means independent self-production; instead, it has to do with self-maintenance, with systems that *re*-produce and sustain their components (Buchanan 2018). This confusion notwithstanding, there are other problems with autopoiesis, having to do with whether it applies only to living systems and if it *sufficiently* explains such systems.

[9] This might explain why some have called for complementary approaches to the internal dynamics described by autopoiesis. Thompson (2007) says that autopoiesis can be complemented by developmental systems theory, as the latter highlights the couplings with the environment necessary for normal development. My take shares similarities with this proposal but based on the fundamental limitation of DST elucidated in the last chapter, it seems insufficient for the task at hand.

the body itself becomes a constructed niche through its relations to others (Gilbert 2019). In a way, then, we could say that autopoiesis takes organisation as an answer, not as a question. This view entails that the life cycle is maintained by the whole organism *including* its symbionts. It concerns the holobiont—the whole assemblage of an organism plus symbionts it interacts with and depends on for normal development (Gilbert and Tauber 2016)—, a notion that overlaps with closure of constraints, to which I now turn.

The Return of Organisation

> Most biologists would admit that even today no adequate theory of the organism exists, from whatever perspective. (Haraway 2004: 204)

We have seen that a mechanical approach does not discard organisation completely; instead, it takes it as self-evident instead of making it the object of study. In other words, it becomes a tacit background condition. But in recent years, the importance of organisation and fields has again become the subject of study. In the ensuing sections, I relate the views of the earlier organicists to more recent works on organisation and infer why Haraway is sceptical of autopoiesis by comparing it to *closure of constraints* (Montévil and Mossio 2015)—a reinterpretation of the autopoietic outlook informed by the works of Robert Rosen and Stuart Kauffman (Montévil 2018).

Closure of constraints denotes a closure which is specific to biological systems, where the whole organisation maintains itself by the interdependence of constraints. There is a dialectical relation at play, highlighted by Jaeger (2019), between the material processes that compose an organism and system-level constraints acting upon these processes. Constraints canalise processes while conserving themselves on the relevant timescale of the processes on which they act (Montévil and Mossio 2015). Also, there is an asymmetry here because the constraints affect the processes, but the processes do not affect the constraints at the same interval. Of course, at other timescales, constraints are themselves subject to change. Constraints are processes that are slower than the processes they constrain (Meincke 2018). While they are maintained by the processes they constrain, they display relative autonomy and invariance vis-à-vis them. This asymmetry is called *dynamic decoupling*—'by which the regulatory subsystem exhibits degrees of freedom that are *not specified* by the dynamics of the regulated one' (Bich et al. 2020: 7, my emphasis).

As we have seen, constraints act as local boundary conditions on processes, but they do not simply restrict their action. They are also enabling. This point can hardly be stressed enough. By reducing their degrees of freedom, constraints canalise processes and thus make something possible at the expense of something else. This might sound paradoxical, but it simply means that it channels the local flux of energy and matter so that it can be put to work in a non-random fashion. Hence, everything that modifies the default state of a system could be considered a constraint.

Ascription of biological function is tied to the existence of constraints on the behaviour of the parts. Constraints allow for a kind of *recursive specification* (Kauffman 2000), as the system maintains itself through its components—contributes to the determining of its conditions of existence. Furthermore, constraints achieve closure when there is 'mutual dependence between a set of constituents which could not exist in isolation, and which maintain each other through their interactions' (Montévil and Mossio 2015: 2). As such, closure denotes mutual dependence between constraints, where the degrees of freedom at one level are influenced by that of the next above level. Constraints at different levels explain how this works: First-order constraints are acted upon by second-order, regulative constraints, which 'provides the organism with the possibility of acting upon its own dynamics' (Bich et al. 2020: 6). In fact, no first-order constraints are unregulated since the notion of an isolated first-order regime is a theoretical construct. While first-order constraints directly channel metabolic processes, regulatory constraints modulate their activity. They must therefore be context-sensitive and able to qualitatively change the function of other constraints. This dialectic of constraints and processes maintains or reproduces living systems: 'If each constitutive constraint in a livings system is both dependent on and generative for at least one other constraints, then there is a *closure of constraints*' (Jaeger 2024: 163, emphasis original).[10] Again, this is not just a matter of conservative *restraints*. Talk of constraints between levels (and in general) must be thought of more like 'bias': restrictions on freedom contribute to the constructive aspect of canalisation. As such, constraints are productive, not only conservative.

Structural Coupling

We are literally making our world together, we critters. (Kauffman 2000: 75)

There are evident similarities between the autopoietic and organisational approaches. The main issue for Montévil and Mossio (2015) is its vagueness concerning the concrete interaction between thermodynamical openness and organisational closure. They write, that 'no details are given regarding how the two dimensions are interrelated, how closure is actually realised, what constituents are involved, and at what level of description' (Montévil and Mossio 2015: 180). It is thus unclear what is distinctly biological about the model. Take structural coupling, which defines the interdependence between organism and environment within the autopoietic approach. Such couplings may give rise to composite unities, which may themselves not be autopoietic if the relations established between them do not give rise to a self-producing system which maintains itself. A higher order autopoietic system

[10] The current organisation presupposes former closure. This is called organisational continuity, which underscores how continuous change is not optional for living systems, but the only way they can reproduce themselves (DiFrisco and Mossio 2020).

is one 'whose autopoiesis entails the autopoiesis of the coupled autopoietic unities which realize it' (Maturana and Varela 1980: 109).

The parts involved in realising the higher order autopoiesis remain unities unto themselves. They retain the characteristic feature of autopoietic systems: specifying their own boundaries. If not, they would be allopoietic, i.e. producing something other than themselves. This reveals that the structural couplings involved in autopoiesis are external (Dempster 1998). They occur in the cognitive realm—as a matter of correlated mental states, not correlated *operating conditions* (Di Paolo 2018). It means that, according to autopoietic theory, the effect that an organism has on its environment is considered secondary 'by-products [...] not important for describing the nature of the system' (Chemero 2012: 53–54). The move from autopoietic theories to properly *enactive* theories,[11] says Ezequiel Di Paolo, happens when the changes in both states and operating conditions, i.e. when 'coherence [...] operate as constraints and facilitative conditions that shape the structure of the organism and the environment (including other organisms) through time' (Di Paolo 2018: 88).

The founders of the autopoietic approach were aware that such systems depend on defining a world in which it may exist 'in relation to their autopoiesis' (Maturana and Varela 1980: 123). It is always caught up in such interactions with the world, unable to step out a view itself from an observer's perspective. But it seems more like a projection made from inside a self-contained system than something which is co-constructed with it. Some have therefore argued in favour of an extended autopoietic approach, which addresses the problems posed by biotic entrenchment, which is the process of integrating heterogeneous elements into the internal functioning of a biological system (Hernández and Vecchi 2019; Virgo 2020; Virgo et al. 2011). On this account, one may characterise multispecies partnerships in terms of closure of constraints, even if the partners by themselves do not achieve closure. Thus, one must ascribe closure to the extended organism.

But as Hernández and Vecchi (2019) emphasise, there are cases in which such a description is not preferable, as the parts of the partnership can be said to achieve closure by themselves. In other cases, however, the integration is so extensive that the partners have causal capacities which they would not have outside the partnership. Hernàndez and Vecchi argue that the idea of *tendency towards closure*, which I discuss more below, solves the problem of whether the partner, partnership or both should be said to achieve closure. This tendency is tied to the degree of merging between systems of constraints.[12]

---

[11] Enactivism is a theory within the philosophy of mind which accentuates the active involvement that an organism has with its environment, and how cognition is 'an embodied, lived process, based on self-organizing and recurrent sensorimotor patterns' (Read and Szokolszky 2020: 2; see also Di Paolo et al. 2018). In this book, I also draw on insights from ecological psychology. While I cannot do full justice to these topics, related to philosophy of mind, I plan to engage more with them in future works. I hope, however, to indicate the mutual relevance of topics in philosophy of mind, philosophy of science, and philosophy of nature. Consult Sandnes Haukedal (2025) for more on philosophy of mind.

[12] Building on the organisational view, Heras-Escribano and Saborido (2023) have proposed that we should understand parts of the environment as subject to organisational closure, and therefore speak

The limitations of autopoiesis seem tied to the cybernetic logic of feedback loops, which might be an apt metaphor for machines, but inadequate for organisms. Feedback loops are non-specific when it comes to levels of organisation, and the constraints that different levels impose on each other.[13] From the organisational perspective, they are not specifically biological but generic. As more levels become involved in the function, a simple feedback model becomes increasingly deflationary. It flattens them and does not account for their complex interaction and constitution.

Organisational models have greater explanatory power, as they subsume feedback mechanisms under the larger dynamics of interacting constraints (Bich et al. 2020). They disclose an unresolved tension between 'the individual-metabolic and the collective-ecological sides of the phenomenon' (Ruiz-Mirazo et al. 2004: 329). Moreover, organisational closure allows for a 'richer description of the processes and constraints at play [...] by making explicit the different hierarchical orders involved' (Bich et al. 2020: 1). It reveals how the integrity of the organism depends on the environment, and how lower-level processes are constrained, not dominated, by more complex levels.

We could recast the problem as a contradiction between autopoiesis and symbiosis incorporated by sympoiesis. Furthermore, Ruiz-Mirazo and others claim that resolving this tension demands combining the top-down and the bottom-up approaches; the overarching constraints on life must be explained in tandem with the processes that enable such constraints. Haraway's concerns seem to overlap with theirs: Autopoietic models are limited because they do not render the full implication of 'being inserted in a collective-ecological evolutionary dynamics' (Ruiz-Mirazo et al. 2004: en3, 342).

Surely, the operational closure of autopoietic does not mean isolation from the environment. Biological autonomy is achieved through recursive webs of interaction (Kirchhoff 2018). Yet, environmental processes are not constituent parts of the formation of the autopoietic systems. From an embryological perspective, this means that organisms are fully formed adults before they enter symbiotic relations, instead of becoming adults in and through such associations (Gilbert 2002).[14] So, while Maturana and Varela (1980) propose a theory of relations between processes and not

about an integrated organism-environment system. This view, which is aligned with ecological psychology, underscores an overlap between question on biology and psychology which should be researched further.

[13] Bich et al. (2020) enumerate three shortfalls with feedback loops, through the example of glucose regulation: first, it tends to neglect the nature of the relations between the parts and their place within the whole organism; second, it flattens the description of the system by overlooking the various categories of objects in play as well as their hierarchical relations; third, it is built on the relationship between concentrations of glucose, insulin and glucagon, and does not foster the inclusion of further factors involved in the regulation of glucose metabolism such as the nervous system and the gut.

[14] John Collier (2004) claims that the fact that autopoiesis does not imply self-organisation but presupposes an already organised self leads Maturana and Varela into an awkward position. They would say that organisation is spontaneous, *and*—like Haraway—that nothing self-organises. Simultaneously, autopoiesis is supposed to be the basis of individuality. Collier concludes, 'the problem is that they have not properly considered the role of self-organization in autonomous systems, or for that matter, taken self-organization seriously at all. At best, an autopoietic system

'properties of components' (113), they make interspecies relations optional and thus trivial (Kirchhoff 2018). As Moreno and Mossio (2015) state,

> interactions with the environment do not enter into the definition-constitution of the autonomous system; rather, the interactions with the environment [...] follow on from the specific internal identity of each autopoietic system. (xxvii)

Autopoietic systems lack immediate contact with the environments, even if they must remain energetically open to endure. Adding ambiguity, Maturana and Varela conceive of interaction as possibly forming higher-order autopoiesis, in which one system is subordinated (allopoietic) to another, but *not* as an inherent part of producing the autonomous system itself. In my view, Montévil and Mossio (2015) offer a less equivocal way of articulating what Haraway is seeking while abiding by the demands of Maturana and Varela.[15]

In summary, autopoietic theory lacks a clear definition of biological closure. This criticism, first made by Fleischaker (1988), concerns the abstract nature of autopoietic models—how physical conditions of possibility are unspecified.[16] We could say that the model fits the scriptures of physics, which dealt with generic and ahistorical objects, instead of biology, in which objects are specific (Montévil 2020). And even if the claim that nothing makes itself might not go beyond the original meaning of autopoiesis, it is worth noting that nothing *maintains* itself independently either. A biological system might achieve closure—and maintain itself—only by making-with-others, in a constitutive and nonoptional interplay. Sympoiesis, then, combines autopoiesis and allopoiesis, merging the making and maintenance of self and other (Žukauskaitė 2020). Organisms are not units unto themselves, and thus autopoietic, but multispecies consortia (Haraway 2016). While autopoiesis does not deny such interplays, it diminishes their importance by focussing on dynamics internal to the system (see Kirchhoff 2018).[17]

---

can be self-governing, and rearrange its organization, but it cannot produce new organization. This is more akin to a mechanistic device than biological systems or human minds.' (2004: 168)

[15] 'An adequate theory of the biological phenomenon should permit the analysis of the dynamics of the concrete components of a system in order to determine whether or not the participate in processes that integrate a biological phenomenon' (Maturana and Varela 1980: 114).

[16] As Moreno writes: 'Maturana and Varela have not analyzed which type of material organization may satisfy this requirement. Actually, their view on autonomy was explicitly abstract and functionalist. More recently, other authors have developed different approaches on the concept of biological autonomy. Hooker, Collier, Christensen, and Bickhard [...] have stressed both the material-energetic and the interactive dimension of autonomous systems, derived from the fact that the systems are in far-from-equilibrium conditions: since the cohesive organization of these systems exists only in far-from-equilibrium conditions, they must maintain an adequate interchange of matter and energy with their environment in order to keep this cohesion. Otherwise, these systems will disintegrate. Therefore, an autonomous system needs an internal mechanism able to organize and channel energy flows for the system's self-maintenance' (Moreno 2013: 61).

[17] Godfrey-Smith (1996) used autopoiesis to exemplify what he meant by internalism. Compare to Kirchhoff (2018) 'What I am *not* saying is that the AT [theory of autopoiesis] is committed to *metaphysical internalism* when it comes to expressing the relationship between mind and its material realizers. The point is merely that when explaining the organization of a living system, the classical formulation of the AT focuses on dynamics *internal* to a system' (2528).

And even if the autopoietic model had incorporated such conditions it would nevertheless not count as evolutionary, since '[t]he ability to produce diversity and growth in complexity as a result of a selective evolutionary process [...] is not at reach for autopoietic systems' (Ruiz-Mirazo et al. 2004: 329). Having omitted questions of the material and thermodynamical conditions underlying biological organisation, they are, as Marques and Brito (2014) write, 'struggling to give a diachronic account of how material systems came to express this peculiar circular organization'; furthermore, 'they offer a bottom up approach, which builds from physical self-organizing process up to biological, and cognitive, complexity' (98). Because of these reasons, theories of autopoiesis are not up for the task of formulating a more complex systems biology.

Holobiont Organisation

> I did not write "smallest possible units of analysis" because the word *unit* misleads us to think that there is an ultimate atom made up of internal differential relatings, which is a premise of autopoiesis and other theories of organic form [...]. I see only prehensile turtles all the way up and down. (Haraway 2008: en31, 313)

The unit of selection may be the organism plus its symbionts, not the organism as an individual. Haraway seems aligned with those who argue that the holobiont constitutes a level of selection, such as Gilbert et al. (2018). Suárez and Triviño (2019) regard this as a question of downward causation, a view that fits with our overall organisational approach. Insofar as a holobiont exerts downward causation on its components, they argue, it should be conceived of as an organism in an extended sense. In terms of closure of constraints, the different organisms act as constraints on each other, achieving higher-order closure as an interdependent consortium (Nunes-Neto et al. 2014; Mossio and Pontarotti 2020).[18] Similarly, Montévil and Mossio reveal that there are cases where it does not make sense to say that one organism alone achieves closure. This is precisely the kind of constitutive intra-actions Haraway is concerned with, and thus worth quoting at length:

> Let us now turn to all those cases in which two or more biological organisms establish a form of mutual dependence due to stable interactions between them, such that each of them can be said to rely on the other(s) for its own maintenance. In these situations, in which a fundamental organisational continuity exists between the interacting organisms, the upper boundaries of closure ascription seem to extend beyond each organism, insofar as the notion of maximally closed system applies only to the encompassing system which contains all (known) constraints subject to closure. If we were to limit ourselves to this analysis, it would be impossible to describe systems including different nested levels of organisational closure and systems belonging to closed systems (and specifically mutually dependent organisms) would not themselves realise closure as discussed above. Moreover, *since biological organisms are systematically involved in such interactions it would follow*

[18] Whether they are best seen as individuals or communities is an ongoing debate that I will not attempt to settle but see Skillings (2016), Suárez and Triviño (2019), and Gilbert (2014).

> *that most of the time individual organisms cannot be said to realise closure.* The main theoretical upshot would be a serious weakness for any account based on closure, which could not be considered a distinctive property of organisms in many biologically relevant cases. (Montévil and Mossio 2015: 10, my emphasis)

Interdependent systems of closure achieve higher-order closure *without* the closure of each system breaking down. The boundaries of the individual organism do not coincide with the boundaries of the self-maintaining organisation (Moreno and Mossio 2015; Jaeger 2024). Had they coincided, closure of constraints would not be able to account for constitutive intra-action between organisms, through which they maintain themselves.

Montévil and Mossio (2015) mention mutualistic symbioses as cases where you can delineate boundaries between partners, but no strict discontinuity exists. The holobiont realises an organisation that extends beyond the organism and displays simultaneous differentiation and integration (Mossio and Pontarotti 2020). In these cases, they propose the notion of *tendency towards closure*, which is 'a measure of the degree of organisational integration of organisms and, as well as, an operational tool for drawing the boundaries between them, even when they establish functional dependence' (Montévil and Mossio 2015: 22).[19] Closure is never complete—as the organism depends on external constraints—and it does not need to be. It only needs to be sufficiently closed for the system to reproduce itself through its components.[20]

Insofar as traits are reconstructed over generations, through a host of interactions with members of the microbiota, a recurrent unit of selection is produced. The organisms which fail to reconstruct these relations, on the other hand, are less likely to survive and reproduce. This indicates why closure of constraints has to do with evolutionary systems—not the development of one organism but the recursive specification of the whole web of constraints needed to maintain and reproduce that organism. We find a similar claim in *When Species Meet*:

> Gilbert's approach is not a holistic systems theory in the sense that Margulis and Sagan lean toward, and his fractal "turtles all the way down" arguments do not posit a self-referential unit of differentiation. [...] Rusten Hogness suggests the term *turtling all the way down* might better express Gilbert's kind of recursivity. I think that for Gilbert the noun *differentiation* is permanently a verb, within which mortal knots of partly structured difference are in play. In my view, Margulis and Sagan's symbiogenesis is not really compatible with their theory of autopoiesis, and the alternative is not an additive mechanistic theory but a going even more deeply into differentiation. (Haraway 2008: 33, emphasis original)

---

[19] They continue: 'It is worth emphasising, in this respect, that such a measure comes in degrees. For example, one can conjecture that the tendency to closure is higher for a unicellular eukaryote than for a cell in a metazoan. Similarly, the tendency to closure of a biofilm is arguably weaker than that of an individual bacterium, or a metazoan. The same differences might also emerge when comparing closed systems located at various nested levels of organisation (see the following subsection), such as, for instance, in the case of the ant and its colony' (Montévil and Mossio 2015: 22).

[20] When I say 'external' here, I use it like Di Paolo et al. (2018: 25): an external process is one that either (1) affects the system but is not itself enabled by it, or (2) a process that is enabled by a system but does not enable it in return. These processes do not enter the *closure* of the network, even if they interact with it.

Unlike dissipative systems, a biological system must differentiate the functions of the parts to maintain itself. Viewing differentiation as a verb underscores the ongoing and permanently unfinished reproduction of organisation. A living system must integrate new components and differentiate their functions constantly to survive.

Gilbert's 'interspecies epigenesis' indicates a contradiction between symbiogenesis and autopoiesis that Margulis, who adopted Maturana and Varela's (1980) idea of 'composite unity', seems unaware of. In *Staying with the Trouble* (Haraway 2016), Haraway explains that the problem is not systems theory *per se*, but a restricted form of it which is too tied up with the logic of autopoiesis and bounded individualism. Again, nothing truly *self*-organises: 'Symbiosis makes trouble for autopoiesis, and symbiogenesis is an even bigger troublemaker for self-organizing individual units' (Haraway 2016: 61). While autopoiesis is an important aspect of complex systems, self-organisation is outright misleading if it is considered the workings of an individual unit. The challenge is 'finding an idiom for the paradoxical and indispensable linkages of openness and closure, [...] repeated recursively' (Haraway 2008: 46). Arguably, the organisational perspective provides this idiom.

Conclusion

The space of possibilities of a biological system is not given in advance, and thus cannot abide by cybernetic principles. It is not a closed system, where the 'mode and range being defined in advance in accordance to a set design' (Nicholson 2018: 147). By contrast, as they tend towards closure, biological systems become functionally differentiated and integrated, as well as self-maintaining. As such, 'the activity of the whole system plays a role in producing and maintaining its parts over time' (Bich et al. 2020: 2). Explanations of homeostasis in terms of feedback loops fail to account for such processes or their constitution. Like autopoiesis, they do not explain how a new organisation is brought about, only how an organisation which is already produced—like a device—can be modified (Collier 2004).[21]

According to Bich, Mossio, and Soto, such models tend to (1) take neat localization of functions for granted, a strategy that fails insofar as many functions are distributed over the whole system; (2) it represents systems in a flat manner, without any hierarchies, which means that there is 'only one kind of "causal relation"' (Bich et al. 2020: 4); (3) searching for additional components and variables is not promoted, and (4) 'description in terms of feedback assumes the existence of a value (or, more precisely, an interval of values) to be kept stable—a set point—without providing an

[21] Nunes-Neto et al. (2014) state: '[C]losure of constraints is not synonymous with homeostasis. The notion of homeostasis implies that the organization of the system is maintained by homeostatic mechanisms, which are not necessarily produced by the system itself; instead, the idea of closure of constraints means that the system is maintained by constraints generated by the very organization of the system. In sum, the notion of closure of constraints is a more detailed representation (or elaboration) of self-maintenance than homeostasis' (129).

explanation of how it is established or how it can be modified' (Bich et al. 2020: 4). In short, it fails to account for the historical constitution of causal relations.

From our perspective, the fourth point enumerated is most relevant, as it accentuates the processual dimension that is encapsulated by *worlding*, for which homeostasis and feedback loops are limited theoretical tools. For living organisms, there exists no pregiven 'normal range' by which to measure a variable. The normal range is a function of the ongoing activity of the whole, and there are no criteria of normality for comparison. In a word, a biological system is historical, in that its dynamics change through time. The same cannot be said for physical systems, whose objects display invariance instead of ongoing change. Conversely, sympoietic systems maintain themselves through continuous change (Dempster 1998). This is called *extended criticality* in the organisational approach and denotes how biological systems undergo constant changes as their structures are inherently unstable (see Chap. 6). These changes happen at thresholds which are constructed by the system itself in an ongoing manner. It explains why biological systems are homeorhetic, viz. nonstationary in the sense that they return to a specific trajectory (Bechtel and Bich 2024). They do not maintain specific states, as homeostatic systems, but the whole dynamics and trajectory of the system. They maintain their identity through continuous change.

Some systems outsource the production of components they need to maintain themselves to other organisms and undergo genome reduction through this outsourcing (see West-Eberhard 2003). This indicates the opportunism of organisms, taking the most energy-efficient path afforded to them. There are systems within systems, *turtling all the way down*:

> We conjecture that a relationship between two closures of constraints which involves both separation and a nested hierarchy provides the theoretical basis for characterising, in our framework, a distinction between levels of organisation. Two closed regimes constitute two different levels of organisation if they are both separated and hierarchically nested; accordingly, cells and multicellular organisms constitute two different levels of organisation. (Montévil and Mossio 2015: 25)

This overlaps with Dempster's definition of sympoiesis, as well as models of extended autopoiesis. It does not mean erasing all boundaries between systems. Instead, there is a kind of centrifugal organisation of different closures, in which one organisation might be dependent on another without therefore being swallowed by it. As Bich (2019) argues, one should not confuse 'self-specification of functional boundaries [...] with functional self-sufficiency' (300). There is nothing in the definition of closure which precludes cross-control and higher levels of closure—without thereby undermining the functional closure of the systems involved. This means that while the organism as a holobiont consists of many life cycles, each of these cycles achieves functional closure, while also contributing to the overall closure of the partnership between different species. Sympoiesis could then mean extending autopoiesis to encompass not only the production of membranes but 'processes in the ecological context' (Froese and Stewart 2012: 72). This means that autopoiesis derives from organisational closure, which involves multiple agents.

When we thematise closure, then, we are not speaking of something that is completely closed, but *ajar*, as Dempster says—marked by an ongoing tendency towards closure which is never completed. By dealing with Dempster's approach, Haraway sheds light on the obligatory relations between species, outside of which they could not survive. She also makes the case for the teleological capacity of organisms not only to react in a prescribed fashion to perturbations but to respond, which involves channelling the impacts of whatever befalls it. By reference to Gilbert et al. (1996), Haraway points towards a new synthesis, which assumes the field metaphor from her dissertation. 'In this new synthesis',

> the developmental or morphogenetic field is proposed to mediate between genotype and phenotype. Just as the cell (and not its genome) functions as the unit of organic structure and function, so the morphogenetic field (and not the genes or the cells) is seen as a major unit of ontogeny, whose changes bring about changes in evolution. (Gilbert et al. 1996: 357, quoted in Haraway 1997: en9, 325)[22]

The evolutionary impact of development and the need for a diachronic perspective is evident here, in opposition to the ahistorical method of molecular genetics: 'Embodied information with a complex time structure is reduced to a linear code in an archive outside time' (Haraway 1997: 245). The field metaphor is the most useful as a mediator between genotype and phenotype, underscoring that the path from one to the other is not linear.

To return to systems theory: Like Montévil and Mossio (2015), Haraway's main issue with systems theory seems to lie in its definition of closures. She is sceptical 'because nothing self-organizes' (Gane and Haraway 2006: 141). The notion of closure of constraint was proposed to explicate what is meant by closure for biological systems since the relation between material openness and organisational closure—and what kind of causal regime it would constitute—is unclear in autopoietic theories (Montévil and Mossio 2015). What is added by the notion of sympoiesis is an emphasis on the multispecies dimension of such closures, and their ajar nature, as ongoing processes of differentiation and integration. But such dimensions are already implicit in the notion of closure of constraints, as I have tried to show.

In summary, Haraway's philosophy provides fertile ground for diachronic biology of constitutive interaction. It offers a model of specificity not as stemming from lower-level determinism but from cyclical processes of interactions (Ruiz-Mirazo et al. 2008). While reduction is part of the scientific process, the reductive approach works only for synchronic entities and even then, it is never complete. Also, while it might enumerate the parts which are necessary for a specific function or mechanism, it cannot explain sufficient conditions, as that requires keeping the whole system in view. In other words, reduction is useful to test the limits of an analytic procedure, but it 'reaches a terminus at the watershed of life' (Longo et al. 2012: 1380).

---

[22] Coincidentally, Ochoa refers to the same article in his suggestion that a new synthesis, which he calls 'Unmodern'. This is not a synthesis of genetics of natural selection but instead a synthesis between development and evolution, a hypothetical "relegated Unmodern Synthesis" that would have linked embryology with evolution' (Ochoa 2024: 414). It undermines natural selection as creative, gene-centrism and the 'extrapolationist argument'.

This indicates a dialectical model where nature is underdetermined and thus cannot rigidly determine its products. Like Needham proposed, then, we must accept that there are emergent integrative levels in nature, which are irreducible to physics or chemistry. In the following chapter, I will examine this view further in relation to Hegel's philosophy, to show how it allows us to reconsider the interaction between actuality and possibility by introducing the notion that there is no given potential at the outset as possibilities are generated through the history of the system.

Bibliography

Bechtel, William, and Leonardo Bich. 2024. Situating Homeostasis in Organisms: Maintaining Organization through Time. *The Journal of Physiology* 602: 6003–6020. https://doi.org/10.1113/JP286883.

Bich, Leonardo. 2019. The Problem of Functional Boundaries in Prebiotic and Inter-Biological Systems. In *Systemics of Incompleteness and Quasi-Systems. Contemporary Systems Thinking*, ed. Gianfranco Minati, Mario R. Abram, and Eliano Pessa, 295–302. Cham: Springer. https://doi.org/10.1007/978-3-030-15277-2_23.

Bich, Leonardo, Matteo Mossio, and Ana M. Soto. 2020. Glycemia Regulation: From Feedback Loops to Organizational Closure. *Frontiers in Physiology* 11. https://doi.org/10.3389/fphys.2020.00069.

Buchanan, I. 2018. *A Dictionary of Critical Theory*. London, UK: Bloomsbury Publishing.

Buiatti, Marcello, and Giuseppe Longo. 2013. Randomness and Multilevel Interactions in Biology. *Theory in Biosciences* 132 (3): 139–158. https://doi.org/10.1007/s12064-013-0179-2.

Chemero, Anthony. 2012. Modeling Self-Organization with Nonwellfounded Set Theory. *Ecological Psychology* 24 (1): 46–59. https://doi.org/10.1080/10407413.2012.643729.

Collier, John. 2004. Self-Organization, Individuation and Identity. *Revue Internationale de Philosophie* 228 (2): 151–172. https://doi.org/10.3917/rip.228.0151.

Dempster, M. Beth. 1998. *A Self-Organizing Systems Perspective on Planning for Sustainability* (MA thesis, Environmental Studies, University of Waterloo).

Dempster, Beth. 2000. Sympoietic and Autopoietic Systems: A New Distinction for Self-Organizing Systems. In *Proceedings of the World Congress of the Systems Sciences and ISSS 2000*, ed. J. K. Allen and J. Wilby.

Depew, D. J., and B. H. Weber. 1995. *Darwinism Evolving: Systems Dynamics and the Genealogy of Natural Selection*. Cambridge, MA: MIT Press.

Di Paolo, Ezequiel. 2009. Overcoming Autopoiesis: An Enactive Detour on the Way from Life to Society, in *Autopoiesis in Organization Theory and Practice* (Advanced Series in Management) ed. Rodrigo Magalhães and Ron Sanchez, VI, 43–68. Bingley, UK: Emerald Publishing. https://doi.org/10.1108/S1877-6361(2009)0000006004.

Di Paolo, Ezequiel A. 2018. The Enactive Conception of Life. In *The Oxford Handbook of 4E Cognition*, ed. Albert Newen, Loen De Bruin, and Shaun Gallagher, 71–94. Oxford, UK: Oxford University Press.

Di Paolo, Ezequiel A., Elena Clare Cuffari, and Hanne De Jaegher. 2018. *Linguistic Bodies: The Continuity between Life and Language*. Cambridge, MA: MIT Press. https://doi.org/10.7551/mitpress/11244.001.0001.

DiFrisco, James, and Matteo Mossio. 2020. Diachronic Identity in Complex Life Cycles: An Organizational Perspective. In *Biological Identity: Perspectives from Metaphysics and the Philosophy of Biology*, ed. Anne Sophie Meincke and John Dupré, 177–199. London, UK: Routledge. https://doi.org/10.4324/9781351066389-10.

Emmeche, Claus. 2004. Organicism and Qualitative Aspects of Self-Organization. *Revue Internationale de Philosophie* 58 (228): 205–217.

Fleischaker, Gail Raney. 1988. Autopoiesis: The Status of Its System Logic. *Biosystems* 22 (1): 37–49. https://doi.org/10.1016/0303-2647(88)90048-2.

Froese, Tom, and John Stewart. 2012. Enactive Cognitive Science and Biology of Cognition: A Response to Humberto Maturana. *Cybernetics & Human Knowing* 19 (4): 61–74.

Gane, Nicholas, and Donna Haraway. 2006. When We Have Never Been Human, What Is to Be Done? Interview with Donna Haraway. *Theory, Culture & Society* 23:135–158. https://doi.org/10.1177/0263276406069228.

Gilbert, Scott F. 2002. The Genome in Its Ecological Context: Philosophical Perspectives on Interspecies Epigenesis. *Annals Of The New York Academy Of Sciences* CMLXXXI: 202–218. https://doi.org/10.1111/j.1749-6632.2002.tb04919.x.

Gilbert, Scott. 2014. A Holobiont Birth Narrative: The Epigenetic Transmission of the Human Microbiome. *Frontiers in Genetics* 5: 282. https://doi.org/10.3389/fgene.2014.00282.

Gilbert, Scott. 2019. Evolutionary Transitions Revisited: Holobiont Evo-devo. *Journal of Experimental Zoology Part B: Molecular and Developmental Evolution* 332: 307–314. https://doi.org/10.1002/jez.b.22903.

Gilbert, Scott F., and David Epel. 2015. *Ecological Developmental Biology: The Environmental Regulation of Development, Health, and Evolution.* 2nd ed. Sunderland, MA: Sinauer Associates, Inc. Publishers.

Gilbert, Scott F., and Alfred I. Tauber. 2016. Rethinking Individuality: The Dialectics of the Holobiont. *Biology & Philosophy* 31 (6): 839–853. https://doi.org/10.1007/s10539-016-9541-3.

Gilbert, Scott F., John M. Opitz, and Rudolf A. Raff. 1996. Resynthesizing Evolutionary and Developmental Biology. *Developmental Biology* 173 (2): 357–372. https://doi.org/10.1006/dbio.1996.0032.

Gilbert, Scott, Emily McDonald, Nicole Boyle, Nicholas Buttino, Lin Gyi, et al. 2010. Symbiosis as a Source of Selectable Epigenetic Variation: Taking the Heat for the Big Guy. *Philosophical Transactions of the Royal Society B: Biological Sciences* 365: 671–678. https://doi.org/10.1098/rstb.2009.0245.

Gilbert, Scott F., Jan Sapp, and Alfred I. Tauber. 2012. A Symbiotic View of Life: We Have Never Been Individuals. *The Quarterly Review of Biology* 87 (4): 325–341. https://doi.org/10.1086/668166.

Gilbert, Scott F., Eugen Rosenberg, and Ilana Zilber-Rosenberg. 2018. The Holobiont with Its Hologenome Is a Level of Selection in Evolution. In *Landscapes of Collectivity in the Life Sciences.*, ed. Snait B. Gissis, Ehud Lamm, and Ayelet Shavit, 305–324. Cambridge, MA: MIT Press.

Godfrey-Smith, P. 1996. *Complexity and the Function of Mind in Nature.* Cambridge, UK: Cambridge University Press.

Griffiths, P. E., & Stotz, K. 2013. *Genetics and Philosophy: An Introduction.* Cambridge, UK: Cambridge University Press.

Haraway, Donna J. 1991. *Simians, Cyborgs and Women: The Reinvention of Nature.* New York, NY: Routledge.

Haraway, Donna J. 1997. *Modest_Witness@Second_Millennium.FemaleMan_Meets_OncoMouse: Feminism and Technoscience.* New York, NY: Routledge.

Haraway, Donna J. 2004. *Crystals, Fabrics, and Fields: Metaphors That Shape Embryos.* Berkeley, CA: North Atlantic Books.

Haraway, Donna J. 2008. *When Species Meet.* Minneapolis. MN: University of Minnesota Press.

Haraway, Donna J. 2016. *Staying with the Trouble: Making Kin in the Chthulucene.* Durham, NC: Duke University Press Books.

Haukedal, R. S. 2025. The Concrete Universal of Embodied Concept. *Verifiche* 53 (The Concrete Universal: Relevance, Meanings and Perspectives): 97–122.

Heras-Escribano, Manuel, and Cristian Saborido. 2023. Affordances and Organizational Functions. *Biology and Philosophy* 38 (1). https://doi.org/10.1007/s10539-023-09891-4.

Hernández, Isaac, and Davide Vecchi. 2019. The Interactive Construction of Biological Individuality Through Biotic Entrenchment. *Frontiers in Psychology* 10. https://doi.org/10.3389/fpsyg.2019.02578.

Hooker, Cliff. 2011. Introduction to Philosophy of Complex Systems: B. In *Philosophy of Complex Systems* (Handbook of the Philosophy of Science) ed. Cliff Hooker, 841–909. Amsterdam: North Holland. https://doi.org/10.1016/B978-0-444-52076-0.50027-4.

Jaeger, Johannes. 2019. Dynamic Structures in Evo-Devo: From Morphogenetic Fields to Evolving Organisms. In *Perspectives on Evolutionary and Developmental Biology*, ed. Giuseppe Fusco. Padova: Padova University Press. https://doi.org/10.31219/osf.io/mprc8.

Jaeger, Johannes. 2024. The Fourth Perspective: Evolution and Organismal Agency. In *Organization in Biology*, ed. Matteo Mossio, 159–186. Cham: Springer. https://doi.org/10.1007/978-3-031-38968-9_8.

Juarrero, A. 1999. *Dynamics in Action: Intentional Behavior as a Complex System.* Cambridge, MA: MIT Press.

Kirchhoff, Michael D. 2018. Autopoiesis, Free Energy, and the Life–Mind Continuity Thesis. *Synthese* 195 (6): 2519–2540. https://doi.org/10.1007/s11229-016-1100-6.

Kauffman, S. 1993. *The Origins of Order: Self-organization and Selection in Evolution.* Oxford, UK: Oxford University Press.

Kauffman, S. 2000. *Investigations.* Oxford and New York: Oxford University Press.

Longo, G., M. Montévil, and S. Kauffman. 2012. No Entailing Laws, but Enablement in the Evolution of the Biosphere. In *Proceedings of the 14th Annual Conference Companion on Genetic and Evolutionary Computation* (GECCO '12), 1379–1392. New York, NY: Association for Computing Machinery. https://doi.org/10.1145/2330784.2330946.

Margulis, Lynn. 1993. *Symbiosis in Cell Evolution: Microbial Communities in the Archean and Proterozoic Eons.* New York, NY: W.H. Freeman & Co. Ltd.

Marques, Victor, and Carlos Brito. 2014. The Rise and Fall of the Machine Metaphor: Organizational Similarities and Differences between Machines and Living Beings. *Verifiche* 43 (1–4): 77–111.

Maturana, H. R., & Varela, F. J. 1980. *Autopoiesis and Cognition (Vol. 42).* Dordrecht: Springer Netherlands. https://doi.org/10.1007/978-94-009-8947-4.

McFall-Ngai, Margaret. 2014. The Importance of Microbes in Animal Development: Lessons from the Squid-Vibrio Symbiosis. *Annual Review of Microbiology* 68: 177–194. https://doi.org/10.1146/annurev-micro-091313-103654.

Meincke, Anne Sophie. 2018. Autopoiesis, Biological Autonomy and the Process View of Life. *European Journal for Philosophy of Science* 9 (1): 1–16. https://doi.org/10.1007/s13194-018-0228-2.

Montévil M. 2018. A Few Pending Challenges from the Perspective of a Theory of Organisms. *Constructivist Foundations* 13 (3): 377–379. http://constructivist.info/13/3/377.

Montévil, M. 2020. Historicity at the heart of biology. *Theory in Biosciences.* https://doi.org/10.1007/s12064-020-00320-8.

Montévil, M., and M. Mossio. 2015. Biological Organisation as Closure of Constraints. *Journal of Theoretical Biology* 372: 179–191. https://doi.org/10.1016/j.jtbi.2015.02.029.

Moreno, Alvaro. 2013. Autonomy. In *Encyclopedia of Systems Biology*, ed. Werner Dubitzky, Olaf Wolkenhauer, Kwang-Hyun Cho, and Hiroki Yokota, 60–62. New York, NY: Springer. https://doi.org/10.1007/978-1-4419-9863-7_51.

Moreno, Alvaro, and Matteo Mossio. 2015. *Biological Autonomy* (History, Philosophy and Theory of the Life Sciences) XII. Dordrecht: Springer. https://doi.org/10.1007/978-94-017-9837-2.

Mossio, Matteo, and Gaëlle Pontarotti. 2020. Conserving Functions across Generations: Heredity in Light of Biological Organization. *The British Journal for the Philosophy of Science* 73 (1): 249–278. https://doi.org/10.1093/bjps/axz031.

Needham, Joseph. 1936. *Order and Life*, Terry Lectures. New Haven, CT: Yale University Press.

Needham, Joseph. 1942. *Biochemistry and Morphogenesis*, Biochemistry and Morphogenesis. London: Cambridge University Press.

Needham, Joseph. 1943. *Time: The Refreshing River. Essays and Addresses 1932–1942.* New York, NY: Macmillan.
Nicholson, Daniel J. 2018. Reconceptualizing the Organism: From Complex Machine to Flowing Stream. In *Everything Flows: Towards a Processual Philosophy of Biology*, ed. John Dupré and Daniel J. Nicholson, 139–166. Oxford, UK: Oxford University Press. https://doi.org/10.1093/oso/9780198779636.003.0007.
Noble, D. 2006. *The Music of Life: Biology Beyond Genes.* Oxford, UK: Oxford University Press.
Nunes-Neto, Nei, Alvaro Moreno, and Charbel N. El-Hani. 2014. Function in Ecology: An Organizational Approach. *Biology and Philosophy* 29 (1): 123–141.
Ochoa, C. 2024. Deconstructing the Extended Evolutionary Synthesis: Do We Need a New Theory of Evolution? In *Unity and Disunity in Evolutionary Biology: Deconstructing Darwinism*, ed. R. G. Delisle, M. Esposito, and D. Ceccarelli, 403–439. Cham: Springer International Publishing. https://doi.org/10.1007/978-3-031-42629-2_14.
O'Malley, Maureen. 2015. Endosymbiosis and Its Implications for Evolutionary Theory. *Proceedings of the National Academy of Sciences of the United States of America* 112: 10270–10277. https://doi.org/10.1073/pnas.1421389112.
O'Malley, Maureen. 2017. From Endosymbiosis to Holobionts: Evaluating a Conceptual Legacy. *Journal of Theoretical Biology* 434. https://doi.org/10.1016/j.jtbi.2017.03.008.
Peterson, Eric L. 2017. *The Life Organic: The Theoretical Biology Club and the Roots of Epigenetics.* Pittsburgh, PA: University of Pittsburgh Press.
Polanyi, Michael. 1968. Life's Irreducible Structure. Science 160 (3834): 1308–1312. https://doi.org/10.1126/science.160.3834.1308.
Pradeu, Thomas. 2012. *The Limits of the Self: Immunology and Biological Identity.* New York, NY: Oxford University Press.
Raimondi, V. 2021. Autopoiesis and Evolution: The Role of Organisms in Natural Drift. *Adaptive Behavior*. https://doi.org/10.1177/10597123211030694.
Read, Catherine, and Agnes Szokolszky. 2020. Ecological Psychology and Enactivism: Perceptually-Guided Action vs. Sensation-Based Enaction. *Frontiers in Psychology* 11: 1270. https://doi.org/10.3389/fpsyg.2020.01270.
Ruiz-Mirazo, Kepa, Juli Peretó, and Alvaro Moreno. 2004. A Universal Definition of Life: Autonomy and Open-Ended Evolution. *Origins of Life and Evolution of the Biosphere* 34 (3): 323–346. https://doi.org/10.1023/B:ORIG.0000016440.53346.dc.
Ruiz-Mirazo, Kepa, Jon Umerez, and Alvaro Moreno. 2008. Enabling conditions for 'open-ended evolution'. *Biology & Philosophy* 23: 67–85. https://doi.org/10.1007/s10539-007-9076-8.
Saetzler, Kurt, Carlos Sonnenschein, and Ana Soto. 2011. Systems Biology beyond Networks: Generating Order from Disorder through Self-Organization. *Seminars in Cancer Biology* 21: 165–174. https://doi.org/10.1016/j.semcancer.2011.04.004.
Skillings, Derek. 2016. Holobionts and the Ecology of Organisms: Multi-Species Communities or Integrated Individuals? *Biology & Philosophy* 31 (6): 875–892. https://doi.org/10.1007/s10539-016-9544-0.
Sperry, Roger W. 1982. Bridging Science and Values: A Unifying View of Mind and Brain. In *Mind and Brain: The Many-Faceted Problem*, ed. John Eccles, 255–269. Washington, D.C.: Paragon House.
Stewart, J. 2019. *Breathing Life Into Biology.* Newcastle upon Tyne, UK: Cambridge Scholars Publishing.
Suárez, Javier, and Vanessa Triviño. 2019. A Metaphysical Approach to Holobiont Individuality: Holobionts as Emergent Individuals. *Quaderns de Filosofia* 6 (1): 59. https://doi.org/10.7203/qfia.6.1.14825.
Thompson, Evan. 2007. *Mind in Life: Biology, Phenomenology, and the Sciences of Mind.* Cambridge, MA: Belknap Press of Harvard University Press.
Varela, Francisco J., Humberto R. Maturana, and R. B. Uribe. 1974. Autopoiesis: The Organization of Living Systems, Its Characterization and a Model. *Currents in Modern Biology* 5: 187–196. https://doi.org/10.1016/0303-2647(74)90031-8.

Virgo, Nathaniel. 2020. The Necessity of Extended Autopoiesis. *Adaptive Behavior* 28 (1): 23–26. https://doi.org/10.1177/1059712319841557.

Virgo, Nathaniel, Matthew D. Egbert, and Tom Froese. 2011. The Role of the Spatial Boundary in Autopoiesis. In *Lecture Notes in Computer Science*, ed. G. Kampis, I. Karsai, and E. Szathmáry, MMMMMMDCCLXXVII, 240–247. Berlin and Heidelberg: Springer. https://doi.org/10.1007/978-3-642-21283-3_30.

Weber, Andreas, and Francisco Varela. 2002. Life after Kant: Natural Purposes and the Autopoietic Foundations of Biological Individuality. *Phenomenology and the Cognitive Sciences* 1: 97–125. https://doi.org/10.1023/A:1020368120174.

Weiss, Paul. 1960. Ross Granville Harrison, 1870–1959, Memorial Minute. *Rockefeller Institute QuarterLy* 6.

Weiss, Paul. 1968. *Dynamics of Development: Experiments and Inferences* (Selected Papers on Developmental Biology). New York, NY: Academic Press.

West-Eberhard, M. J. 2003. *Developmental Plasticity and Evolution. Developmental Plasticity and Evolution.* Oxford, UK: Oxford University Press. https://doi.org/10.1093/oso/9780195122343.001.0001.

Whitehead, Alfred North. 1967. *Science and the Modern World*. New York, NY: The Free Press.

Žukauskaitė, Audronė. 2020. Gaia Theory: Between Autopoiesis and Sympoiesis. *Problemos* 98. https://doi.org/10.15388/Problemos.98.13.

Chapter 5
Self-Reference and Enablement

Abstract This chapter advances a dialectical approach to biological organisation, synthesizing insights from Hegel, Žižek, and Stuart Kauffman to challenge mechanistic and reductionist frameworks. Central to the argument is Kauffman's concept of *the adjacent possible*, which illustrates how biological systems generate novel possibilities through their own activity, rather than unfolding pre-given potentials. To make this argument, key Hegelian concepts, such as negativity, retroactive positing, and true infinity, illuminate how contingency and necessity are internally related in evolution, with biological systems actively shaping their own conditions of existence. Quantum indeterminacy further underscores the ontological openness of nature, enabling downward causation and bio-resonance across levels of organisation. By integrating dialectical philosophy with systems biology, the chapter proposes a non-reductionist, processual ontology of life, where emergence, historicity, and self-determination redefine our understanding of biological agency. This framework not only bridges philosophy and science but also calls for a re-evaluation of teleology, moving beyond Kantian regulative ideals to a robust, materialist account of immanent organisation.

Introduction

> [T]he entire system is totally, intensely conservative, locked into itself, utterly impervious to any 'hints' from the outside world. Through its properties, by the microscopic clockwork function that establishes between DNA and protein, as between organism and medium, an entirely one-way relationship, this system obviously defies any 'dialectical' description. It is not Hegelian at all, but thoroughly Cartesian: the cell is indeed a *machine*. (Monod 1972: 110–11)

The epigraph from Jacques Monod epitomises the philosophical backdrop of the MS, a view that has been challenged in previous chapters and whose opposition to dialectical thinking has been indicated. In my view, Monod gets things backwards: By assigning the role of specifying the complete ontogenesis to genetic information alone, any circular or reciprocal determination is rejected. Causation is bottom-up,

R. Sandnes Haukedal, *Agency and Organisation: The Dialectics of Nature and Life*, Library of Ethics and Applied Philosophy 44,
https://doi.org/10.1007/978-981-95-1924-8_5

unilinear, from part to whole, and what arises from these processes has no influence on the base it emerges from. This ignores the downward organisation of chance and the fact that biological randomness is not randomness *within* a pre-established space of possibilities but has to do with the construction of this space (Longo and Mossio 2020; Longo and Montévil 2017).

Against Monod, I have sketched an alternative where the development of an organism cannot be codified because its potential cannot be stated in advance. It has a specific ontogeny, as does the information it carries. Also, Monod considers chance to be physical, whereas natural selection acts on the biological level to order this chance. This interplay of chance and necessity must be reconceptualised: Chance and necessity cannot be reduced to the interplay between microscopic events and natural selection, in which mutations are contingent but made necessary by selection. Systems biology requires a more encompassing model of this interaction.

In the previous chapter, we touched upon the role of the machine metaphor, producing a situation where organisms are considered passive objects. I continue this critique while expanding on the role of contingency and possibility in evolution. In line with Monod, I will emphasise how evolution is the interplay of chance and necessity, but I add that there is no way to predefine the objects of evolution, and thus no way to localise randomness at a specific level. So, while Monod is not wrong to focus on the interplay of chance and necessity, he reduces their interplay to one between two externally opposed processes—one random, the other deterministic. By contrast, I show how contingency and necessity are internally related to each other.

In Žižek's view, the hallmark of great thinkers is that they misrecognise the 'basic dimension of their breakthrough' (Žižek 2014: 34). Before, we applied sympoiesis to conceptualise the higher-order closure between organisms. I now develop this perspective further to get at the status of actuality and possibility in biological systems and the structural similarities between the organisational and dialectical view.[1] What might seem a mere transcription of notions from the organisational approach into Hegelian terminology is equally excavating the misrecognised impact of his notions, as 'an archive of lost theoretical alternatives' (Gambarotto and Illetterati 2020). And since the organicist tradition rejuvenated by the organisational approach is heavily indebted to dialectics (inadvertently or not), one might argue that I am retracing the forgotten importance of this perspective instead of merely transcribing. Perhaps we should apply Žižek's idea to the organisational approach and say that they too misrecognise the scope of their insights.

To understand contingency and how it interacts with notions of organisation and emergence, I introduce the dimensions of Žižek and Hegel's thinking that are most relevant to our concerns.[2] I aim to show how their perspectives may inform the

---

[1] Kauffman (2016) names Hegel as the first modern thinker of becoming, inaugurating the worldview that Kauffman advocates. Unfortunately, he does not elaborate much on this topic but evidently subscribes to a textbook version of Hegel since he claims that Hegel believed there 'would be an end to emergent history' (Kauffman 2016: 248).

[2] Some may take issue with the fact that I leave many parts of Žižek's philosophy, like Jacques Lacan's theory, out of the discussion but, again, this is because I am led by the overall aim of this book, which is not to reconstruct the theories of the philosophers I engage with in a systematic

renewed interest for teleology and organisation within biology and philosophy and be informed in return. Furthermore, I discuss Žižek's take on quantum physics as indicating the incompleteness of nature, which I relate to the larger debate concerning the lack of causal completeness that has surfaced within biology, and the role of quantum mechanics therein. This points to another commonality between the organisational and dialectical approaches: Questioning the completeness of the physical realm to make sense of ontological emergence.

This is the longest chapter of the book and contains a lot of different theories. I have therefore divided it into two parts, proceeding as follows: Part one introduces some of the core principles of dialectics and what we might learn from quantum physics. In the first two sections of this part, which prepare the ground for the rest of the chapter, I outline some of the basic motifs that Žižek takes from Hegel's thinking: negativity, positing the presuppositions, retroactivity, and true infinity. I spend the most time on the first concept, as it is important for the rest of what I am saying and sheds light on the others. I then move to a discussion of the ambiguous status of external reality in Hegel's theory, and how this may inform our understanding of the interaction between organism and environment. As we shall see more clearly in later chapters, Hegel treads a fine line between dualism and realism, where it is not that nature is outside us or that we can access it in a completely unmediated manner, but that our mediation is part of nature itself (Zambrana 2015).

Next, I turn to Žižek and quantum physics, which sets the stage for a broader debate about emergence and the lack of ground that it implies.[3] This indicates why biology must change its relation to physics. This rather long section is followed by an interlude on the notions of levels and scales, where I introduce the notion of *bio-resonance* to get at the constitutive interaction between ontological levels. I then link what I have said on quantum physics to *ergodicity* and the question of whether the universe can visit all its possible states or not. Kauffman says 'no' and draws the implications of this notion: Since the universe is non-ergodic, only a small fraction of possible complex things will never exist. We are thus dealing with a canalised universe, where what exists presently constrains what may exist in the future. This ends the first part of the chapter.

In the second part, I return to Hegel and go more deeply into his notion of contingency and necessity, as well as possibility and actuality, to see how these processes are dialectically entangled. I argue that the absolute relation that Hegel proposes between these notions challenges the modern view which considers possibilities as given prior to their actualisation. After this analysis, I get back to Kauffman and his analysis of possibility and actuality in living systems, formalised in what is called the *Adjacent Possible*, hiding on the fringes of what is actual. Combining the insights from Kauffman and Hegel, we find an overlap in their thinking around contingency

---

manner but to gauge how they can be mobilised scientifically. To be sure, this is not to say that Lacan is useless for such a project, simply that I did not find a way to integrate him in this book.

[3] I hold that Hegel is a strong emergentist, and that dialectical materialism is the philosophical background from which emergence arises, with its emphasis on the discontinuity arising from within continuity—or the shift from quantitative to qualitative change (Johnston 2019).

and necessity. This is extended into a more scientific debate concerning emergence and the question of ontological ground. Here, I underline the overlaps between the dialectical perspective and notions of emergence and downward causation in other domains. In the final two sections, I aim to draw implications of these similarities and conclude that the organisational approach would benefit from adopting a Hegelian view instead of the Kantian approach that it has tended towards thus far, and that such a view is already implicit in it.

Part I: The Incompleteness of Nature

> This relationship between the complementary dynamics of organization and chance is akin to a Hegelian dialectic. They remain antagonistic within the immediate domain, but they become mutually dependent over the larger realm. (Ulanowicz 2009: 7–8)

Friedrich Engels (1959) held that Darwin's theory demonstrated the *inner connection* between chance and necessity proposed by Hegel. Evolution proves that what seems given is the product of natural history, how it has *become necessary* through a process of becoming which is permanently unfinished. This 'reintroduce[s] the openness of the future into the past' (Žižek 2012: 464), reinscribing potentiality into actuality. Below, I argue that the notion of contingency indicated here concerns how the current state is not given but mediated through the negation of possibilities which continue to haunt actuality. Focussing on the dialectics of contingency and necessity, I aim to demonstrate that dialectics is not a method that resolves into an ideal unity, quite the opposite: Any self-contained unity is illusory and negates itself.

This does not mean that Hegel abandons unity in favour of dualism. It is rather that any unity is differentiated, a speculative identity of different parts that depend on each other. In the above epigraph, Ulanowicz articulates such an identity and a crucial dimension of Hegelian thought: the complementarity of organisation and chance, also a central facet of the organisational approach. The relation that is akin to dialectics is the 'paradoxical role that chance and disarray play in the persistence of complex systems, because, without them, a system lacks the flexibility necessary to adapt and becomes defenseless in the face of novel perturbation' (Ulanowicz 2009: 7). So, how is chance harnessed to enable the maintenance of life? This chapter will take us part of the way, but we cannot get there until we discuss plasticity and variation in later chapters.

In this section and the next, I introduce some of the core concepts that we will elaborate on later, such as negativity, positing, and mediation. Žižek undermines the caricature of Hegel as a panlogician attempting to deduce everything logically. Simultaneously, he avoids the opposite caricature, wherein Hegel simply adds a historical dimension to Kant's epistemological categories (Pippin 1989). Žižek does not espouse an evolutionary narrative in which the (explicit) 'for-itself' simply grows out of the (implicit) 'in-itself' in a progressive manner, triggering a latent potential.

Instead, he holds that there is no dialectical synthesis awaiting us at the end of the progress, no final sublation of contradictions.

The basic idea of this version of dialectical materialism is that the self-referential process that characterises living organisms is enabled by ontological lack. This takes us to negativity, both as a lack of ontological grounding, and the process through which this ground is negated and made positive. Some claim that Hegel never developed what he meant by negativity (Henrich 2008; Bhaskar 2008). But this is not coincidental—as negativity is not *something* that can be pinned down conclusively but is developed through a process of becoming. First, negativity is not simply nothingness or absence. There is no affirmation in pure nothingness; it is indeterminate, an 'absence of determination' (Hegel 2010: 59), which is why Hegel claims that it is identical to being, as the latter is also completely indeterminate. Thus, he arrives at the notion that being and nothing are equally *becoming*. As such, he shows that when we try to establish an unmediated foundation, we fail to make sense of it as anything other than pure nothing which we cannot say anything determinate about. Being is therefore a false start for his system (Houlgate 2006), and Hegel sets out searching for another ground which is not simply presupposed and abstract but mediated and concrete. Hegel finds this ground in the living activity of the organism (as argued more fully in Chap. 7).

One aspect of negativity is absence, opposition, and non-existence. This sheer negativity of *nothingness*, a limit to knowledge refers to that which is hidden, outside, unconscious, or dead. It is the abstract negativity of Kant's things in themselves, outside possible experience. Hegel does not stop at this notion of negativity but instead shows how negation is historical and mobilised by the organism. He demonstrates how this abstract negativity, qua being, fails to stabilise itself as something positively given and is therefore negated. In this view, negation is not the same as cancellation but has a positive dimension, as *determinate negation*. Actualisation is a process of negating determinate negations; not releasing a given potential but (co-)producing it. Negativity, then, points to an absence of given potentials outside history. Negation of this ground is also a positing and construction of a potential that did not predate the activity of positing itself. Anything positively 'given' is the consequence of this positing activity or negation.

This might be the most challenging and important thing to comprehend. Hegel makes the case that possibilities are historically produced, that there is no stable ground where possibilities reside but instead a virtual ground that is in becoming and never settles. It is easier to think of negativity as pure nothingness forever outside our comprehension than to consider how it operates in reality. But this is what Hegel forces us to consider. This is a movement from the abstract to the concrete, where negativity is a principle of differentiation.

Fundamentally, then, Hegel is showing how any self-relation is equally a relation to the other—first as abstract negativity, then differentiated into more concrete forms of negativity (Bowman 2013). Hence, negativity has a duality of process-product as its formative opposition. It is a determinate product of an absolute process of negativity, which is an ongoing response to negativity as a product (Bhaskar 2008). In this way, it opposes itself, says Hegel, through the products it posits. These products, positive determinations, fail to stabilise themselves as self-identical since their boundaries cannot be established without the other. They immediately transgress their boundaries to remain what they are. In this view, unity is a process of differentiation and integration. It reveals 'the connection of the *differentiated* [...] to *that from which it is differentiated*' (Hegel 2010: 746). The negative, says Hegel, holds these moments together, and mediates both. In this way, any finite and determinate thing negates itself and gives way to further permutations of negativity, which are positively given but unstable since they are not self-identical. As such, negativity is a category of construction as much as destruction (we see this more clearly in Chap. 6).

Finite things are infinite in the sense that they are preserved through their constant transformations. They are unities that contain their internal difference. Through the process of actualisation, things themselves sublate the contingent conditions that produced them to make them necessary for their becoming. Hegel argues that this historical becoming provides them with subjectivity since they do not simply unfold a given potential but construct it. Moreover, they have a certain kind of rationality, understood in retrospect, as the process of their becoming has become necessary for the things they currently are. Their historicity and self-organisation impose constraints on how we make sense of them.

Ng (2013) claims that negativity can be understood in connection with the *double constitution* of self-consciousness, how it is both natural and spiritual:

> That life is necessary and constitutive means that self-consciousness in fact has a *double constitution*: self-consciousness is always both a living object and a self-conscious subject, and its identity and non-identity with life—that it *is* life and yet distinguishes itself from and opposes itself to mere life, that it *is* purposive form but also has *knowledge* of that form and can determine it in different ways—constitutes its negativity. (35)

This principle applies to all organisms. Insofar as they can relate to their natural conditions of becoming and sublate them into their own functioning, they display subjectivity as the capacity to self-determine through negation. Negativity is observed in the purposiveness of organisms, and in the lack that leads organisms to seek nurture outside themselves and thus overturn their external determinations. What distinguishes humans is that we can consciously know about this purposiveness, while other organisms embody it unwittingly.

We thus confront the materialist dimension of Hegel's thinking, undermining the charge that his philosophy is spinning in a void. Instead, 'empirical content [is] acting as a constraint on the spontaneity of our conceptual capacities' (Ng 2013: fn26, 51), through their internal principles of becoming. In other words, their intelligibility depends on the objective conditions that produced them. Otherwise, our categories would be external to what they make intelligible, and we would have no way to know whether they are adequate or merely subjective projections. The double constitution of self-consciousness is a way to overcome this problem. Through it, the subject becomes self-conscious and aware of itself as both natural and spiritual. What I later describe as the problem of nature is based on this unity and eternal opposition, how nature is never left behind by spirit, and how either can be reduced to the other (Ng

2020). As we will also see later, experience is enabled through the friction between the organism and its environment.

The organism becomes self-aware, able to make itself an object for itself by being faced with another living organism.[4] We should be careful here, however, as it is too easy to say that the negative *is* the other and place it outside the organism. Instead, the disparity between the organism and the outside object is internal to both, since they are subject *and* object, constrained by external determination but also able to transgress them, even if they are never completely overcome. It is a disparity within substance itself. This underscores the close link between negativity and subjectivity, as the latter is the empty point of self-relation that negates determinate negations. This process of negation is also the construction of the subject as inseparable yet distinct from substance.

I am not arguing that nature as such is subjective, but that nature enables different forms of subjectivity or self-determination and that it has an evolutionary impact. Negativity is an activity that constantly returns to itself, through the process of actualisation that it enables. It is a principle of movement. Hegel writes that the negative

> belongs to the content itself and is the *positive*, both as its *immanent* movement and determination and as the *totality* of these. Taken as a result, it is the *determinate* negative which emerges out of this movement and is likewise thereby a positive content. (Hegel 2018: 37, emphases original)

The negative is not merely negative, but part of what is positively produced. It is the formative movement of content. It is what 'processes the dialectical process and is therefore also "positive" in that very respect' (Zaslawski 2018: 47). Zaslawski claims that we perceive this negativity empirically as the instability that allows for emergence to occur, which is like the lack of specification in the initial conditions. I develop this insight below. In summary, negativity concerns a process of self-limitation and specification from which emerge new forms of negativity. There is nothing outside this process—no noumenal realm or substance outside the subject, only the negativity of its becoming. Thus, a core principle of Hegel's logic is that the negative is also positive. Anything which is positively given is a precarious product of a constructive process, where the possibility did not predate the actualisation. This means that we should not take negativity, *as form*, as externally related to matter, but understand it as part of the becoming of any specific content.

---

[4] This is also prior to intersubjective recognition of the other, as in the lord and bondsman-dialectic, a topic I cannot explore here, but see Ng (2013), who writes that 'the living object is a condition inserted behind the intersubjective relation, a condition that determines the very form that any possible social-historical normative authority can assume' (fn28, 52–53).

Retroactive Positing

We could say that negativity is a process of differentiation that lacks an ontological ground and therefore must produce it in an ongoing manner. Had such a ground existed, the whole process would simply unfold a given potential without the capacity to determine it. Negativity transforms as Hegel moves from abstract being towards concrete actuality, i.e. toward living beings able to act in the world. When we get to this point, says Zambrana, negativity is 'understood as externalisation [...] and self-relation is the result of recollection' (2015: 68). Negativity is embodied as a movement outward to the environment whereas recollection is Hegel's term for the opposite movement, a return of the self from this movement whereby it gathers itself. This could be considered a normative process—an internalisation of constraints through life activities such as habit formation (Tahar 2022). I will assess this form of negativity (and negativity as form), as an 'activity of deficiency', more in the next chapters when discussing how organisms recreate their surroundings by relating to and reconstructing them. For now, we could think of it in terms of organicism: as indicative of why form is an ongoing process of giving and receiving form.

Retroactivity or positing the presuppositions, integral to Žižek's version of dialectics, is tied to negativity (Fuchs 2014). Positing, the process of mediation and actualisation, is the way the subject reflexively relates to the contingent conditions of its activity and hereby makes them necessary for its current state. Crucially, the presuppositions of the positing are not positively given prior to the positing. If they were, they would be outside the positing activity of the organism, or abstract nothingness. This explains why presuppositions can only be established in retrospect. Also, this is not to universalise contingency, in the sense that everything may be otherwise; we are not replacing the ordered universe with empirical chaos (Zupančič 2017). Instead, positing the presupposition entails that necessity is an activity that posits contingent presuppositions as part of its constitution. In other words, necessity must actualise itself through contingent phenomena. As Žižek writes: 'the very process through which *necessity arises out of necessity is a contingent process*' (Žižek 2012: 467). As such, necessity is relative, qua 'contingent upon the peculiar circumstance of a distinctive situation' (Erkan 2021: 184).

As we will see, Hegel introduces possibility into actuality by challenging a purely actual notion of reality and the view that logical categories only concern *possible* experience, which is Kant's understanding. This shift implies retroactivity, which demands that the past is open to repetition because it is unfinished in the sense that possibilities are not positively given at the outset. Hence, retroactivity is not merely about registering the necessity of the development; instead, it 'activates and posits the development as necessary' (Radnik 2024: 65). Žižek writes that this is a 'movement of restoring the dimension of potentiality to mere actuality, of unearthing, in the very heart of actuality, a secret striving towards potentiality' (Žižek 2012: 464). This means recognising that reality is always in a state of becoming, charged with virtual possibilities, and that any necessity is an emergent product of contingent processes (more below). It points to 'the paradox of a contingent actual emergency which retroactively creates its own possibility: only when the thing takes place can we 'see' how it was possible' (Žižek 2008: 180). It also means that retroactivity is closely tied to reciprocity since what is posited as the precondition is precisely the perceived otherness that an organism relies on to maintain itself.

I will argue that the circular nature of biological determination is comparable to this notion of retroactive positing, which is how contingency is made necessary, i.e. how necessity is not given but produced through a contingent process. The openness of the future is hereby introduced into the past. Retroactivity means that the beginning is negated. But instead of regarding this as a loss of *something* present at the outset, we must transpose this loss back into the initial conditions. Žižek states that there is no substantial unity prior to the loss; the loss itself retroactively created this illusory unity, purportedly outside any experience (Žižek 2014). This means that the positing always depends on 'previous sublation that has left something behind' (Fuchs 2014: 851). In other words, sublation does not mean cancellation but is rather an expansion that includes the contradiction as 'the necessary connection between opposites' (Collins 2000: 775). It expands the concept by acknowledging the opposite that is implicitly part of it.[5]

A final concept before moving on: Hegel's concept of 'true infinity', which I have hinted at. Against the quantitative infinity of more of the same, good infinity concerns the infinite self-relation that a system has to itself. Thus, it is qualitative: it is not about limitless expansion but reflexive self-determination. By relating to each other, determinate existences 'form a *world* of reciprocal dependency and an infinite connection of grounds and grounded entities' (Hegel 2015: 190–91, emphasis original). This 'is only the dialectical movement, this course of self-engendering, advancing, and then returning into itself' (Hegel 2018: 40). This self-relation is not isolated from its other but dependent on it—which is what true infinity means: that the finite and the infinite are not outside each other but identical. Hegel says that we arrive at this notion of infinity simply by identifying what is implicit in the notion of infinitude as the opposite of finitude. It makes the infinite into another finite *thing* transcending possible experience, and hence a pure nothingness. Against this notion, the infinite must be included in the notion of finite things, as they never simply disappear but are preserved in their vanishing. This is similar to how biological systems keep their identity in the midst of continuous material turnover.

We find this principle in determinate negation, which Hegel (1982) describes as *absolute unrest*. Finite things are not at peace with themselves but are involved in processes that go beyond them and cannot be exhausted. As a consequence, we are not dealing with a given reality of finite things but rather an ongoing process of construction. Zambrana (2015) argues that this reveals how Hegel introduces ideality into reality. Kant is again helpful: when he says that the negative is outside our ideal representation of reality, he says that the truth is on the side of this negative that we

---

[5] For example, the mediation of immediacy is externalised and made into external immediacy; then, the contradiction between these immediacies is negated, and we arrive at the 'posited' immediacy, which is a unity-in-difference between self and other. At this stage, we acknowledge that the external world is permeated by the mediating activities of organisms. I develop this further in Chap. 7.

cannot reach (Kant 1998). Hegel, on the other hand, says that the ideal partakes in 'real' being, that is it not external to it.

True, self-relating, infinity concerns how the organism relates to and maintains itself through its other. This entanglement of self and other, says Žižek, is displayed by the self-determination at the basic level of life, when a living cell produces a membrane which sets it apart from its environment and allows it to self-determine, like in autopoietic models.[6] A minimal form of subjectivity allows the organism to sense its environment as distinct from itself. The infinity involved implies an organisation that regulates the processes from which it emerged. It diverges from spurious infinity since it remains self-identical by differentiating itself. A flame, for instance, exemplifies bad infinity since there is no internal differentiation of the processes involved (Montévil and Mossio 2015).[7] It does not display the organisation that requires differentiation of parts with separate functions. Physical objects do not display the process of self-differentiation and reunification that characterise the living. After this introduction to some of the concepts that I concretise in this chapter, we turn to the problem of nature in Hegel's philosophy.

The Problem of Nature

As we move towards the problem of nature, I will spend some time on Hegel's view on living organisms and the question of intelligibility. In Hegel's view, subjectivity is embodied by any object that manages to set itself apart from and interiorise its surroundings, thus isolating itself from efficient causality by relating to itself. The fundamental question then becomes (see Žižek 1993, 2012, 2016): How must nature itself be for something like subjectivity to emerge? How is our cognition of reality enabled by the structure of reality itself? The short answer is that life and subjectivity are possible because nature is underdetermined because it lacks the causal specificity needed for self-sufficiency. In other words, nature is necessary—it cannot be overcome—but insufficient for the development of autonomous life. This excess in nature is the space where subjectivity, the empty point of self-relating, emerges.

Multiple authors have characterised life as a circular organisation, but few have traced this notion back to Hegel (Marques 2016). But before we get to the dialectics of organisation, the problem of nature must be addressed. As I argue below, Hegel denaturalises nature by highlighting how it lacks self-consistency. Nature is no longer

---

[6] Godfrey-Smith (2016) argues that subjectivity only demands a point of view and an agenda. While 'point of view' does not imply consciousness, the agenda, from an organisational view, would be the reproduction of organisation. I should also note the similarity—or 'conceptual compatibility' with autopoiesis (Zaslawski 2018).

[7] Such differentiation is the criterion that distinguishes between organisational principles pertaining to living systems from more generic physical cycles. This seems to be overlooked by Gambarotto and Nahas (2022).

a self-enclosed whole, a stable background condition for our activities. Hegel undermines reductionism, and 'conceiving it [nature] instead in terms of goal-directedness and, ultimately, freedom' (Gambarotto and Illetterati 2020: 15).

We see this in the *Phenomenology of Spirit*, where Hegel claims that the attempt made by consciousness to explain living organism according to laws collapses on itself, as it cannot explain the constitution of the laws themselves except by referencing further laws, introduction an infinite regress. It fails to account for living organisms because they are not composed of isolated elements. A living being maintains itself in and through its relation to its other. As such, it is a concrete totality, meaning that its parts are not just aggregated but realise a higher (emergent) order (Gambarotto 2020). As such, they are *members*, not *parts*, circumscribed by a structuring principle that emerges from the whole nexus of constraints that tends towards closure:

> For Hegel, the organism is essentially the processual unity of different functions that are articulated in different members. The organism is, in Hegel's language, a 'concrete unity', i.e., a unit that is articulated in difference, a unity of specialized parts that act as such only within the organic whole. (Gambarotto and Illetterati 2020: 10)

A mechanical and chemical explanation can identify the components of an organism as parts, but not as members that are circumscribed by the system they enable. Apart from this unity, members 'would stop being what they are' (Illetterati 2016: 194). They are not permanent but constantly destroyed and engendered (Hegel 2004). To be more precise, members are means that make use of materials to reproduce themselves (and vice versa).[8]

This distinction, as Luca Corti argues, is like the difference between constraints and processes discussed before, where members are constraints that 'act upon material in a causal way, without themselves being affected (in a relevant time scale) by the process' (Corti 2023: 236). Unlike the material (processes) they constrain, members are functional components—reproduced through their reciprocal interplay and relation to the whole (Bich 2021). Like the organisation approach, Hegel underscores how the differentiation and mutual dependence of such components enable the self-maintenance of a biological system. He writes that the unity of the self is preserved by the movement outwards (Hegel 1991). This is a crucial point for the conceptual and historical overlaps I aim to articulate between these perspectives.

It also indicates a distinction between existence and actuality: For *existing* things, inanimate objects, there is an external relation between form and content. But for something to be *actual* the relation to the other is internal and necessary. A stone exists but does not act on its own. It lacks actuality because it cannot posit itself by negating its generic possibilities. If we explain a living thing through the principles we apply to a stone, we obfuscate how the parts are not indifferent to the whole but are gathered as concrete determinations which contribute differentially to the

[8] To be sure, Hegel also says that the celestial bodies of the solar system are members because they too are part of a systematically organised whole. He calls this the 'organism of mechanism'. It is not yet the explicit form of the organism but it does display similar principles, in-themselves, however nascently (see Ferrini 2024).

organisation. Again, think of a flame: while it consists of various chemicals, they all contribute to the same operation. There is no differentiation of functions.

As a processual unity, the organicist notion of totality does not mean closedness or rigid control. Instead, it concerns emergence through the interplay of closure and openness. Likewise, Hegel dispute self-enclosed notions of totality, as there are oppositions that are never resolved and any totality must include its opposite. He says that life—as its own end—is a 'rounded-out totality' (Pinkard 2012). This totality must be both open and closed to maintain itself. The organism as a life cycle (or a cycle of cycles) exemplifies an open and dynamic totality (Gilbert 2019), as does the closure of constraints specific to a biological organisation.

Totality is a processual category (Bhaskar 2008). It is about the ongoing activity of the organism and the relations through which it maintains itself within a range of viability. It is a process of differentiation through which the organism produces itself as a 'unity of multiplicity' (Hegel 2004). A closed totality is dead, whereas an open totality is enabled by and maintained through its persistent contradiction. Not only does it have to integrate the interaction of disparate parts; but it must also sustain far-from-equilibrium conditions in its metabolic exchange with the environment to remain alive. To achieve this, the organism negates and repurposes the internal and external processes that both sustain and threaten to undermine it (Longo and Montévil 2014).

The contingency of nature frustrates any attempt to elucidate its unfolding deductively. As such, nature is the precondition for spirit and something it can never overcome. It thus poses a 'perpetual problem for spirit's self-actualisation' (Furlotte 2018). This indicates that nature is open-ended and underdetermined, not the stable background of our activity and that organisms are 'irretrievability entangled [...] with the materials of their factical environment' (Furlotte 2018: 26). The self-referential dimension is part of the constitution of both inside and outside; the self-unity of the organism can only maintain itself 'in what *it* makes to be *its own externality*' (Ferrini 2020: 254). Anything that counts as immediate is always already mediated, related to the history of the organism and the way it has moulded it. Hegel calls this *the immediacy of mediation*. What appears immediate is the congealed version of the conditions—and the mediation of these conditions—that make up a concrete phenomenon. Again, Hegel is dealing with existing conditions of actuality, not just formal conditions of possibility (Zambrana 2015).[9]

Take perception: Whatever the organism experiences in its environment, it is not only tied to its activity—and thus self-referential—, but also shaped and enabled by neuronal firing, which is related to what happens in the rest of the organism; and this, in turn, is moulded by previous engagement with the outside world, memories, state of mind etc. The idea of pure immediacy obfuscates the situatedness and historicity of experience (Hegel 2010). It gives the impression that form is imposed on indeterminate matter from without. If this was the case, form would both be self-sufficient

[9] But we should clearly not take him to be a pre-critical thinker, as he does not deny that our access to reality is mediated, which is the position called naïve realism. Hegel moves beyond Kant to understand how mediation is ontologically possible.

and depend on matter. Against this, Hegel holds that form is internal to matter, which means that not just form but also matter can determine itself, by taking on multiple historically constituted forms (Zambrana 2015).

The contradiction between nature and reason is generative and problematic. Living organisms embody nature's externality to itself since its failure to contain itself is concretised in finite beings. This could be understood through the double constitution spoken of before and implies that nature is not a completely coherent and rational structure. Rather, we account for the emergence of spirit through nature's incoherence. The subjective and objective sides are distinct moments of such a unity, not indistinguishable but existing *through* their relation. The externality of nature cannot be sublated. Yet, it is not outside the dialectic, as that would reintroduce the idea of a permanent nothingness outside our reach. Hegel's notion is more troubling: We are always caught up with the externality we try to grasp. Externality remains a problem for the organism because it can never establish itself once and for all as independent from it.

An organism *is* nature's externality to its own conceptual scheme, and yet not completely outside it. Furlotte (2018) claims that this follows from an understanding of Hegel's model as 'radically autopoietic'. There is never complete overlap between our conceptual apparatuses and the natural processes it makes intelligible:

> Nature is something that is open as well as closed with respect to its intelligibility, since its essence is both logical and non-logical, rational and non-rational. The externality of the idea marks the specific logical structure of nature. If this is so—and this is the crucial point—nature according to Hegel is external not just with respect to a mind that considers and analyses it; nature is external with respect to its own logical structure. (Illetterati 2020: 62)

As such, nature is neither outside nor inside conceptual determinations. It cannot be reduced to a logical discourse but still evinces a historically contingent organisation. When Illetterati says that it is external to its own logical structure, it is related to the notion that nature is the *idea*—the unity of the ideal and the real—in its externalised form. Nature is the breaking of the unity of the real and ideal in and through its contingent products. The paradox is that the shattering of the unity of the idea is also the only way that it can be actualised. Thus, the concrete instances of nature simultaneously manifest and fragment the idea. They display the idea in a differentiated manner, as a process which contains its negation. The broader implications of this become more evident later when a discussion of the relation between concept and idea is undertaken.

For now, the point is to understand why nature is an unresolved contradiction for Hegel. It has inherent rationality, but this rationality is incomplete. Put differently, the rationality of nature is manifested in a patchy way. This impotence, the contingency inherent in nature, makes conceptual deduction of its products futile. Yet, nature still has an inchoate conceptual structure, revealed through the natural sciences, since this 'rationality can be grasped only by exploring the detail, assuming it, and recognizing it for what it is' (Illetterati 2020: 64). Nature displays fragile rationality since it is shaped and embodied through contingent and singular events. It does not evince a necessary and logical becoming that can be universalised in abstract terms.

As Illetterati explains, since every natural entity is concrete, it 'never immediately coincides with the conceptual structure that makes it intelligible' (Illetterati 2020: 61). There are 'traces of notional determination' in the concrete products, according to Hegel, but these traces do not exhaust the instantiations. This underscores the importance of a detailed study of nature, from which universal principles are distilled but these grasp the phenomenon at hand only imperfectly (Stone 2013). As noted initially, there is shared ground between empiricism and philosophy of nature, as the latter must make use of the results of the former.[10]

Since the subject is also natural and concrete, it embodies the impotence of nature. Subjects are external to themselves—dependent yet autonomous from externality. Again, we see the redoubled split, where each side is split internally. What is external to the subject is internal to the substance, and vice versa. There are not simply two sides since each side includes the division within itself. So, it is not merely that the environment is outside the organism, but that the split between organism and environment is internal to both and marks their mutual incompletion. Thus, their limitation is also their condition of existence. As Hegel says, 'each is what it is in itself, the totality of the relation, precisely through its other' (Hegel 2010: 463). The reciprocity between self and other is crucial for understanding the agency of living organisms. We see, then, how an altered conception of nature coincides with a specific notion of subjectivity, as *autonomous from yet dependent on* natural determinations (Johnston 2013). Below, I relate this to strong emergence.

Not Only as Particle but Also as Wave

> The true problem is thus not how an organism adapts to its environs, but how it is that there is something, a distinct entity, which must adapt itself in the first place. And, it is here, at this crucial point, that today's biological language starts to resemble, quite uncannily, the language of Hegel. (Gabriel and Žižek 2009: 105)

In the epigraph, Žižek questions the premise of adaptation, the living organism—and how it emerges and maintains itself. This premise is unthought in adaptationist reasoning, and a crucial part of integrating development into evolutionary explanations. By contrast to such reasoning, Žižek stresses the radical contingency at the heart of nature—how it is constrained, taken up and made useful by biological systems while constraining such systems in return, and he underscores a point that I have been trying to develop further; namely, how this way of thinking is fundamentally dialectical.

Hegel's notion: the impotence of nature (*Ohnmacht der Nature*) is the failure to attain conceptual stability, or to determine itself perfectly. Nature is weak because it

---

[10] Ferrini emphasises this point on several occasions. She writes that philosophy of nature must make use of scientific results, 'because empirical physics, although it is not comprehending (*begreifende*) speculative cognition, is nevertheless thinking (*denkende*) cognition of nature' (Ferrini 2024: fn4, 145).

does not embody universal categories in a flawless manner. As Adrian Johnston (2019) says, nature displays 'indefinite contingent proliferations' that cannot be predicted. This ontological weakness translates into an epistemological weakness. Nature's contingent process of becoming lacks internal stability which means that natural processes are not completely 'intelligible on their own terms' (Di Giovanni 2010: xliv). Science must complement them to make them intelligible but cannot attain perfection (Johnston 2019). Yet, the limits of natural determination must be set by the sciences, not philosophy in isolation.

This impotence is concretised as a lack of causal specificity, as natural processes do not determine their own becoming in a rigid or isolated manner. Science cannot be completed because it also grapples with the incompleteness of nature (Bowman 2013). It is this indeterminacy that enables the transformation of nature into second nature through habit formation (Renault 2013), as discussed in the next chapter. But this does not mean that nature is completely circumscribed by subjective determinations, making objective knowledge impossible. Natural processes have internal principles of becoming, however inchoate, which place constraints on what emerges from them and their intelligibility. These are similar to the principles of self-organisation we have already discussed.

Saying that nature is unintelligible in itself—completely outside reason—would be saying more than is warranted, as it would be saying that the world as such is unintelligible.[11] And it would undercut that very statement itself, as its own preconditions would be inaccessible. Against Kant, Hegel holds that his categories have a basis in empirical findings. The fuzziness of categories is not coincidental because nature is itself riddled with contingency. It is thus a question of how and to what degree natural processes are intelligible to us, without presupposing an answer in advance. To get clearer about this ontological contingency, I now turn to quantum physics.

In Žižek's philosophy, quantum physics is a *concrete universal*—a concrete instance that embodies a universal principle.[12] Similarly, Bitbol (2012) calls it 'a reinforced version of a universal situation' (247). It thus provides lessons for other sciences. Not as an analogy, but as a characteristic shared across the sciences. Insofar as it reveals something about the interaction between levels and the causality involved in this interaction, its relevance to our discussion is evident. As Longo and Montévil (2014) claim, it is not that quantum physics tells us anything directly about biological systems, but that it might 'inspire our attempts in system biology' (164). Similarly,

---

[11] Illetterati (2020) writes: 'By recognizing nature as a way of being characterized primarily as externality, Hegel aims at a rational understanding of nature without assuming that nature is itself the transparent expression of this rationality, while at the same time refusing to think that rationality is simply a subjective network superimposed on nature to make it *rational*—as if nature in its legality were nothing more than a construct of this subjective rationality' (63).

[12] Here, it is useful to mention Stern and Weiss' shorthand, which is that 'abstract universals are merely common to many instances, whereas concrete ones determine their own variations' (Stern and Weiss 2024: fn9, 24). It indicates a point which is similar to the one I make about quantum physics—which is that it does not have one specific role to play within different instances but instead operate different at different levels of organisation.

I attempt to show how quantum physics indicates the preconditions for autonomous action, without making strong statements about how quantum events affect biological systems—just to say that they cannot be ignored. This follows from the view that we cannot distinguish sharply between levels of organisations since there is constitutive bio-resonance between them.

Quantum physics concerns how the smallest units in the universe behave. Its fundamental ambiguity is that quantum events may occupy a specific point in space as a particle or be distributed as waves. A wave function is a mathematical description of this wave, which defines the probability of the particle having specific properties. The wave does not, however, have a specific location. But when a detector is used to check, the spatial distribution of waves collapses to a point. This is called the collapse of the superposition. The behaviour depends on measurement: If we observe what happens, it behaves like a particle; if we do not, it behaves like a wave. Measurement introduces such indeterminism, but this may only be demonstrated indirectly since we cannot know how it would behave without measurement. In the double-slit experiment, the wave distribution is shown by marks left around the edges of the slits that indicate a pattern produced by the inference of waves. A particle would not leave such marks since it has a definite location (Hobson 2017). Thus, the experiment indirectly shows the dual nature of quantum phenomena as both wave and particle. Importantly, the measurement apparatus is entangled with the entity it uncovers; it partakes in the production of either a wave or a particle. Exactly why this is the case remains a mystery, but the duality has been robustly confirmed.

The multiple interpretations of this situation are not our concern here. But Žižek holds that it takes measurement to produce a stable quantum result; before measurement, there is only a virtual superposition. After, a symmetry breaking has occurred, which means that the 'quantum state space assumes privileged directions' (Longo and Montévil 2014: 135). In other words, the equiprobability of different states is cancelled as certain states are actualised instead of others. They are no longer symmetrical, viz. not equally probable.

The details and multiple interpretations of such experiments are not our concern now. What matters is how quantum phenomena shed light on other levels of organisation and their interaction. Žižek claims that quantum physics indicates that our incapacity to understand nature completely is a symptom of ontological incompleteness. In his view, it is not enough to say that life is characterised by absences. Instead, the absences found at the level of life are enabled by absences at a physical level or the absence *of* a bottom level:

> [I]nanimate nature cannot be the zero-level out of which higher ontological levels emerge, there must be some absentials at work already there, prior to inanimate materiality, to render possible the emergence of life out of matter and of awareness out of life. (Žižek 2016: 39)

In the same vein, he states that 'the basic epistemological lesson of quantum physics' is that 'we cannot get to know reality the way it is independently of us because we are part of reality' (Žižek 2012: fn31, 546). This could be cast as a minimal definition of his version of dialectical materialism: The observer is not outside but caught up in what is observed. Assuming reality to be completely independent of the subject is a non-starter because it fails to account for our access to objective reality.

Quantum events are also the precondition for retroaction, in the sense that the atom is '*always open to future and past reworkings!*' (Barad 2011: 143, emphasis original). The past was never fully constituted and therefore open to being iteratively reworked. Likewise, retroaction points to 'an 'open' ontology of not yet fully constituted reality' (Žižek 2012: 925). It does not determine a specific future but enables it (Fuchs 2014). This incompleteness—a prerequisite for downward causation (explained below)—does not mean that the quantum realm is all-encompassing and linearly overdetermines other levels of reality. Rather, it indicates that there is no ultimate reality, no stable physical ground. Focussing on the quantum realm, then, would only be reductive if it is understood in terms of a foundational level that is completely distinct from the classical realm—which implies that these levels 'influence each other by way of efficient causality' (Bitbol 2012: 234). We should instead understand how the quantum level affects living systems through enablement since the dynamic constraints of physical processes depend on how they are integrated and regulated by organisms (Longo and Montévil 2013). These events are not efficient causes but form attractors, tendencies in phase space (Emmeche et al. 2000). They provide a wellspring of stochasticity that the organism may harness but does not act directly on other levels; instead—and since levels display radically different rates of activity—quantum events may only affect adjacent levels by transformative steps.

By questioning the possibility of reduction, and revealing the instability of any ground, quantum physics may stimulate the articulation of a systems biology which is not bottom-up. It challenges the view we have surveyed, in which biology has been an extension of classical physics concerned with isolated particles (Longo 2020). Had the MS modelled itself on quantum physics, which questions the very separability and identity of particles, another biology would probably have been articulated. As I argue later, this biology cannot assume pre-given state spaces and a stable base level from where everything else derives—other levels at best 'releasing' the causal impact of the bottom level. Rather, if biology is an extension of physics, it must consider the various kinds of interaction between quantum and classical levels—namely their bio-resonance, i.e. the constitutive interaction between levels of organisation. Different levels must thus be read through each other, as we move from a physics of determination to one of enablement and differential causation—a physics that deconstructs itself (Barad 2007, 2011).

In a sense, quantum physics is about bio-resonance or *bio-entanglement* (Buiatti and Longo 2013). Observed values are co-constituted with the observer, underscoring the interaction between the quantum and the classical level (Longo 2020). As Longo says, 'properties of life would contribute to establish the properties of the inert' (Longo 2020: fn5, 138). Similarly, Pattee (2013) highlights how quantum measurement is only completed when a classical result is obtained. By reference to Niels Bohr, Žižek also draws attention to this paradox: 'it is not only that there is no classical reality which is not sustained by fuzzy quantum fluctuations; it is also that there is no quantum universe which is not always already hooked onto some bit of classical reality' (Žižek 2012: 938). Quantum events are only made intelligible in classical terms, but this renders them incorrect from a quantum perspective. When described in quantum formalisms, however, they are nonsensical (Krips 1990). Žižek takes this to mean that the quantum realm depends on downward causation to become stable. By itself, it is not fully constituted, but virtual. It does not even constitute an 'autonomous sphere of being' (Žižek 2012: 918). This reminds us of the notion that no level is unconstrained.

In summary, quantum physics demonstrates how our failure to grasp reality in itself means that '*there is no (balanced, self-enclosed) Nature* to be thrown out of joint by man's *hubris*' (Žižek 1996: 235, emphases original). Nature is always already imbalanced. No neutral position is conceivable, as we are entangled with what we attempt to grasp. This undermines the very idea of ultimate reality—transposing the lack attributed to our cognitive faculties onto objective reality. Žižek's reason for appealing to quantum physics is therefore clear: He needs a notion of incomplete physical reality to make ontological retroactivity plausible. Thus, we should understand it as a reinforced version of the impotence of nature described above. Žižek hereby places more emphasis on emergence than cumulative evolution (Marques 2016). The importance of this view becomes more evident when we relate it to downward causation.

Interlude: Levels and Scales

> [T]he very phase space of evolution changes in unprestatable ways. In consequence [...] we can write no equations of motion for the evolving biosphere, nor know ahead of time the niche boundary conditions so cannot integrate the equations of motion which we do not have. No law entails the evolution of the biosphere. (Longo et al. 2012: 7)

Longo (2018) states that '[t]he least critique one may address to [the] gene-centric approach, largely based on classical dynamics, is a reference to quantum indetermination which seems present at the molecular level, both in genetic and epigenetic phenomena' (456). Since random events may prompt phenotypic changes, quantum physics undermines genetic determinism. These changes are not mere noise but have nonlinear effects. For example, a 'change in the hormonal cascade may seriously damage a tissue's coherence and, years later, cause or enable cancer' or a 'quantum event at the molecular level (a mutation) may be amplified by cell to cell interaction and affect the organism, whose changes may downwards affect tissues, cells, metabolism' (Longo and Montévil 2013: 6).[13] We should think of this in terms of bio-entanglement.

Here, we could distil a general principle: Causation does not only work by summation or aggregation of local events. From local interactions emerges new phenomena and levels that cannot be predicted based on the parts in isolation. Because of this, and

[13] The analogy between biological and quantum entanglement is correct in the sense that the states involved cannot be understood apart from each other. But should not be taken too far, as entangled states in biology are more robust than quantum states (Noble et al. 2019).

because downward causation becomes viable, we need a global view of the system involved (Longo and Montévil 2011). And we need an adequate notion of levels and their interaction.

As in Haraway's discussion of organicism, there is no reason to assume that the elementary components are simple. Also, we may only understand them by their reciprocal interaction with other levels of organisation (Longo 2019). Changes at a lower level become causally effective not only by propagating upward but also by affecting the overall organisation which constrains the action of the lower levels themselves. This retroaction exemplifies the contingency of necessity, as seen in how the quantum and classical levels interact:

> [T]he effects of the classical/quantum blend may show up at a different level of observability and may retroact. First, a mutation or a random difference in the genome, may contribute to the construction of a new phenotype. Second, this phenotype may retroact downwards, to the molecular (or quantum) level. (Kauffman et al. 2012: 5)

Depending on the context, the random quantum effect may become causally efficacious at the level of the whole system; but this is not specified by the quantum event alone. For instance, a mutation is not efficacious per se but depends on the context that supports it. And since the rate of mutation can be changed in stress responses and certain mutations may even be targeted, it is inaccurate to say that mutations *originate* in genes or that they are simply stochastic events (Miquel 2011). Moreover, a mutation might never reach a threshold at which its effect is noticeable even if it has indirect effects by its interaction with other processes, as seen in epistasis, where the expression of a gene depends on its interaction with other genes. Hence, the effect is not additive by default. Also, there is no principled cut-off scale at which molecular stochasticity, quantal or not, becomes irrelevant to the system. This indicates the fuzzy boundary between scales and levels, lacking anatomical location (Noble et al. 2019).

This ambiguity might explain the lack of consensus in defining levels of organisation and why the notion changes depending on context. It constitutes an open empirical question (Brooks et al. 2021), and I can only sketch how levels are understood from the organisational perspective. It contrasts with the compositional definition of levels, which considers it a simple part-whole phenomenon (Bich and Bechtel 2022). Here, the whole exerts influence on its part in a direct (linear) manner, and the levels are simply assumed to exist, based on their composition, without any explanation of their constitution. This direct connection between part and whole is questioned by the notion of organisation, as it allows for a more dynamic conception of levels (Baedke 2021). In this view, 'the interrelation among elements that goes well beyond mere composition: such systems manifest integrated global emergent properties, capable of regulating the behavior (dynamics) of their constituents' (Umerez 2016: 75). It is not the precise structure of levels that matters, but how the interaction between levels enables their individuation (Umerez 2021). We return to this when discussing downward causation.

This view casts light on the dialectical nature of part-whole relations. The parts are only parts through the whole and vice versa; they are not indifferent to each other, as quantitative differences are. Passing from a mechanical to an organic conception of totality means seeing it as a synthetic unity—not '*composed of parts* but *organically formed* of limbs, organs and functions' (Sève 2008: 88), which is to say that it is an emergent reality. Lucien Sève argues that Hegel's concept of emergence offers a third position that neither negates nor trivialises emergent novelty. It is neither a rearrangement of pre-existing components (which is to deny novelty) nor created ex nihilo. 'Something' is needed to account for this: 'This something is its *organization* as a whole, the *overall connection of its elements* and its *logic*, which together constitute precisely what is novel about the new measure that has emerged at the tipping point' (Sève 2008: 94). The elements are necessary but insufficient for this as they rely on the whole organisation.

Notions of upward and downward causation are informed by our notion of levels. The organisational approach holds that scales and levels are interdependent. Any sustained level undergoes constant changes on an interval of viability (Longo et al. 2012). As I discuss later, a criterion for such level transitions is symmetry breaking, where the failure of determination at one level gives rise to another, with the 'functional unity of the organism' as the highest level. In this view, a change of level coincides with a break in determination at another. Or, as Sève argues, the disorganisation of the old is the prerequisite for reorganisation. In other words, the weakness of nature is the disorganisation that enables the process of organisation.

Symmetry breakings constitute and explain the historicity of a living organism. The challenge, as we saw in the case of quantum physics, is to formulate the circumstances in which physical processes are relevant to biological ones—and vice versa—, and the historically specific form they take at different levels (qua concrete universals). Therefore, we should not confuse levels with scales (Longo et al. 2012). While the latter concerns qualitative measures of magnitude, such as time and space, the former concerns qualitative features, such as molecules, cells, tissues, organs etc… In short, scales are continuous, while levels are discrete. This means that a level of organisation is the combination of different scales. In a sense, then, a level might be said to emerge from scales that interact with each other in a fractal-like manner, corresponding to a 'local maxima of regularity and predictability' (Wimsatt 1976). We identify levels based on symmetry breaking occurring at multiple spatial scales.[14]

Bio-resonance denotes how levels of organisation interact constitutively. It is analogous to Poincaré's observation of how the interaction of three gravitational bodies (that by themselves behave deterministically) produces random behaviours (deterministic chaos). However, it differs in that the resonances are operative *between* levels of organisation, not just bodies at the same level (see Longo 2018). While some levels are more stable than others and might display law-like behaviour—as the causal relationships between its phenomena are stable—none is completely

[14] 'Certainly, all causes can be traced back to the constraint of physical forces acting among all the particles present, these constraints being the geometric configuration of the particles in space and time. But realising this can give us a radically integrative view telling us that material objects composed in a hierarchy of levels of organisation and interacting with one another via physical forces are in fact all parts of one pattern of elementary particles with symmetry breaking occurring at multiple spatial scales, from which we identify the levels' (Farnsworth 2022: 12).

isolated and unperturbed by others (Bizzarri et al. 2020). With increased size and complexity comes more degrees of freedom, which also translates into increased leakage between levels. This makes higher levels less well-defined than lower ones (Wimsatt 1994).

Finally, bio-resonance between levels follows from *and* yields variation; it both stabilises and destabilises the organism, as it produces randomness which the organism makes use of (Longo and Montévil 2017). The task is to account for dependence between levels without thereby casting their relation as one of domination. Closure of constraints allows us to sidestep the problem of breaking with the arrow of time of efficient causation since the relationship between levels is understood in terms of constraints that enable and regulate but *do not cause* events at lower levels (Juarrero 2023). There are not two separate streams of efficient causation acting in opposite directions, but entangled webs of enablement. This notion of dynamic levels integrating scales instead of components allows us to surpass the *direct* whole-part understanding of the interplay between levels and adopt a more complex model, where no level is complete (or self-sufficient) but reciprocally constrained and enabled by interaction with others.

Ergodicity Versus History

If the space of possibilities is larger than that of actualities, and the universe—above the atomic level—is vastly non-ergodic, *history enters* (see Longo et al. 2012).[15] Let us unpack this: non-ergodicity means that evolution cannot explore all possible forms and come up with the most optimal solution to every problem. Had there been enough time in the universe to do this, any change would be reversible—and no canalisation would be possible.[16] An analysis assuming the universe is ergodic thus undermines the unique trajectory of evolution (Longo and Montévil 2013). In other words,

> evolution is both the result of random events at all levels of organization of life and of constraints that canalize it, [...] by excluding, by selection, incompatible random explorations. So, ergodic explorations are also restricted or prevented both by selection and the history of the organism. For example, the presence and the structure of a membrane, or a nucleus, in a cell canalizes also the whole cellular activities along a restricted form of possible dynamics. (Longo et al. 2012: 5)

---

[15] 'Vastly' is not an overstatement here: 'Now the universe is 13.7 billion years old and has about 10^{80} particles. The fastest time scale in the universe is the Planck time scale of 10^{-43} s. If the universe were doing nothing but using all 10^{80} particles in parallel to make proteins the length of 200 amino acids, each in a single Planck moment, it would take 10^{39} repetitions of the history of the universe to make all the possible proteins the length of 200 amino acids just *once*!' (Kauffman 2016: 43).

[16] If the universe were ergodic this would undermine natural selection, as 'it would mean that a negatively selected phenotype would "come back" anyway' (Longo et al. 2012: 6).

Instead of an abstract state space, biology must consider the concrete history of the system in question. In contrast to physics, even if biological objects display stability, they are not invariant (generic) but undergo constant changes to maintain their trajectory (homeorhesis). This places normative constraints on which transformations are possible.

Like any process, a quantum event is path dependent and must be propagated by events which do not belong to the state space of quantum physics, but are described by classical physics; thus, quantum events and classical events combine in evolution. But exactly how they combine cannot be stated in advance. Mathematical or algorithmic predictions of what may come to exist in the biosphere are futile (Longo et al. 2012). Instead, the 'challenge is then to guess a right 'mesoscopic' level where random variations constructively affect the dynamics and are neither corrected nor averaged out' (Longo and Mossio 2020: 392). Again, the quantum level is not isolated from the level of classical objects; instead, it might interfere with it, happen simultaneously, or be amplified by it. Levels interact in unpredictable ways. As such,

> evolution is both strongly canalized (or far from ergodic) and yet indeterminate, random and acausal. Our key point is then that random events, in biology, do not "just" modify the (numerical) values of an observable in a pre-given phase space, like in physics. They modify the very phase space, or space of pertinent biological (evolutionary) observables, the phenotypes. (Longo and Montévil 2013: 7)

This explains why quantum physics is within the Newtonian framework: Even if it shows that there is no strict determinism operating at the physical level, it still defines a state space in advance to give a probability distribution (see Kauffman 2020). The list of possibilities is predefined—viz. not affected by the history of the system, only its current state (see Longo 2018). If we want to escape the Newtonian hold on science, we cannot rely on quantum physics alone.

The construction and maintenance of biological organisation are strongly irreversible processes, which is why you cannot replay the tape of the universe and expect the same result. Soto et al. (2008) draw the consequences of this view: 'In order to predict an outcome, one must first reproduce a situation, and not the contrary' (259). The non-ergodicity of the universe, which means that it is canalised, is comparable to the unstable and virtual status of the assumed ground levels. Organisms do not eliminate this contingency but make use of it. They constrain the processes they are moulded by and thus self-determine by making them necessary for their becoming. This notion of qualitative change dovetails with true infinity as the process of reflexive self-determination. I now move to the close intertwinement of actuality and possibility there involved in this process.

Part II: The Organisation of Possibility

> To successfully study the role of development in evolution, one needs to replace the orthodox school of thought, dating back to Aristotle, of nature as a continuum, which has become entrenched into the Darwinian theory of evolution [...], with an alternative philosophy, whose roots can be found perhaps in Hegel—a theory based on a more dialectical and discontinuous view of nature. The latter approach emphasizes the interactive aspects of natural processes and studies them on the basis of the constraints emerging from such interactions. (Alberch 1989: 48)

Above, Pere Alberch, one of the founders of evolutionary developmental biology, underscores the interactive aspects of natural processes through Hegel. One such interaction is that between actuality and possibility. As I will argue, these modalities are internally related, in that something become possible through its actualisation. Below, I make this point more precise by reconstructing Hegel's argument. It will be somewhat technical, but the general point should be familiar since it concerns the historical construction of possibilities. As I aim to show in this rather long section: Had possibilities existed independently from actuality, it would imply that possibilities are prior and external to actuality. Instead, we must consider their co-primacy (Erkan 2021). The ongoing dialectics between possibility and actuality combats the notion that actualisation is a simple release of pregiven possibilities.

In Hegel's view, actuality and possibility have three modalities: formal, real, and absolute. They display different forms of necessity and contingency. Let us start with the latter. Hegel maintains that any possibility is contingent in the double sense of having and lacking a ground. It has a ground and is contingent upon it, but this ground is itself contingent (Longuenesse 2007). This absence of a stable ground enables and necessitates the activity of grounding. By dealing only with what is actual, treating it as external to what is possible, we deny this contingency. This would leave us in a position like Kant's, where we could speak only of appearances, not things themselves. Instead, Hegel distinguishes between (1) contingency as dependence on the totality of conditions that make something necessary, and (2) a more radical form of contingency tied to the impotence of nature, its lack of stable ground.

In the first sense, contingency means dependence on prior conditions. As we shall see, this local form of contingency can co-exist with global necessity. Any occurrence is dependent on the totality of conditions and thus made necessary by them. But these conditions are themselves contingent. Padui writes: 'Given the totality of conditions of what exists, the world as it exists would necessarily follow. But the very givenness of that totality of conditions would itself have to be contingent, that is, "without a why"' (2010: 247). While contingency as dependency is a logical form of contingency, it also has a more radical form, as the irrationality of nature. This is what Furlotte (2016) speaks of as the reticent facticity that frustrates all attempts at complete conceptualisation.

Hegel argues that the contingency we find in nature is not just an expression of our ignorance. In this view, necessity is a product of the totality of conditions that are contingent, in the sense that it lacks a ground.[17] The relation is transitive. Otherwise, the totality of conditions would depend on something else—and would not be absolute but relative. To perceive necessity as absolute is to acknowledge that it contains contingency. This contingency of necessity is still within reason in the sense that it has to do with a logical limit to what reason may know. But there is also a limit imposed on reason by nature's externality to itself. This is the impotence of nature as a 'without why' that reason can never domesticate. Padui explains:

> [I]t appears that there are two different types of limits for philosophy corresponding to these two senses of contingency, one set by reason itself in determining its own limits, and one set by nature's irrationality thereby limiting Reason's comprehension of natural products. Reason cannot ground the irrational in nature without undermining precisely what makes it truly irrational. So while Hegel does *include* the category of contingency within Reason, he also must *exclude* the radical contingency of real natural objects from rational comprehension. (Padui 2010: 250)

Hegel says that the contingent 'has no ground because it is contingent; and for the same reason it has a ground, because it is contingent' (Hegel 2010: 481). In other words, the contingent ground must be grounded, it depends on organismic activity, as it has no *actuality* apart from its positing. There is no stable ground, being or essence, on which the organism can rely, and precisely because of this lack, it *has* ground as the posited totality of conditions. Stated differently, the activity of grounding is constrained/enabled by the lack of ground. On the one hand, any immediate actuality and possibility could have equally not existed and thus are contingent; on the other, since any contingent being is actual, it depends on the non-actuality of what could have been.

To show why this is the case, Hegel emphasises the dual nature of possibility: first, it is abstract essence which relates only to itself. This formal possibility is a 'limitless manifoldness', without determinations and actuality. It can be whatever we may think of as possible, insofar as it is not contradictory. As such, it is an empty determination. On the other hand, therefore, the possibility is deficient, something that is not yet. It must be posited as actual to become a real possibility. Hence, formal possibilities are contradictory—as much impossible as possible since they lack the conditions of actualisation. Saying something is possible in this sense is also saying it is impossible, that it could fail to happen—and the relation between these options 'determines both as possible' (Hegel 2010: 480).

The formal possibility is contingent since it lacks a ground which guarantees its stability. There is no way to know whether it is possible except by reference to what is actual. We shift from formal to real actuality and possibility—sublating formal possibility—when it is shown that every possibility depends on the 'totality

[17] Padui relates this to the misunderstood notion that whatever is actual is rational: 'When Hegel famously identifies the actual with the rational he cannot be claiming that all entities and events are necessarily determined in a rationally necessary whole, since any relative totality within that whole must be contingent on another ground and the necessity of the whole itself must be radically contingent (in the sense of "without ground"). Rather, Hegel claims that any systematic unity of Reason necessarily implies an internal space of that which is not necessary, a place for contingency as the internal limit and negation of necessity.' (2010: 247)

of relations of determinacy' (Ng 2009: 158)—that actuality is the '*unity of itself and possibility*' (Hegel 2010: 480). We can imagine all sorts of formal possibilities, but real possibilities are always constrained by actuality. The formal possibility is contingent, merely possible since its opposite could also be. It is when the contradiction of contingency—that it is equally grounded and ungrounded—is sublated that we arrive at real actuality. Here, the contingency is not only seen as groundless, as one of many possibilities but also as *grounded in actuality*, in the sense that it emerges as an actual possibility only when the totality of (contingent) relations are in place (Ng 2020). Thus, Hegel replaces the abstract ground of being and essence with ground as the totality of existent conditions.

In this modality, when possibilities are negated to produce new actualities, the actual is mediated by the possible. Thus, possibilities are contained within actuality, influencing what can be actualised. But the relation between them is still contingent, as actuality simply 'delimits the range of what is possible' (Ng 2009: 161). In other words, whatever exists depends on this totality (it is *necessary* for its existence) but this necessity is not absolute, since it depends on a totality of conditions. Ng accentuates that when actual conditions of a thing are considered, what is in question is real necessity. They are the same since 'under the given conditions and circumstances, *nothing else can follow*' (Hegel 2010: 484, my emphasis). We thus move to a materialist conception of possibility, conditioned by empirical conditions. Whatever is possible is constrained/enabled by empirical reality.[18]

Real possibility is the totality of *actual* circumstances. The speculative identity between real actuality and possibility is always already presupposed in the way we speak of these modalities, according to Hegel. It reveals the 'unity of the thing with its conditions'. Ng concludes:

> As with the determinations of formal modality, the oppositions between real actuality and possibility find their identity in real necessity: what is really actual is the really possible, and what is really possible is the real totality of conditions which *is* actuality. Therefore, real actuality and real possibility turn out to be the same, and what is determined in the thought of real modality is real necessity. (2009: 166–167, emphasis original)

Finally, we arrive at *absolute* necessity when we grasp that the becoming of necessity is itself contingent in the radical sense. Contingency is necessary because 'there is nothing that exists that is not determined by a totality of conditions' (Ng 2009: 169), and necessity is contingent because this totality is itself groundless. Absolute necessity is the process wherein necessity and contingency are unified in the way they (re-)determine the relation between a thing and its conditions. Any totality of condition is 'without why' as an expression of the weakness of nature. Unless this was the case, it would depend on a ground outside itself, and we would return to relative necessity (Padui 2010). Necessity is not decided from without but contains contingency as part of its becoming. And since necessity does not have an external

[18] According to Ng (2009), this means that there is no clear distinction between what is a priori and what is empirical in Hegel's philosophy. More on this view in Chap. 7.

ground but is a process of *self-development*, it an expression of freedom (Marcuse 1986).[19]

There is always a remainder in this process of actualisation, but it should not (only) be understood as what is inaccessible to thought. It marks ontological incompleteness, which demands ongoing reconfiguration of conditions and circumstances. It might seem, however, that Hegel says that insofar as something is a real possibility, it *must* happen. This would be a determinist argument: Given certain conditions, there is no room for alternative outcomes. We see why this view is limited through the modality of *absolute* actuality and possibility (Zambrana 2019). It reveals the necessity to be relative to the contingent set of conditions, which can only be specified in retrospect after they have acted, viz. become actual. Again, the notion of positing is helpful: it is only when the totality of conditions is posited *as conditions of possibility*, and not just as a 'manifoldness of dispersed circumstances' (Erkan 2021: 85), that we see how possibility and actuality are united in the process of self-determination.

Real necessity adds conditions to how possibilities are actualised but 'does not capture the role of contingency in actuality's self-manifestation' (Zambrana 2019: 86). As such, it is still abstract, as it does not consider the contingency of the totality of conditions that produce actuality. Absolute actuality, on the other hand, posits its precondition and produces a 'matrix of alternate possibilities' (Yeomans 2012: 161). Here, the fit between actuality and possibility is so tight that alternative possibilities cannot be decided as either possible or actual.

I admit, this might be difficult to comprehend, but the main point to grasp is that possibilities must be *organised* to count as either possible or actual. As Zambrana writes:

> What absolute modality shows is that possibilities [...] [are] included in the internal organization of the totality of conditions that produce actuality in the first place. I suggest that this means as well that unactualized and indeed frustrated possibilities are so determined given such organization. Unactualized, indeed frustrated, possibilities are, too, constitutive of what was expressed, hence intimately related to the articulation of necessity. (2019: 87)

It is not the case, then, that given certain conditions, possibilities are actualised without failure. This is the relative notion of contingency and necessity, where contingency simply means dependence on prior conditions. It is rather that actuality and possibility are so entangled that we cannot distinguish them clearly, and that even non-actualised possibilities contribute to the formation of what is actualised. Based on this, we should say that when the organism posits the conditions of its activity, it also posits unactualised possibilities.

Absolute necessity is the 'absolute conversion of its actuality into its possibility and its possibility into actuality' (Hegel 2010: 487). It reveals the unity of a thing with its conditions, reciprocity, which is founded on the lack of ground outside this unity. Of course, the space of formal possibilities will always be larger than that of

---

[19] If, by contrast, we say that necessity is necessary, we face the problem of explaining where this necessity stems from. It would open another infinite regress, as no permanent ground can be found. It parallels the problem of empiricism, where laws must be grounded in further laws, but the becoming of the laws themselves is left in the dark.

real possibilities, as it depends on the contingent totality that makes them viable. A formal possibility is unconstrained by context, and completely abstract. But a formal possibility can become real if the totality of conditions changes. Ng writes that the

> real modality determines the identity and difference of the actual and the possible by taking up actuality as a set of concrete, dispersed potentialities, as containing *within itself* the possibility of becoming a new configuration of conditions, and hence, a new actuality. What is actual can act, or has the power and potential to produce effects, insofar as a new configuration of circumstances and conditions can arise from the existing multiplicity of conditions. (Ng 2020: 145–146)

We saw that formal actuality coincides with what is merely possible. By contrast, real actuality is tied to the capacity to act since it is related to concrete conditions; it *is* the dynamic process of organising possibilities. We may elucidate this by reference to our previous discussion: Biological constraints produce the emergence of preferred (actual) states from possible ones. This differs from physics, where 'all possible behaviors of a system are considered equally possible' (Bizzarri et al. 2020: 5). As such, physics deals only with formal possibilities. It also exemplifies the difference between spurious and true infinity, as formal possibilities follow the logic of spurious infinite potentiality, or what is called totipotency, since it considers possibilities abstractly, outside actuality. Against this, real possibilities are constrained by context—emerge from the self-delimitation of the whole system—and are thus historical. We can imagine all sorts of things that may emerge, but only those that are *mutually compatible* can be actualised. This principle applies to biological organisation (Gilbert and Barresi 2016).[20] The production of variation in biology is also a process organising possibilities, in which regulatory networks are the contingent totality of conditions:

> The diverse effects of genetic change, sometimes leading to changes in phenotype and other times not, arise because regulatory networks process genetic inputs, *organizing them into functional phenotypic outputs*. (Lala et al. 2024: 97, my emphasis)

But the real modality only encapsulates contingency as dependence on prior conditions *within a state space*, and not the contingency of this state space itself. In conclusion, therefore, we need the absolute modality to get at how possibilities are themselves part of an actual totality—how something is *actualised as possibility* by exerting influence on the activities of living beings' process of actualisation. We also need it to grasp the internal relation between possibility and actuality, due to their lack of ground.

It seems that we could understand this through enablement: while real possibilities are determined by the conditions in place, absolute possibilities are enabled by and constrained by the positing that produces a new state space. Contingency is not external but internal to necessity and enables the autonomous development of the thing in question. In other words, the boundary conditions of a system are not decided

---

[20] Another way of describing this is through Leibniz's distinctions between possible and *compossible*, which captures the difference between all imaginable possibilities and possibilities which 'makes sense', are compatible/viable, with a larger whole (Gilbert and Bard 2014).

by an external agent but are internally generated (Juarrero 2013). Much more could be said about this, but I must pause the discussion here and pick it up when we discuss how organisms relate to their environments in Chap. 7.

From Entailment to Enablement

> [T]he point of Hegelian dialectical analysis is not to reduce the chaotic flow of events to a deeper necessity, but to unearth the contingency of the rise of necessity itself—this is what it means to grasp things "in their becoming." (Žižek 2012: 575–576)

To concretise the implications of the above exegesis I now relate it to adjacent debates. I start with strong emergence and the question of whether it demands us to abandon the notion of physical closure, as suggested by Žižek. I relate this to Kauffman's notion of *the adjacent possible* and how it can be informed by the notion of actuality and possibility presented above, as well as the notion of retroactive positing. I continue this debate in the next section, to underscore the virtual status of the presuppositions.

I have argued that a material dimension of Hegel's philosophy is found in the openness and incompleteness of nature, which enables subjectivity to emerge from and retroact on substance (Furlotte 2018).[21] Similarly, strong emergence involves both dependence and autonomy (Wilson 2015). Wilson states that strong (ontological) emergence demand abandoning *Physical Causal Closure*, *viz.* 'that every lower-level physically acceptable effect has a purely lower-level physically acceptable cause' (2015: 353), which is not only necessary but sufficient for the effect. This principle is often defined in terms of causal inherence, in which all higher levels follow from the most basal level and are thus already present there, latently. But, as Wilson (2021) remarks: Even if many endorse this principle, it is 'not a principle of contemporary physics' (124). Hence, there is no reason to assume that it is correct by default.

The notion of physical closure cannot be maintained insofar as we argue in favour of strong emergence that involves novel forms of causation that are both distinct from *and* relevant to the physical domain (Gibb 2019). This criterion points to downward causation, the view that emergent properties or processes has a causal influence on lower-level properties and processes. It is a form of macro-determinism that canalises the activity of the processes that it depends on. Without it, emergence would lack causal autonomy. Absent this notion, emergence is epistemological or weak—which means that higher-level descriptions are optional and that the behaviour of emergent levels can be reduced to processes at lower levels (Tabaczek 2019). From this perspective, emergent novelties are only apparent. This leaves us in an ontological limbo which is not conducive to systems biology.[22]

---

[21] As Andreas Arndt asks: 'how is the materialism/idealism dichotomy supposed to apply to Hegel if Hegel already declares it obsolete with regard to his philosophy?' (Arndt 2023: 9).

[22] We see the intertwinement of these notions through multiple realisability: Multiple realisability means that multiple lower-level processes may realise the same higher-level function while still

If we opt for the causal autonomy of emergent levels, we need a viable concept of downward causation. Like Wilson, proponents of the organisation view have opined that we should abandon 'the completeness and the causal closure of the physical world' (Soto et al. 2008: 257). In these authors' view, this allows for a concept of diachronic emergence as the default state. If this is so, we must forsake the conception of a stable basic level (Bitbol 2012). To account for internally generated constraints, 'a plural vision of causality' is needed (Moreno and Umerez 2000). From this view, we might understand why Žižek believes that quantum physics reveals the impotence of nature, its 'inability to hold fast to the realization of the Notion'. The wager appears simple: Either subjectivity is epiphenomenal, or nature is incomplete. You cannot have it both ways—provide a total description of the universe that permits the agency and subjectivity of organisms.[23]

If the mind is epiphenomenal, if it cannot enact any changes in the world, it becomes a mystery why it evolved at all, as it lacks a plausible function (Kauffman and Radin 2021). The same goes for rudimentary forms of agency, without intentionality nor consciousness (Sultan et al. 2022). If reductionism is true, it would be redundant to postulate such phenomena because they are only apparent. An adequate theory of organisation, on the other hand, might provide the missing link in the debate concerning emergence and reductionism (Kauffman and Clayton 2006). The maintenance of life happens through engendering constraints that achieve closure and thus maintain the overall organisation by constraining the release of energy—both organising and producing entropy, a wellspring of randomness that it can make use of, but which also threatens to undermine its organisation.[24]

The dialectical unity of the logical and material is found in the self-determining concept, which is not merely formal, as for Kant, but actual.[25] The concept *is* the purposive activity of organisms, directed towards their self-maintenance. It thus refers to the self-determining articulation of concrete conditions that maintain the

maintaining that the lower-level base is the ultimate cause. Higher levels lack autonomy and do not affect the lower level in a downward fashion. In other words, the emergent properties are distinct (ontologically autonomous) but lack causal autonomy (see Wilson 2015).

[23] Similarly, Tom Froese has proposed what he calls *Irruption Theory* as a way to understand how the actions of organisms are not merely epiphenomena because the actions of an agent is 'underdetermined by its material basis'. Underdetermination and contingency opens up a space for the efficacy of normative action. He moreover states that 'there are compelling arguments that the notion of a clockwork universe has been invalidated by the most successful natural science, quantum physics' (Froese 2023: 10). I plan to explore such issues, related to issues in philosophy of mind, such as how we should understand consciousness, in future works.

[24] In what follows, I cannot thematise the specific conception of entropy involved here, which is implicit in what I say. I should note, however, that organisation is understood as anti-entropy, by construction and reproduction of biological complexity. At the same time, since there is energy dispersal in each iteration of the cycle, and this process is irreversible, there is also entropy production (Longo and Montévil 2014). Organisation and chaos are not opposites but dialectically entangled in the reproduction of biological organisation. This notion of entropy could elucidate negativity as the cancelling of virtual possibilities through the production of disorganisation. This is a topic I plan to cover in future works.

[25] I cannot cover this topic here, but see Ng (2020), Khurana (2013), and Sedgwick (2012), among others.

organism (Zambrana 2015). I cannot expand on this yet, but we should note this shared commitment of the dialectical and the organisational approach. Let us instead turn to Kauffman, who indicates something like Hegel by way of example: Could you tell me all the possible uses of a screwdriver apart from its actual use? Of course, you cannot. This is not because there are infinite ways of using it, as that would mean that it could be used for whatever. Instead, there are *indefinite* ways of making use of it, which is to say that you cannot state ahead of time all the possible uses, just like you cannot state how a trait can be re-functionalised in advance (Kauffman 2020). A screwdriver can be repurposed but within limits defined by its history and context. Thus,

> the Arrival of the Fitter does not depend in any specific way on any specific predefinable set of cellular or molecular screwdrivers. Although any specific instantiation of an adaptive step to a fitter function or new function will in fact utilize some specific cellular or molecular screwdrivers, which ones may happen to be the ones so selected cannot be algorithmically prestated.

This takes us beyond weak emergence:

> [N]ot only is there the familiar "multiple realizability" philosophers are used to, the situation is more radical—the multiple realizations for a given new use cannot be algorithmically listed, so are unprestatable. In this sense, the adaptive change of the organism is beyond entailing law and beyond statable mechanism. (Kauffman 2013: 11)

'Beyond entailment' means that emergence is not just epistemological, viz. about the way we make the world intelligible. Rather, it is an ontological claim. It is not enough to state that a phenomenon at one level may be realised by many different configurations at a lower level. This would still imply that the emergent level supervenes on a base, so that any change in the former is predicated on the latter. Kauffman not only overturns the mechanical approach but also the quantum mechanical one—since both rely on defining the state space in advance to make calculations. Had he said 'infinite' he would be speaking of an abstract manifold (formal possibility) without actuality. It would be an empty determination of contingency, of whatever is conceivable (Malabou 2016). But Kauffman does not deal with imaginable ways of using a screwdriver in the abstract sense; his examples are concrete, constrained by the history of the object, and therefore indefinite or virtual. Also, since the relation between actuality and possibility is not one of determination but enablement, he seems aligned with the absolute conception of their interplay, where they are reciprocal.

Kauffman's adjacent possible formalises the opportunism of nature, and how actuals do not primarily cause but enable new *possibles*. The context does not enable all formal possibilities, only a subset (Gilbert and Bard 2014). Adjacent possibles exemplify the contingency of necessity: As soon as something contingent occurs, it *could* become a necessary premise for whatever else might occur. What currently exists does not determine what may occur; rather, it enables something else which is also contingent, in the dual sense of dependent and unpredictable. Contingency is primary and internal to necessity, and the becoming-necessary of the contingent only appears necessary after the fact. The past becomes necessary in and through the present it enables; but it remains open, containing virtual possibilities that may or may not be actualised.

Only through its consequences does an event *become necessary* for whatever comes after. As Žižek remarks, while the necessity of contingency simply means that any necessity must realise itself through contingent phenomena, the contingency of necessity is a more potent concept, which states that necessity emerges from a contingent process:

> when Hegel describes the progress from "external" contingent appearance to "inner" necessary essence, the appearance's "self-internalization" through self-reflection, he is not thereby describing the discovery of some preexisting inner Essence. something that was already there ([...] a "reification" of the Essence), but a "performative" process of constructing (forming) that which is "discovered". (Žižek 2012: 467)

Note the historicity involved, and take Kauffman's example, the internet: The internet is an actual that stemmed from earlier actuals through which it was an adjacent possible. From it, we have gotten things like Facebook. But the internet did not determine Facebook, in the sense that it *had* to occur as soon as the internet did. Instead, the internet was an enabling boundary condition for Facebook to arise; and as soon as Facebook emerged it became (retroactively) a necessity for something else—like Instagram—to occur. But you could not prestate the emergence of either of these inventions from the emergence of the internet.

This dovetails with positing the presuppositions: Facebook became the presupposition for Instagram after the fact. It acted as the niche in which something like Instagram could emerge, but no one could have said that until after it occurred. Before it happened, there were an *indefinite* number of things which the internet might lead to but whatever becomes actualised is rational in the sense that we can trace its internal necessity. This does not mean that whatever *exists* in this world is rational (or adaptive) but that it becomes rational (or intelligible) only from the perspective of what enabled it, in this case what became actual due to the internet. We should thus not take it as a normative claim that states that what exists is good, but rather as a claim about how reason is able to make nature intelligible, however imperfectly because there is already an inherent rationality there, in the making (Stern 2006). It is the inchoate rationality of nature we spoke of before.

Downward Positing

Actuals beget new possibilities that pave the way for new actuals. But unactualised possibilities have an actuality of their own; they produce real effects as virtual points of reference. This is evident both in quantum physics and the sphere of language, says Žižek. He mentions paternal authority as an example: the virtual threat of punishment affects the actions of subjects even if this threat is never actualised (Carew 2014). (I discuss virtuality more in relation to habit in the next chapter.) The possibility is negatively present but not actual, and yet it influences actuality. Furthermore, the adjacent possible of an actual is not cancelled wholesale when another actual comes about. Žižek says that 'the actualization doesn't simply abolish the previous panoply of possibilities: what might have happened continues to echo in what actually happens as its virtual background' (2016: 49). It is not only a new actual reality that is produced; the conditions for this actuality also change (Žižek 2012). We could link this to the idea of nature as a bricolage, which puts old elements to new uses in an unprestatable manner. Žižek takes an example from biology:

> [T]he fact that our DNA overlaps around 98 percent with that of some apes means something: it sets some coordinates. [...] Nature is not a complete determinist order: it is in some sense ontologically incomplete, full of improvisations; it develops like French cuisine. Is the origin of many of its famous dishes or drinks not that, when they wanted to produce a standard piece of food or drink, something went wrong, but then they realized that this failure can be resold as success? (Žižek 2020: 282)

Here, Žižek is reiterating the point we made above: that DNA is a constraint that enables but does not determine how it is put to work. Nature improvises in a manner that upends the rules of the game; even if what exists now constraints what may exist in the future, we cannot know with any certainty what might emerge. It is jury-rigged from whatever evolution finds lying around. This is a radicalisation of the unpredictable nature of Poincaré's three-body problem of deterministic chaos, where the interaction between three gravitational bodies (planets) produces unpredictable behaviour.

Another example: If you throw a coin 100 times, you know the alternatives even if you do not know what the result of each toss will be. They are equally probable, and thus we are simply epistemically ignorant of what will happen because the exact ordering cannot be predicted. But while the outcome is random, the sample space is given in advance, and we know that it tends towards an equal distribution. By contrast, the biosphere flows into an adjacent possible that it constructs—a construction that cannot be known in advance: 'We not only do not know what *will* happen, we do not even know what *can* happen' (Kauffman 2020: 19) since the possibility space itself is changing, like the groundless imposition of a new necessity:

> When a chaotic period of gestation culminates in the explosive eruption of a new Form which reorganizes the entire field, this very imposition of the new Necessity/Order is in itself thoroughly contingent, an act of abyssal/ungrounded subjective decision. [...] While "substance" already stands for the encompassing unity of opposites, for the medium within which particular forces reproduce themselves through their struggle, in a "substantial" relationship the two aspects, synthesis and splitting, are not yet brought to self-relating, so that splitting *as such* would be that which brings about a synthesis, so that imposing a new Necessity would be the highest gesture of contingency. (Žižek 2012: 196)

The epigraph brings out the notion of self-relating negativity, which overcomes the bad infinity of two vacillating poles. It is only when synthesis and splitting are seen as internally related that we get to true infinity, which produces a space of possibilities. The reflexive moment of self-determination that generates a new state space is not limited to human subjectivity since it emerges from more basic forms of subjectivity (as I argue more fully in the chapter on niche construction).

Within a new form, a determinate negation, there are relations of contingency (as dependence on conditions), but this is not enough to get at the radical contingency at work in instituting a new network of enabling relations. Anything that emerges depends on the current space of possibilities, where the interaction between agencies 'can create possibilities that did not exist before' (Fuchs 2014: 862). Again, witness the dual nature of contingency: any chance event is contingent upon the pre-conditions of the current field and the history that shaped it, which is what it challenges at the same time.

Hegel says that contingency as immediate actuality, 'is at the same time the possibility of something else' (Hegel 2015: 217). In the same paragraph, Hegel states that actuality is a process that 'contains the seed of something completely other'. This process is an organisation of chance, wherein actuality is shaped through the dialectical of necessity and chance (Hörz 2009). There is therefore no a priori way of distinguishing a system from its elements, since 'to identify the elements, we need to understand first how the system has been modified by its history' (Soto et al. 2008: 267). Linear part-whole explanations fail since a part (process) has a function only by reference to the historical development of the system in which it operates.

This underscores why there are 'strange entanglements' in biology: for example, a gene might produce proteins through a process which requires the influence of the very proteins that it produces. It is impossible to apply simplistic cause-and-effect models to such processes' (Soto et al. 2008). Or take Hegel's example: He speaks of someone having a talent at music who started playing after his father was shot in battle. Hegel is not satisfied in stating that this event was the *cause* of the abilities. The gun that shot the father, and the war that led to the gun being fired could equally be seen as causes, as could the steel of the gun or the monarch that started the war (Hegel 2010; Ng 2020). Neither of these events could be said to single out the cause of the musical ability to develop. Hegel says that any event we are enumerating 'is not a cause at all but only a single *moment* which belonged to the *circumstances of the possibility*' (Hegel 2010: 496, emphases original). Misunderstandings arise when actuality is understood in terms of efficient causality and external mechanisms, not as enabling constraints that produce adjacent possibilities—that may or may not be actualised based on the activity of the organism. The totality of conditions cannot cause anything except through being taken up and reworked by the individual; by themselves, mechanisms over- and underdetermine what happens.[26]

The releasing cause should not be confused for the enabling nexus of causes and their constitution, and how they are integrated and actualised through purposive activity. Here, we observe the overlap between reciprocity and enablement. It also underscores Hegel's anti-reductionistic stance. In living things, causes are sublated; living organisms do 'not let a cause continue to work its causality in it undisturbed but will rather interrupt and transmute it' (Hegel 2010: 496). The same goes for 'the

---

[26] 'Mechanism underdetermines because there is an infinite regress of causes, so the set of conditions is never complete; mechanism overdetermines because there are too many causes and provides no criteria of its own with which to distinguish genuine, essential causes from accidental, non-essential ones' (Ng 2020: 154) As per Johnston (2015), from such overdetermination emerges underdetermination, as the infinite regress is short-circuited to produce emergent reality.

rise of the great out of the small': it is not due to an ultimate cause but due to the way such causes are taken up and sublated, by 'the conversion that spirit works on the external' (Hegel 2010: 497). Apart from relations, an external circumstance lacks causal specificity. It gains actuality by being sublated. Further, it is not enough—or even possible—to look at one cause in isolation; instead, it is the conjunction of causes that produces effects, which then reveals their causes backwards: 'the cause does not just *have* an effect but, in the effect, refers *as cause* back to itself' (Hegel 2010: 504).

The two moments are not separate, as that would mean that 'it is not only *something posited* but something that rather *exists in itself*' (Hegel 2010: 701, emphases original). To exist is to be a condition for something else, and something is only a condition in retrospect. We could tie this to the notion that we cannot prestate what may become an evolutionary factor. An infinite regress is avoided by assuming that systems cannot be exhaustively described at the outset and that they evince properties which emerge in and through their history. Being historical, is distinct from simply temporal, means that organisms 'embody in their very structure the conditions under which they were created' (Juarrero 1999: 8). While initial conditions leave their mark on what emerges from them, they are also custom to change in a nonarbitrary downward manner as the organism imposes new boundary conditions on their functioning. This process provides the emergent level with agency and causal force (see Soto et al. 2008).

Becoming Initial

> Matter is a causa sui, it has the capacity to organise itself and produce new forms and levels of organisation of matter. The self-organisation of matter is the ultimate absolute recoil: In every transition from one form of the organisation of matter to another (e.g. from inanimate to animate nature, from the animal to the human, from capitalism to communism, etc.), matter posits its own presuppositions as the ultimate absolute recoil, namely the capacity to produce forms of matter and to thereby reproduce itself. (Fuchs 2014: 856)

To return to biology, the lack of information at the outset means that a mother system—the initial conditions of a system—gives rise to daughter systems with unpredictably changed boundary conditions. The initial boundary conditions change as the system evolves (Kauffman 2016). What is lost, therefore, was never fully articulated. It only determined basic or generic features, 'a range of possibilities that need to be supplemented with an additional causal explanation at a different level' (Moreno and Umerez 2000: 105). Also, it is not completely lost, in the sense that the initial state does not affect what emerges from it. Instead, it reverberates in its successors:

> Some older qualities may cease to exist so that parts of something turn into nothing, but the old continues to exist and shape the new. And finally, given that there are mere potentialities that have not been realised and constitute non-being (or not-yet being) in the old and the new,

> also and old nothing turns in a sublation into a new nothing: A new field of possibilities, of non-existing realities that are pure potentialities, emerges. (Fuchs 2014: 854)

In other words, the past is integrated into a new *relational field*, that shapes the potentialities that were not realised in the old. The past and future are reorganised as 'horizons of possibilities' (Bich and Bocchi 2012). This is another way of formulating why we cannot expect the bottom level to contain enough information and causal specificity to determine other levels. It also rehashes Denis Noble's middle-out approach. If the bottom level is incomplete, it means that initial conditions are constrained by what emerges from them. In a sense, then, initial conditions *become* initial only by reference to the system in which they are actualised as such. Similarly, Soto, Sonnenschein, and Miquel write that processes of morphogenesis is not just the differential expression of genes, that

> in addition to this upward causation, cellular and tissue events occurring before the expression of a particular set of genes takes place may act downwardly modifying the expression of these genes at a later time (diachronic emergence). (2008: 271)

As we saw before: no level is completely isolated and unconstrained, as organising principles are operative everywhere. Also, as Hegel's virtual beginning, the mother system cancels itself out but never completely so. He says that

> each step of the *advance* in the process of further determination, while getting away from the indeterminate beginning, I also a *getting back closer* to it; consequently, that what may at first appear to be different, the *retrogressive grounding* of the beginning and the *progressive further determination* of it, run into one another and are the same. (Hegel 2010: 750)

The grounding and the development of the ground coincide, as the ground is itself only a process of grounding. Moreover, the process reveals how 'material being, as incomplete and inconsistent, contains within itself the potentials for the creative genesis of modes of subjectivity exceeding this same ontological foundation' (Johnston 2014: 121–22). Subjectivity emerges from the incompleteness of substance and frustrates attempts to understand it completely. Substance self-differentiates as subject, as a negative process that concretises substance through a historical becoming. The indeterminate beginning is concretised but never cancelled through the formative negations it embodies as a processual unity. Hence:

> The past itself is open towards the future, that is, it is not simply "what really happened" but full of cracks, of alternate possibilities—the past is also what failed to happen, what was crushed so that "what really happened" could have happened. (Žižek 2023: 427)

As we have seen, since an event at one time changes what might happen at a later stage, the potential for change is itself a product of the history of the organism, and possibilities are therefore not positively given at the outset. Take genetic information: The *potential* information found in DNA only becomes actual when it is realised in an environment. You cannot speak of the informational content of genes apart from their actualisation as phenotype:

> Genetic information constitutes potential information [...] that is only realized when confronted by environmental information. More generally, development necessarily entails interactions between the various sources of information possessed by individuals, be they inclusively heritable [...] or not, genetic or nongenetic [...]. (Danchin 2013: 352)

Current genetic information interacts with previous activity, which imposes constraints on its current meaning. Again, DNA is not an efficient but a formal cause in the sense that it constrains the synthesis of molecules it depends on (Moreno and Umerez 2000). What is inherited by a fertilised egg is not 'genes' but DNA, which is made into genetic information in the cell through the relationship between the DNA, transcription factors, RNA-splicing factors' (Gilbert and Bard 2014: 130). As the information potential of the DNA is actualised by higher-order structures such as cells, causal chains may emerge in unpredictable ways; a molecule, might, for instance, be afforded new properties through interaction with other molecules, affected by the constraints of the overall organisation, which are themselves processes: 'In the context of new systemic properties, basic properties are changing. However, this new type of constraint is not fixed at the beginning. This is the meaning of downward causation' (Soto et al. 2008: 272). In this version of organicism, top-down and bottom-up combined, allowing for a third option, between reductionism and holism (Noble 2006), in which no privileged level of causality exists. Against the critique of organicism as holism, then, where the whole has primacy over the parts, we should instead speak of a *relational organicism* (Quadrio 2012), where there is no privileged level.[27] This view adheres to the 'principle of biological relativity' proposed by Noble (2006, 2011), where no level of organisation has a priori primacy. It is also what is meant by heterarchy, in contrast to hierarchy.

Res Potentia

> I never pretended that one can insert reality into the past and thus work backwards in time. However, one can without any doubt insert there the possible, or, rather, at every moment, the possible insert itself there. Insofar as inpredictable and new reality creates itself, its image reflects itself behind itself in the indefinite past: this new reality finds itself all the time having been possible; but it is only at the precise moment of its actual emergence that it begins to always have been, and this is why I say that its possibility, which does not precede its reality, will have preceded it once this reality emerges. (Bergson 1991: 1340)

The principle formulated by Bergson (often reiterated by Žižek) might initially seem counterintuitive but it is vital for the position I am favouring: Possibilities do not predate their actualisation. As biological necessity is a function of contingent events being taken up and maintained by a concrete organisation, natural history makes a comeback. The way genes splice and recombine for instance, 'seem to become instantiated as "biological laws," even though they are entirely historically contingent' (Kauffman 2000: 137). The 'laws' of biology are subject to retroactive change;

[27] In line with Paul Burkett (2014), we could also talk about *relational holism*.

they are 'historicised invariants' (Longo 2018). They are, in a strong sense, the product of historical and contextual changes, and thus contingent in both senses above.

In contrast to classical physics, where invariants such as laws are postulated to explain changes in physical objects, biology cannot make such postulations but must take the construction of invariants themselves as their subject matter (Montévil 2020). This indicates a novel kind of randomness:

> It is our contention that biology requires a new form of randomness different from the one used in physics. The new form can be loosely described as the emergence of new possibilities [...] i.e., as the appearance of new dimensions in the space of possibilities assuming that these dimensions are associated with qualitatively new behaviors. Our principle of variation states that in each biological organism such new possibilities can emerge. (Montévil 2018: 379)

This randomness at the level of form points to the contingent emergence of actuality and possibility. It also takes us back to the infinitude inherent in finite beings—their circular determination, which cannot have a limit outside its own movement. In other words, their indefinite potential changes as new forms emerge. As Hegel says, the true infinite is the turning back of any finite thing on itself (Hegel 2010). It is the process of negativity relating to itself through its positive products.[28]

The view that the world is self-enclosed and complete excludes potentiality and deals only with actuality, says Kauffman. He combats it through the notion *res potentia*: ontologically real possibilities, the counterpart but not the opposite of res extensa. This aligns with the idea that the past is not fully constituted. The realm of possibilities is the proto reality through which actuals emerge. It is the shadowy realm of res potentia, a realm we only visit indirectly, through what is actualised. Here, what *could have happened* at earlier stages echoes in the new constellation (Žižek 2016). In contrast to the main thrust of modern science, this means that we do not have a dualism between consciousness as the realm of possibilities and hard reality on the other side. Reality is ontologically open before consciousness intervenes and negates its multiplicity, as seen in quantum measurement (Žižek 2006).

Negativity, then, signifies the impossibility of reducing reality to actuality. It is an excess of possibility over actuality, which influences what is actualised.[29] In a similar vein, Montévil (2018) speaks of pre-possibilities as opposed to real possibilities. The set of everything that might happen to occur is an ahistorical idealisation concerned with pre-possibilities, not real possibilities, which are always specific to the context which makes them viable. Possibility and compatibility are therefore intimately linked (Longo and Montévil 2011). Montévil relates this to music and how novelty emerges there. A musical score is treated in an abstract manner when none of the individual elements are explicated. In such a model, every interval of

[28] For a view that links this to the notion of entropy and anti-entropy (organisation), see Sandnes Haukedal and Mylonas (2024).

[29] As Russon (2014) writes, it is 'an absence, not in the sense of something "not there", but in the sense of something that is there as something that can never be present, something that is recognized through physical actuality but is irreducible to it' (52).

music is equiprobable. But as soon as any elements are made explicit, they constrain the possible paths that may be taken.

The main point—elucidated by the adjacent possible—is that in biology you must contextualise possibilities for them to count as real. While physics deals with the generic properties of systems, biology cannot make such an abstraction without distortion. This means, as Montévil explains, that Bergson is correct to say that the possibility of a symphony does not pre-exist its actualisation. The composer cannot have a clear sense of the symphony before it is made. If she did, the symphony would be made beforehand and not truly be novel. As Campanella writes: 'Bergson argues that the possible should not be conceived as something that awaits to be realized on the basis of given conditions' (2018: 199). It does not wait to be actualised since it only becomes a possibility through its dialectical interaction with actuality.

While you may generically produce the set of all possible symphonies, in more specific terms it should be evident that you cannot assume that all musical scores would 'make sense as symphonies' (Montévil 2018). It constitutes a formal possibility, unconstrained by context. The emergent and specific set is much smaller than the generic set of all possible symphonies. Also, enabling constraints must be compatible with the context already in place. Equally, we may define all possible sequences from DNA base pairs generically, but the biologically feasible sequences depend on their context. Again, biological systems are not generic—no organism has an identical organisation. In contrast to physics, the objects of study are not interchangeable (see for instance Longo and Montévil 2011). Thus, you cannot suppose that models adopted from physics apply in other contexts without remainder.

Conclusion

> The possibility of locating oneself within one's reality has to remain a possibility: however, and herein resides the crucial point, this possibility itself has to actualize itself *qua possibility*, to be active, to exert influence, *qua possibility*. (Gabriel and Žižek 2009: 109)

Žižek elucidates the dialectical approach to biology by reference to exaptation. He notes that even hardliners like Daniel Dennett accept how the variation which evolution uses for its purposes are produced in a radically contingent manner. Yet, they minimise the importance of such randomness, as they assume that natural selection alone makes it functional. Only a dialectical approach to nature reveals the full implication of contingency:

> First, the dialectical approach proper is *structural:* the New emerges not as an element, but as a structure. In an aleatory way, all of a sudden, a new Order, new harmony, emerges out of Chaos Although we can (retroactively) ascertain a long gestation period, one last element triggers the swift shift from Chaos to new Order—"hard" Darwinists do not talk in terms of such a structural "totality." Second difference: this new Order cannot be accounted for in terms of "adaptation"—it is not only that a univocal *ad quem* is missing here (adaptation to what?), one also cannot presuppose a univocal agent of it (adaptation of what?). A vicious cycle is inescapable here: we cannot explain the very emergence of an organism in the terms

> of a strategy of adaptation. If an organism is to adapt in order to survive, it must be there in the first place. An organism evolves to survive, but it cannot emerge in order to survive: it is meaningless to say that I live *in order* to adapt myself. In short, a newly emerging Order "creates" ("posits") its environs—in relating to its other, it relates to itself. (Žižek 2006: 199)

First, the idea that novelty is to do with the emergence, not of an element but a structure has obvious overlaps with how changes in the overall organisation alter the identity of an organism. What emerges is a new space of possibilities, a qualitative change. The switch is what dynamical systems theory calls a critical transition, which is when a threshold is reached, and nonlinear and qualitative changes ensue.

But the change in question is more radical: What changes is not an element; rather, the element and its context change simultaneously. This is the negation of negation, a change of relations between elements (Sève 2008). The new organisation which arises is an open and dynamic totality, constantly reconstructing itself through its infinite self-relation. It works by negation, which is to say that it both sublates and retains parts of its past structure (Tahar 2022). While individual parts constantly degrade and must be reproduced, a diachronic identity is maintained through the overall organisation which subsists throughout the degeneration of the parts (DiFrisco and Mossio 2020).

Stated differently, the organism sublates the presuppositions of its becoming into moments of its constitution, as in the case of the musician above. Such a move is retroactive since the moments only became the necessary steps of the production after the organisation emerged. There was no fully formed bottom level of *determination* involved; at most, this postulated level is a inchoate field that enables the emergence of new fields of enablement relations. By such a movement, retroaction affects its own conditions of possibility.[30]—The second point made by Žižek touches on the question of whether natural selection can be said to produce the variation it acts on. Whatever emerges through self-organising processes will be maintained by natural selection if it manages to survive, but the variation itself is not produced by natural selection *in order for* the organism to survive. To avoid this tautology, natural selection must be conceived as less powerful than the MS presumed. It is mostly concerned with excluding the incompatible or deleterious. Life cannot be explained as a process of adaptation but as a process of self-maintenance through organisation. As such, it is the precondition of adaptation both as a process and result.

The emergence of the new from the old is the subject matter of dialectics, but not as an unfolding of a given potential. Instead, what happens in the turning back is that the gap is redoubled. There is a gap between the old and the new, but also 'the Old "in-itself" (as it was before the New) and the Old retroactively posited by the New from this view' (Žižek 2012: 273). What escapes us is not the before or the after but the 'very birth of the New'. This self-relating negativity is also a way of

[30] Of course, this does not mean that there are not occasional cases where DNA drives the change. As Longo (2018) notes apropos the aetiology of cancer: in most cases, DNA changes are not the causes but the products of changes in tissue organisation. But one cannot assume a priori that DNA changes can never drive phenotypic changes, even if 99.9 of mutations in cancerous cells are passage mutations, i.e. driven by other changes.

conceptualising how evolution delimits itself. It also hints at how quantum events enable internal teleology:

> The lesson of quantum physics is that, beneath solid material reality, there is a quantum level at which determinism breaks down. Hence the claim that the indeterminacy discovered by quantum physics opens up a space within which the "higher level" teleological causality can determine the "lower level" material events, without relying on any spiritualist notion of the power of our minds to magically suspend natural causality. (Žižek 2012: 744)

Of course, indeterminism or contingency is not itself enough to speak of self-determination; it is just a prerequisite. It must be made use of by the organism in its positing activity. Positing the presupposition concerns through which the organism 'generates the very material conditions that engender and sustain it' (Žižek 2006: 119–120). Arguably, we find the same kind of bootstrapping in closure of constraints. As Longo and Soto (2016) put it: 'the conditions of possibility for the emergence of new objects are generated along the way' (8). Again, we can think of negativity as an ongoing production of possibilities—as the unfinished production of an ontological ground.

An emergent totality is 'is itself riddled with 'unpredictable, contingent future presents promising to retroactively alter or overturn altogether such teleologies through positing their own new teleologies' (Johnston 2018: 123). Such positing is, as we have seen, a matter of constructing a self-maintaining organisation, which actively keeps itself within a range of viability—from which '*teleological properties are a direct and necessary consequence*' (Jaeger 2024: 165, emphasis original). This positing replaces the external teleology of linear growth, often associated with Hegelian thought:

> Hegel's dialectic itself is not yet another grand teleological narrative, but precisely an effort to avoid the narrative illusion of a continuous process of organic growth of the New out of the Old; the historical forms which follow one another are not successive figures within the same teleological frame, but successive re-totalizations, each of them creating ("positing") its own past (as well as projecting its own future). (Žižek 2012: 272–273)

We might understand the appeal of quantum mechanics, as Žižek seeks to conceptualise emergence without a magical suspension of reality. Furthermore, positing the presupposition concerns the same kind of bootstrapping as that found in closure of constraints, through which the organism 'generates the very material conditions that engender and sustain it' (Žižek 2006: 119–120), producing its conditions along the way.

Similarly, Victor Marques (2016) remarks: 'for [...] downward determination to have any material efficacy we need to presuppose the noncompleteness of physical causality' (127). We have seen that proponents of the organisational view apply principles from quantum physics in an analogous manner (Soto et al. 2008; Wilson 2015). This underscores the structural similarities I am arguing for. If you want to speak about subjectivity as anything other than an epiphenomenon, you must pay the ontological price and accept that material reality is unfinished.[31] Yet, this 'in no way

[31] This dovetails with the enactive approach, where some favour Hegel's perspective on teleology (see Di Paolo et al. 2018).

implies that all we can do is to reconcile ourselves with this impossibility, i.e. accept reality as imperfect' (Žižek 2012: 477). More importantly, it means that organisms have the capacity to change their reality and hereby change themselves.

So, what is the relevance of the Hegelian perspective to the organisational approach? It is not so much that Hegel anticipates the organisational approach or that he may improve it, but rather that 'Hegel's logical analysis of organization is assumed by the causal account of closure and differentiation outlined by the organizational view' (Cooper 2020: 10). Thus, his philosophy provides the logical justification for the organisational perspective. The latter complements Hegel's view of the organism and underscores the normativity involved in organismic action. This homology is imperfect but suggestive.

Žižek mentions but does not develop the link between Hegelian *good infinity*, active self-limitation, and self-organisation. By contrast, authors like Varela, Rosen, and Kauffman make explicit references to Kantian wholes—as self-organising cause and effect of themselves—but seem unaware of how Hegel draws the full implications of this idea. We could even say that this view is already implicit in Kauffman, as he is clearly a realist about purposive behaviour. Insofar as should regard Kantian wholes—where the parts 'exist for and by means of the whole' (Kant 2000)—as actual, viz. something we *can* have positive knowledge of, we must make a move to Hegel and *dialectical wholes* (Marques and Brito 2014; Cooper 2020). Marques writes:

> the challenge here is to show how a properly teleological, normative being can emerge from an indifferent material reality by means of natural evolution. But isn't that, in a sense, already the aim of dialectical materialism: to explain how freedom itself can take place in a completely natural world, to give a nontranscendent account of the emergence of the autonomous, self-determined subject by the nonteleological (self)organization of matter? Isn't dialectical materialism, from its beginning, a synthesis of Hegel and Darwin? We would like to say that what is really at issue in dialectical materialism is a conceptual articulation of a synchronic concept of a whole that is cause and effect of itself (an ontologized natural purpose) and a diachronic historicization of nature, animated by purely immanent tensions. (Marques 2016: 114)

In other words, the current identity of the organisation is enabled by the ongoing exchange with the environment. It is a synchronic identity produced through diachronic differentiation. This resonates with the historical aspect of the evolution of the biosphere, characterised by self-reference. The rules by which you might predict the evolution of evolution are immanent; there is no meta-evolution. The space in which evolution operates expands as a function of evolution itself.

Commenting on analyses which fix the space of evolution in advance, Nigel Goldenfeld and Carl Woese (2011) write: 'Such approaches to evolution miss what is to us the central aspect of evolution: it is a process that continually expands the space in which it operates through a dynamic that is essentially self-referential' (386). But self-reference does not mean that we are dealing with closed systems, as in autopoietic theories; instead, as every system fails to contain itself, self-referentiality is never complete. In short, otherness is part of the functioning of the self, not a mere add-on from which it may escape into the secure confines of the self (Kisner 2014; Pradeu 2012). As Furlotte (2018) writes,

> For although autopoietic self-relationality has a self-determining structure of its own, because it still depends upon its preceding conditions (its members are, after all, *also* composed of chemical, physical, and mechanical bodies), it is only possible within, and alongside, the register of nature's overwhelming externality, the lack of interiorized conceptuality that characterizes its previous stages of nature. (199)

Organic life constructs a field of possibilities that emerge from preceding conditions. The lack of conceptuality in nature enables autopoiesis but also troubles it. Again, sympoiesis is more adequate, as it encapsulates the allopoietic ('other-making') dimension of autopoiesis. In the sentence following the passage above, Furlotte touches on an important distinction: On the one hand, the organism is an ideality which sets itself apart from and negatively overrides its mechanical, chemical, and physical presuppositions; yet, on the other, it cannot overcome them. So, when it is said that 'subjects crush the alleged external substantiality of the reality of the world standing before his or her knowing, by making it *ideal*' (Ferrini 2020: 181), this should not be interpreted as saying that we have hereby come to dominate nature. Instead, nature always retains its disruptive dimension, frustrating attempts to dominate it. The organism persists as part of an externality—the totality of conditions—that both sustains and threatens to undermine it, and its activity is spurred by this tension.

On the other hand, therefore, is not merely that the environment is external to the organism, but that the organism—even if always caught up with what is external to it—has the power to retroactively posit whatever was presupposed for its maintenance. It is not affected by *everything* which is outside it (Sultan 2015). It is determined in a way which it must itself posit, thus not determined at all, except on rare occasions (Rosen 1991). Also, organisms are both autopoietic *and* allopoietic. They produce otherness in their dealing with their environment. Later, I tie this to the notion that purposive activity is an embodiment of the concept.

When life emerges, a ground is produced which cannot be reduced to the condition it emerged from and which retroacts on them. The psychochemical field that life emerges from can neither be overcome nor dominated; it is historical and plastic and remains a problem for the organism—while also being the preconditions for its maintenance. Any self-relation, such as autopoietic systems, depends on the weakness of these conditions. A contingent event retroactively creates its conditions of possibility by harnessing other processes. It is not that the conditions were produced ex nihilo, but that they can only count as conditions after the fact—when it is revealed what they *could* be conditions for. This introduces a paradox, which is that something both is and is not a condition until after it is released as such. But that does not make it any less actual. When life emerges from nature, it changes nature in return.

Finally, the lack of ground or abstract negativity is positively charged in the sense that it makes possible the emergence of subjectivity. The 'insubstantial potentiality' that the lack of grounding represents must be embodied to be actualised (Johnston 2008). By itself, it is only the possibility of autonomy. It enables specific and internal 'dynamics of unfolding', constructing boundaries and mediating the organism's interaction with the environment (Jaeger 2024). This minimal agency characterises all life. The organisation evinces autonomy from physical laws and thus cannot be predicted by them. It demonstrates self-determination as 'closure to efficient causation' (Rosen 1991). From the vast realm of formal possibilities, the more limited realm of the actual is produced. This is the increasingly complex, and thus more and more sparsely explored, space of the possible. Granting historicity its due, this is not an exploration of a preformed sphere of possibilities but an ongoing and contingent construction which *becomes* necessary.

Bibliography

Alberch, Pere. 1989. The Logic of Monsters: Evidence for Internal Constraint in Development and Evolution. *Geobios* 22: 21–57. https://doi.org/10.1016/S0016-6995(89)80006-3.

Arndt, Andreas. 2023. *Hegel in Marx: Studien Zur Dialektischen Kritik Und Zur Theorie Der Befreiung*. Berlin: Dietz.

Baedke, Jan. 2021. The Origin of New Levels of Organization. In *Levels of Organization in the Biological Sciences*, 153–173. Cambridge, MA: MIT Press.

Barad, Karen. 2007. *Meeting the Universe Halfway*. Durham, NC: Duke University Press.

Barad, Karen. 2011. Nature's Queer Performativity. *Qui Parle* 19 (2): 121–158. https://doi.org/10.5250/quiparle.19.2.0121.

Bergson, Henri-Louis. 1991. *Oeuvres*. Paris: PUF.

Bhaskar, Roy. 2008. *Dialectic: The Pulse of Freedom*. London and New York: Routledge.

Bich, Leonardo. 2021. Autonomous Systems and the Place of Biology Among Sciences. Perspectives for an Epistemology of Complex Systems. In *Multiplicity and Interdisciplinarity, Essays in Honor of Eliano Pessa*, ed. Gianfranco Minati, 41–57. Cham: Springer. https://doi.org/10.1007/978-3-030-71877-0_4.

Bich, Leonardo, and William Bechtel. 2022. Organization Needs Organization: Understanding Integrated Control in Living Organisms. *Studies in History and Philosophy of Science* 93: 96–106. https://doi.org/10.1016/j.shpsa.2022.03.005.

Bich, Leonardo, and Gianluca Bocchi. 2012. Emergent Processes as Generation of Discontinuities. In *Methods, Models, Simulations and Approaches: Towards a General Theory of Change*, ed. Gianfranco Minati, Mario Abram, and Eliano Pessa Pessa, 135–146. Singapore: World Scientific Publishing Company. https://doi.org/10.1142/9789814383332_0009.

Bitbol, Michel. 2012. Downward Causation Without Foundations. *Synthese* 185 (2): 233–255. https://doi.org/10.1007/s11229-010-9723-5.

Bizzarri, Mariano, Oleg Naimark, José Nieto-Villar, Valeria Fedeli, and Alessandro Giuliani. 2020. Complexity in Biological Organization: Deconstruction (and Subsequent Restating) of Key Concepts. *Entropy* 22 (8). https://doi.org/10.3390/E22080885.

Bowman, B. 2013. *Hegel and the Metaphysics of Absolute Negativity* (Modern European Philosophy). Cambridge, UK: Cambridge University Press.

Brooks, Daniel S., James DiFrisco, and William C. Wimsatt. 2021. Introduction: Levels of Organization: The Architecture of the Scientific Image. In *Levels of Organization in the Biological Sciences*, ed. Daniel S. Brooks, James DiFrisco, and William C. Wimsatt, 1–20. Cambridge, MA: The MIT Press. https://doi.org/10.7551/mitpress/12389.003.0004.

Buiatti, M., and G. Longo. 2013. Randomness and Multilevel Interactions in Biology. *Theory in Biosciences* 132 (3): 139–158. https://doi.org/10.1007/s12064-013-0179-2.

Burkett, P. 2014. *Marx and Nature: A Red and Green Perspective* (G—Reference,Information and Interdisciplinary Subjects Series). Chicago, IL: Haymarket Books.

Campanella, Sara. 2018. "Synthesis" and Behavior: A New Role for Selection. In *Predictability and the Unpredictable: Life, Evolution and Behaviour*, ed. David Ceccarelli and Giulia Frezza, 193–204. Rome: CNR Edizioni.
Carew, Joseph. 2014. *Ontological Catastrophe: Žižek and the Paradoxical Metaphysics of German Idealism.* Ann Arbor, MI: Open Humanities Press.
Collins, Ardis B. 2000. Hegel's Unresolved Contradiction: Experience, Philosophy, and the Irrationality of Nature. *Dialogue (PST)* 39 (4): 771–796.
Cooper, Andrew. 2020. Do Functions Explain? Hegel and the Organizational View. *Hegel Bulletin* 41 (3): 1–18. https://doi.org/10.1017/hgl.2020.14.
Corti, Luca. 2023. The Logical Form of a Living Organism: Hegel, Naturalism, and Biological Autonomy. In *Nature and Naturalism in Classical German Philosophy*, ed. Luca Corti and Johannes-Georg Schülein, 227–245. New York and London: Routledge.
Danchin, Étienne. 2013. Avatars of Information: Towards an Inclusive Evolutionary Synthesis. *Trends in Ecology and Evolution* 28: 351–358. https://doi.org/10.1016/j.tree.2013.02.010.
Di Giovanni, George. 2010. Introduction. In *The Science of Logic*, xi–lxii. Cambridge, UK: Cambridge University Press.
DiFrisco, James, and Matteo Mossio. 2020. Diachronic Identity in Complex Life Cycles: An Organizational Perspective. In *Biological Identity: Perspectives from Metaphysics and the Philosophy of Biology*, ed. Anne Sophie Meincke and John Dupré, 177–199. London, UK: Routledge. https://doi.org/10.4324/9781351066389-10.
Di Paolo, E. A., Cuffari, E. C., & De Jaegher, H. 2018. *Linguistic Bodies: The Continuity between Life and Language.* Cambridge, MA: MIT Press. https://doi.org/10.7551/mitpress/11244.001.0001.
Emmeche, Claus, Simo Køppe, and Frederik Stjernfelt. 2000. Levels, Emergence, and Three Versions of Downward Causation. In *Downward Causation: Minds, Bodies and Matter*, 13–35. Aarhus: Aarhus Universitetsforlag.
Engels, Friedrich. 1959. *Anti-Dühring: Herr Eugen Dühring's Revolution in Science.* 2nd ed. Moscow: Foreign Languages Publishing House.
Erkan, Ekin. 2021. Life and Actuality: On Placing Possibility in Hegel's Modal Metaphysics. *Cosmos And History: The Journal Of Natural And Social Philosophy* 17 (3): 171–195.
Farnsworth, Keith D. 2022. How an Information Perspective Helps Overcome the Challenge of Biology to Physics. *Biosystems* 217: 104683. https://doi.org/10.1016/j.biosystems.2022.104683.
Ferrini, Cinzia. 2020. Hegel's Philosophy of Natural and Human Spaces. In *The Palgrave Hegel Handbook*, ed. Marina F. Bykova and Kenneth R. Westphal, 241–266. Cham: Springer. https://doi.org/10.1007/978-3-030-26597-7_13.
Ferrini, Cinzia. 2024. Hegel's Syllogism of Analogy and Organic Conception of Cosmic Life: A Speculative Kantian Legacy in His Absolute Mechanics? In *Hegel's Philosophy of Nature: A Critical Guide, Cambridge Critical Guides*, ed. Marina F. Bykova, 138–157. Cambridge: Cambridge University Press. https://doi.org/10.1017/9781009023160.010.
Froese, Tom. 2023. Irruption Theory: A Novel Conceptualization of the Enactive Account of Motivated Activity. *Entropy*. https://doi.org/10.3390/e25050748.
Fuchs, Christian. 2014. The Dialectic: Not Just the Absolute Recoil, but the World's Living Fire That Extinguishes and Kindles Itself. Reflections on Slavoj Žižek's Version of Dialectical Philosophy in "Absolute Recoil: Towards a New Foundation of Dialectical Materialism". *TripleC:: Communication, Capitalism & Critique* 12 (2): 848–875. https://doi.org/10.31269/triplec.v12i2.640.
Furlotte, Wes. 2016. Lack and the Spurious Infinite: Towards a New Reading of Hegel's Philosophy of Nature. In *Rethinking German Idealism*, ed. S. J. McGrath and Joseph Carew, 191–214. London, UK: Palgrave Macmillan. https://doi.org/10.1057/978-1-137-53514-6_9.
Furlotte, Wes. 2018. *The Problem of Nature in Hegel's Final System.* Edinburgh: Edinburgh University Press. https://doi.org/10.3366/edinburgh/9781474435536.001.0001.

Gabriel, Markus, and Slavoj Žižek. 2009. *Mythology, Madness, and Laughter: Subjectivity in German Idealism.* London and New York: Continuum Publishing.

Gambarotto, Andrea. 2020. Teleology, Life, and Cognition: Reconsidering Jonas' Legacy for a Theory of the Organism. In *Natural Born Monads: On the Metaphysics of Organisms and Human Individuals*, ed. Andrea Altobrando and Pierfrancesco Biasetti, 243–264. Berlin and Boston: De Gruyter. https://doi.org/10.1515/9783110604665-012.

Gambarotto, Andrea, and Luca Illetterati. 2020. Hegel's Philosophy of Biology? A Programmatic Overview. *Hegel Bulletin* 41 (3): 1–22. https://doi.org/10.1017/hgl.2020.21.

Gambarotto, Andrea, and Auguste Nahas. 2022. Teleology and the Organism: Kant's Controversial Legacy for Contemporary Biology. *Studies in History and Philosophy of Science* 93 (2): 47–56. https://doi.org/10.1016/j.shpsa.2022.02.005.

Gibb, Sophie. 2019. The Causal Closure Principle. In *The Routledge Handbook of Emergence*, 111–120. London and New York: Routledge.

Gilbert, Scott. 2019. Evolutionary Transitions Revisited: Holobiont Evo-Devo. *Journal of Experimental Zoology Part B: Molecular and Developmental Evolution* 332: 307–314. https://doi.org/10.1002/jez.b.22903.

Gilbert, Scott F., and Jonathan Bard. 2014. Formalizing Theories of Development: A Fugue on the Orderliness of Change. In *Towards a Theory of Development*, ed. Alessandro Minelli and Thomas Pradeu, 129–143. Oxford, UK: Oxford University Press. https://doi.org/10.1093/acprof:oso/9780199671427.003.0008.

Gilbert, Scott F., and Michael J. F. Barresi. 2016. *Developmental Biology.* 11th ed. Sunderland, MA: Sinauer Associates, Inc.

Godfrey-Smith, Peter. 2016. *Other Minds: The Octopus and the Evolution of Intelligent Life.* London, UK: William Collins.

Goldenfeld, Nigel, and Carl Woese. 2011. Life Is Physics: Evolution as a Collective Phenomenon Far From Equilibrium. *Annual Review of Condensed Matter Physics* 2 (1): 375–399. https://doi.org/10.1146/annurev-conmatphys-062910-140509.

Hegel, Georg Wilhelm Friedrich. 1982. In *Jenaer Systementwürfe II: Logik, Metaphysik, Naturphilosophie*, ed. Rolf-Peter Horstmann. Hamburg: Felix Meiner.

Hegel, Georg Wilhelm Friedrich. 1991. In *Encyclopaedia of the Philosophical Sciences in Outline and Critical Writings*, ed. Ernst Behler. New York, NY: Bloomsbury.

Hegel, Georg Wilhelm Friedrich. 2004. *Philosophy of Nature: Part Two of the Encyclopaedia of the Philosophical Sciences*, With Foreword by J. N. Findlay, translated by A. V. Miller. Oxford, UK: Oxford University Press.

Hegel, Georg Wilhelm Friedrich. 2010. In *The Science of Logic*, ed. George Di Giovanni. Cambridge, UK: Cambridge University Press.

Hegel, Georg Wilhelm Friedrich. 2015. In *Encyclopedia of the Philosophical Sciences in Basic Outline. Part I: Science of Logic* (Cambridge Hegel Translations) ed. Klaus Brinkmann and Daniel O. Dahlstrom. Cambridge, UK: Cambridge University Press.

Hegel, Georg Wilhelm Friedrich. 2018. *The Phenomenology of Spirit.* Cambridge, UK: Cambridge University Press.

Henrich, D. 2008. *Between Kant and Hegel: Lectures on German Idealism.* Cambridge, MA: Harvard University Press.

Hobson, Art. 2017. *Tales of the Quantum: Understanding Physics' Most Fundamental Theory.* Oxford, UK: Oxford University Press.

Hörz, Herbert. 2009. *Materialistische Dialektik: Aktuelles Denkinstrument Zur Zukunftsgestaltung.* Berlin: trafo.

Houlgate, Stephen. 2006. *The Opening of Hegel's Logic: From Being to Infinity.* West Lafayette, IN: Purdue University Press.

Illetterati, Luca. 2016. Nature, Subjectivity and Freedom: Moving from Hegel's Philosophy of Nature. In *'I That Is We, We That Is I.' Perspectives on Contemporary Hegel: Social Ontology, Recognition, Naturalism, and the Critique of Kantian Constructivism*, ed. Italo Testa, 183–201. Leiden: Brill. https://doi.org/10.1163/9789004322967_012.

Illetterati, L. 2020. Nature's externality: Hegel's non-naturalistic naturalism. *Problemi* 58 (11–12): 51–72.
Jaeger, Johannes. 2024. The Fourth Perspective: Evolution and Organismal Agency. In *Organization in Biology*, ed. Matteo Mossio, 159–186. Cham: Springer. https://doi.org/10.1007/978-3-031-38968-9_8.
Johnston, Adrian. 2008. *Žižek's Ontology: A Transcendental Materialist Theory of Subjectivity*. Evanston, IL: Northwestern University Press.
Johnston, Adrian. 2013. *Prolegomena to Any Future Materialism, Volume One: The Outcome of Contemporary French Philosophy* (Diaeresis). Evanston, IL: Northwestern University Press.
Johnston, Adrian. 2014. *Adventures in Transcendental Materialism: Dialogues with Contemporary Thinkers, Adventures in Transcendental Materialism*. Edinburgh: Edinburgh University Press.
Johnston, Adrian. 2015. Lacking Causes. Privative Causality from Locke and Kant to Lacan and Deacon. In *Speculations VI*, ed. Fabio Gironi, 19–62. New York, NY: Punctum Books.
Johnston, Adrian. 2018. *A New German Idealism: Hegel, ŽiŽek, and Dialectical Materialism*. New York, NY: Columbia University Press.
Johnston, Adrian. 2019. *Prolegomena to Any Future Materialism, Volume Two: A Weak Nature Alone* (Diaeresis). Evanston, IL: Northwestern University Press.
Juarrero, Alicia. 2013. Downward Causation: Polanyi and Prigogine. *Tradition and Discovery: The Polanyi Society Periodical* 40 (3): 4–15.
Juarrero, Alicia. 2015. What Does the Closure of Context-Sensitive Constraints Mean for Determinism, Autonomy, Self-Determination, and Agency? *Progress in Biophysics and Molecular Biology* 119 (3): 510–521. https://doi.org/10.1016/j.pbiomolbio.2015.08.007.
Juarrero, A. 1999. Dynamics in Action: Intentional Behavior as a Complex System. Cambridge, MA: MIT Press.
Juarrero, Alicia 2023. Context Changes Everything: How Constraints Create Coherence. Cambridge, MA: The MIT Press. https://doi.org/10.7551/mitpress/14630.001.0001.
Kant, Immanuel. 1998. *Critique of Pure Reason* (The Cambridge Edition of the Works of Immanuel Kant). Cambridge, UK: Cambridge University Press.
Kant, Immanuel. 2000. *Critique of the Power of Judgment*, translated by Paul Guyer and Eric Matthews. Cambridge, UK: Cambridge University Press.
Kauffman, Stuart. 2000. *Investigations*. Oxford and New York: Oxford University Press.
Kauffman, Stuart. 2013. Evolution beyond Newton, Darwin, and Entailing Law. In *Beyond Mechanism: Putting Life Back into Biology*, ed. Brian G. Henning and Adam C. Scarfe, 1–24. Plymouth, UK: Lexington Books.
Kauffman, Stuart. 2016. *Humanity in a Creative Universe*. New York, NY: Oxford University Press.
Kauffman, Stuart. 2020. Eros and Logos. *Angelaki* 25 (3): 9–23. https://doi.org/10.1080/0969725X.2020.1754011.
Kauffman, Stuart, and Philip Clayton. 2006. On Emergence, Agency, and Organization. *Biology and Philosophy* 21 (4): 501–521. https://doi.org/10.1007/s10539-005-9003-9.
Kauffman, Stuart, and Dean Radin. 2021. Is Brain-Mind Quantum? A Theory and Supporting Evidence. *PsyArXiv*. https://doi.org/10.31234/osf.io/qejzr.
Khurana, T. 2013. *The Freedom of Life: Hegelian Perspectives*. Berlin: August Verlag.
Kisner, Wendell. 2014. *Ecological Ethics and Living Subjectivity in Hegel's Logic*. London: Palgrave Macmillan UK. https://doi.org/10.1057/9781137412119.
Krips, H. 1990. *The Metaphysics of Quantum Theory, Clarendon Paperbacks*. Oxford, UK: Clarendon Press.
Lala, K. N., Uller, T., Feiner, N., Feldman, M. W., & Gilbert, S. F. 2024. *Evolution Evolving: The Developmental Origins of Adaptation and Biodiversity*. Princeton, NJ: Princeton University Press.
Longo, Giuseppe. 2018. How Future Depends on Past and Rare Events in Systems of Life. *Foundations of Science* 23: 443–474. https://doi.org/10.1007/s10699-017-9535-x.

Longo, Giuseppe. 2019. Interfaces of Incompleteness. In *Systemics of Incompleteness and Quasi-Systems*, ed. Gianfranco Minati, Mario R. Abram, and Eliano Pessa, 3–55. New York, NY: Springer. https://doi.org/10.1007/978-3-030-15277-2_1.

Longo, Giuseppe. 2020. Naturalizing Physics or, Embedding Physics in the Historicity and Materiality of the Living. *La Deleuziana—Online Journal Of Philosophy* 132–149.

Longo, G., and M. Montévil. 2011. From Physics to Biology by Extending Criticality and Symmetry Breakings. *Progress in Biophysics and Molecular Biology* 106 (2). https://doi.org/10.1016/j.pbiomolbio.2011.03.005.

Longo, Giuseppe, and Maël Montévil. 2013. Extended Criticality, Phase Spaces and Enablement in Biology. *Chaos, Solitons and Fractals* 55. https://doi.org/10.1016/j.chaos.2013.03.008.

Longo, Giuseppe, and Maël Montévil. 2014. *Perspectives on Organisms: Biological Time, Symmetries and Singularities*. Berlin and Heidelberg: Springer. https://doi.org/10.1007/978-3-642-35938-5.

Longo, Giuseppe, and Maël Montévil. 2017. From Logic to Biology via Physics: A Survey. *Logical Methods in Computer Science* 13 (4). https://doi.org/10.23638/LMCS-13(4:21)2017.

Longo, Giuseppe, and Matteo Mossio. 2020. Geocentrism vs Genocentrism: Theories Without Metaphors, Metaphors Without Theories. *Interdisciplinary Science Reviews* 45 (3). https://doi.org/10.1080/03080188.2020.1798588.

Longo, Giuseppe, and Ana M. Soto. 2016. Why Do We Need Theories? *Progress in Biophysics and Molecular Biology* 122 (1): 4–10. https://doi.org/10.1016/j.pbiomolbio.2016.06.005.

Longo, Giuseppe, Maël Montévil, and Arnaud Pocheville. 2012. From Bottom-up Approaches to Levels of Organization and: Extended Critical Transitions. *Frontiers in Physiology* 3. https://doi.org/10.3389/fphys.2012.00232.

Longo, G., Montévil, M., and S. Kauffman. 2012. No entailing laws, but enablement in the evolution of the biosphere. *ACM Proceedings of GECCO*: 1379–1392.

Longuenesse, Béatrice. 2007. *Hegel's Critique of Metaphysics* (Modern European Philosophy). Cambridge, UK: Cambridge University Press.

Malabou, C. 2016. *Before Tomorrow: Epigenesis and Rationality*. Cambridge, UK: Polity Press.

Marcuse, Herbert. 1986. *Reason and Revolution: Hegel and the Rise of Social Theory*. London and New York: Routledge.

Marques, V., and C. Brito. 2014. The Rise and Fall of the Machine Metaphor: Organizational Similarities and Differences Between Machines and Living Beings. *Verifiche* 43 (1–4): 77–111.

Marques, Victor. 2016. Positing the Presuppositions—Dialectical Biology and the Minimal Structure of Life. In *Slavoj Žižek and Dialectical Materialism*, ed. Agon Hamza and Frank Ruda, 113–132. New York, NY: Palgrave Macmillan US. http://link.springer.com/10.1057/9781137538611_9 [accessed 10 January 2018].

Miquel, Paul-Antoine. 2011. Extended Physics as a Theoretical Framework for Systems Biology? *Progress in Biophysics and Molecular Biology* 106 (2): 348–352. https://doi.org/10.1016/j.pbiomolbio.2011.03.006.

Monod, Jacques. 1972. *Chance and Necessity an Essay on the Natural Philosophy of Modern Biology*. New York, NY: Random House.

Montévil, M. 2020. Historicity at the Heart of Biology. *Theory in Biosciences*. https://doi.org/10.1007/s12064-020-00320-8.

Montévil, Maël. 2018. A Few Pending Challenges from the Perspective of a Theory of Organisms. *Constructivist Foundations* 13: 377–379.

Moreno, A., and Jon Umerez. 2000. Downward Causation at the Core of Living Organization. In *Downward Causation*, 99–117. Aarhus: Aarhus Universitetsforlag.

Ng, Karen. 2009. Hegel's Logic of Actuality. *The Review of Metaphysics* 63 (1): 139–172. http://www.jstor.org/stable/40387730.

Ng, Karen. 2013. Life, Self-Consciousness, Negativity: Understanding Hegel's Speculative Identity Thesis. In *The Freedom of Life: Hegelian Perspectives*, ed. Thomas Khurana, 33–67. Berlin: August Verlag.

Ng, Karen. 2020. *Hegel's Concept of Life: Self-Consciousness, Freedom, Logic*. New York, NY: Oxford University Press. https://doi.org/10.1093/oso/9780190947613.001.0001.
Noble, Denis. 2006. *The Music of Life: Biology Beyond Genes*. Oxford, UK: Oxford University Press.
Noble, D. 2011. Neo-Darwinism, the Modern Synthesis and selfish genes: Are they of use in physiology? *Journal of Physiology*. https://doi.org/10.1113/jphysiol.2010.201384.
Noble, Raymond, Kazuyo Tasaki, Penelope J. Noble, and Denis Noble. 2019. Biological Relativity Requires Circular Causality but Not Symmetry of Causation: So, Where, What and When Are the Boundaries? *Frontiers in Physiology* 10: 827. https://doi.org/10.3389/fphys.2019.00827.
Padui, Raoni. 2010. The Necessity of Contingency and the Powerlessness of Nature. *Idealistic Studies* 40 (3): 243–255. https://doi.org/10.5840/idstudies201040316.
Pattee, H. H. 2013. Epistemic, Evolutionary, and Physical Conditions for Biological Information. *Biosemiotics* 6: 9–31. https://doi.org/10.1007/s12304-012-9150-8.
Pinkard, Terry. 2012. *Hegel's Naturalism Mind, Nature, and the Final Ends of Life*. Oxford, UK: Oxford University Press.
Pippin, Robert B. 1989. *Hegel's Idealism: The Satisfactions of Self-Consciousness, Cambridge Paperback Library*. Cambridge, UK: Cambridge University Press.
Pradeu, Thomas. 2012. *The Limits of the Self: Immunology and Biological Identity*. New York, NY: Oxford University Press.
Quadrio, Philip. 2012. Hegel's Relational Organicism. *Critical Horizons: A Journal of Philosophy & Social Theory* 13 (3): 317–336. https://doi.org/10.1558/crit.v13i3.317.
Radnik, Borna. 2024. *Freedom, in Context: Time, History, and Necessity in Hegel*. London, UK: Bloomsbury Academic.
Renault, Emmanuel. 2013. The Naturalistic Side of Hegel's Pragmatism. *Critical Horizons* 13 (2): 244–274. https://doi.org/10.1558/crit.v13i2.244.
Rosen, R. 1991. *Life Itself: a Comprehensive Inquiry Into the Nature, Origin, and Fabrication of Life*. New York, NY: Columbia University Press.
Russon, J. 2014. Philosophy of Mind. In *G.W.F. Hegel: Key Concepts*, ed. M. Baur, 44–58. London and New York: Routledge.
Sandnes Haukedal, Rasmus, and Nikolaos Mylonas. 2024. Entropy and Negativity: The Ecological Implications of Dialectics. *Technophany, A Journal for Philosophy and Technology* 2 (1): 1–25. https://doi.org/10.54195/technophany.15370.
Sedgwick, S. 2012. *Hegel's Critique of Kant: From Dichotomy to Identity*. Oxford, UK: Oxford University Press.
Sève, Lucien. 2008. Dialectics of Emergence. In *Dialectics for the New Century*, ed. Bertell Ollman and Tony Smith, 85–97. New York, NY: Palgrave Macmillan. https://doi.org/10.1057/9780230583818.
Soto, Ana M., Carlos Sonnenschein, and P. A. Miquel. 2008. On Physicalism and Downward Causation in Developmental and Cancer Biology. *Acta Biotheoretica* 56 (4): 257–274. https://doi.org/10.1007/s10441-008-9052-y.
Stern, Robert. 2006. Hegel's Doppelsatz: A Neutral Reading. *Journal of the History of Philosophy* 44 (2): 235–266. https://doi.org/10.1353/hph.2006.0032.
Stern, Robert, and Leonard Weiss. 2024. The Feebleness of the Concept in Nature: A Challenge to Conceptual Realism? In *Hegel's Philosophy of Nature: A Critical Guide* (Cambridge Critical Guides) ed. Marina F. Bykova, 15–35. Cambridge: Cambridge University Press. https://doi.org/10.1017/9781009023160.003.
Stone, Alison. 2013. Hegel, Naturalism and the Philosophy of Nature. *Hegel Bulletin* 34 (1): 59–78. https://doi.org/10.1017/hgl.2013.2.
Sultan, S. E. 2015. *Organism and Environment: Ecological Development, Niche Construction, and Adaptation*. Oxford University Press.
Sultan, Sonia E., Armin P. Moczek, and Denis Walsh. 2022. Bridging the Explanatory Gaps: What Can We Learn from a Biological Agency Perspective? *BioEssays* 44 (1): 2100185. https://doi.org/10.1002/bies.202100185.

Tabaczek, M. 2019. *Emergence: Towards A New Metaphysics and Philosophy of Science.* Notre Dame, IN: University of Notre Dame Press.
Tahar, Mathilde. 2022. Biological Constraints as Norms in Evolution. *History and Philosophy of the Life Sciences* 44 (9). https://doi.org/10.1007/s40656-022-00483-1.
Ulanowicz, Robert E. 2009. *A Third Window: Natural Life Beyond Newton and Darwin.* West Conshohoken, PA: Templeton Foundation Press.
Umerez, Jon. 2016. Biological Organization from a Hierarchical Perspective: Articulation of Concepts and Interlevel Relation. In *Evolutionary Theory: A Hierarchical Perspective*, ed. Niles Eldredge, Telmo Pievani, Emanuele Serrelli, and Ilya Temkin, 63–85. Chicago, IL: University of Chicago Press. https://doi.org/10.7208/9780226426198-005.
Umerez, Jon. 2021. Control Hierarchies: Pattee's Approach to Function and Control as Time-Dependent Constraints. In *Levels of Organization in the Biological Sciences*, ed. Daniel S. Brooks, James DiFrisco, and William C. Wimsatt, 279–295. Cambridge, MA: The MIT Press. https://doi.org/10.7551/mitpress/12389.003.0018.
Wilson, Jessica. 2015. Metaphysical Emergence: Weak and Strong. In *Metaphysics in Contemporary Physics* (Poznan Studies in the Philosophy of the Sciences and the Humanities) CIV, 345–402. Leiden: Brill. https://doi.org/10.1163/9789004310827_015.
Wilson, Jessica. 2021. *Metaphysical Emergence.* Oxford, UK: Oxford University Press.
Wimsatt, William C. 1976. Reductionism, Levels of Organization, and the Mind-Body Problem. In *Consciousness and the Brain: A Scientific and Philosophical Inquiry*, ed. Gordon G. Globus, Grover Maxwell, and Irwin Savodnik, 205–267. Boston, MA: Springer US. https://doi.org/10.1007/978-1-4684-2196-5_9.
Wimsatt, William C. 1994. Levels, Robustness, Emergence, and Heterogeneous Dynamics: Finding Partial Organization in Causal Thickets. In *Canadian Journal of Philosophy Supplementary*, ed. Daniel S. Brooks, James DiFrisco, and William C. Wimsatt, vol. 20, 207–274. Cambridge, MA: The MIT Press. https://doi.org/10.1080/00455091.1994.10717400.
Yeomans, C. 2012. *Freedom and Reflection: Hegel and the Logic of Agency.* New York, NY: Oxford University Press.
Zambrana, R. 2015. *Hegel's Theory of Intelligibility.* Chicago, IL: University of Chicago Press.
Zambrana, Rocío. 2019. Actuality in Hegel and Marx. *Hegel Bulletin* 40 (1): 74–91. https://doi.org/10.1017/hgl.2018.11.
Zaslawski, Nicolas. 2018. Neurodialectics: A Proposal for Philosophy of Cognitive and Social Sciences. *Constructivist Foundations* 14 (1): 42–55. http://constructivist.info/14/1/042.
Žižek, S. 1993. *Tarrying with the Negative: Kant, Hegel, and the Critique of Ideology.* Durham, NC: Duke University Press.
Žižek, Slavoj. 1996. *The Indivisible Remainder: On Schelling and Related Matters.* London, UK: Verso.
Žižek, Slavoj. 2006. *The Parallax View.* Cambridge, MA: MIT Press.
Žižek, Slavoj. 2008. *In Defense of Lost Causes.* London, UK: Verso.
Žižek, Slavoj. 2012. *Less than Nothing: Hegel and the Shadow of Dialectical Materialism.* London, UK: Verso.
Žižek, Slavoj. 2014. *Absolute Recoil: Towards a New Foundation of Dialectical Materialism.* London, UK: Verso.
Žižek, Slavoj. 2016. *Disparities.* London, UK: Bloomsbury Publishing.
Žižek, S. 2020. *Hegel in A Wired Brain.* London, UK: Bloomsbury Academic.
Žižek, Slavoj. 2023. From Quantum Mechanics to Quantum Reality. *Filozofia* 78: 409–428. https://doi.org/10.31577/filozofia.2023.78.6.1.
Zupančič, A. 2017. *What Is Sex?* (Short Circuits). Cambridge, MA: MIT Press.

Chapter 6
Plasticity, Canalisation, and Habit

Abstract This chapter explores Catherine Malabou's concept of plasticity, situating it within contemporary theoretical biology and philosophy. Plasticity is examined as a dual process of giving and receiving form, which challenges traditional oppositions between canalisation and flexibility. The chapter argues that plasticity is not merely a passive response to environmental changes but an active, organisational principle that enables organisms to self-determine through habit formation. By integrating Malabou's ideas with scientific perspectives, the analysis highlights the dialectical interplay between stability and transformation, emphasising how organisms maintain viability through continuous, albeit constrained, change. The discussion also introduces the notion of *explosive plasticity*, where radical ruptures reconfigure the space of possibilities, underscoring the inherent unpredictability in biological systems. Furthermore, the chapter critiques reductionist views in neuroscience, advocating for an ecological understanding of plasticity as a distributed, multi-level phenomenon. Ultimately, plasticity is framed as a generative and destructive force, central to the dynamic, historical nature of living systems. The chapter concludes by linking these insights to broader debates in evolutionary theory and cognitive science, positioning plasticity as a key concept for rethinking agency and organisation in biology.

Introduction

> The biological perspective is brimming with a potential that is both post-metaphysical and post-critical, one that philosophers have been wrong to ignore. (Malabou 2016: 148)

In previous chapters, I have compared works in theoretical biology to developments in contemporary philosophy. I now relate these insights to Catherine Malabou's concept of plasticity, while elaborating on some of the philosophical motifs introduced already. It should come as no surprise that we find considerable overlap between these positions. The notion of emergent levels of organisation, for instance, is tangible in how Malabou depicts the transitions from the mental to the neuronal as a negation, where there is both continuity and tension. Mind is continuous with life but not in a seamless way (Ng 2013). There is a disruption of one from the other,

R. Sandnes Haukedal, *Agency and Organisation: The Dialectics of Nature and Life*, Library of Ethics and Applied Philosophy 44,
https://doi.org/10.1007/978-981-95-1924-8_6

but this does not mean that they are outside each other, as negation is not the same as cancellation.[1]

Dialectics could be defined as the evolutionary emergence of reciprocal levels of organisation (Foster 2020, 2022). This is tangible in the perspectives we have already discussed. To get a clearer understanding of the role of plasticity in this process, I focus mostly on Malabou's first two books, *The Future of Hegel* and *What Should We Do with Our Brain?*, even if I am informed by her other works as well. One reason for this restriction, besides lack of space, is that these works provide the most fertile potential for bringing her approach in dialogue with the organisational view. Second, I believe that her position has not changed significantly, at least not regarding the original definition of plasticity.

To frame the debate, recall one of the theoretical shortcomings of the MS: It failed to explain the generation of variation as anything else than random, which is not an explanation at all. By assuming that there is always enough variation in a population for natural selection to act upon, the generation of variation becomes trivial—not the subject of study except as a molecular event. This makes the generation of, and possible biases introduced by, variation incomprehensible, and completely discards the role of the organism in producing variation, viz. how it is canalised by the structural integrity of the organism. Against this, the question of plasticity takes us to the centrality of form, which is not a static thing but a process.

In Malabou's view, plasticity is the dual process of giving and taking form. I elucidate this through her conception of habit formation. The latter is the process whereby the organism gains leeway from its surroundings by familiarising and extending itself into them. This makes the environment more predictable, but there is a limit to such predictions. To indicate these limits, I examine explosive plasticity as a kind of 'negative possibility' that is permanently there. I make sense of this idea through the notion that organisms undergo constant changes to survive. This is similar to *extended criticality*, a principle which is also operative in the interaction between levels of organisation (Farnsworth 2022). By these means, I try to elucidate Malabou's notion of explosive plasticity through the organisational approach.

I also touch on similarities between Malabou and Žižek's versions of Hegel, especially their emphasis on groundlessness and contingency (Crockett 2010). Unlike Žižek, Malabou deals extensively with biology. This makes our task both easier and more challenging. Easier, as it frees us of the task of teasing out the biological implications of her work; more challenging, as Malabou deals with multiple disciplines, but often does not relate them to her philosophical project in a systematic manner—making it difficult to grasp the full scope of her intervention. She has also been accused of missing the heterogeneous nature of the scientific disciplines she employs, due to a limited intake of scientific work.[2] There might be some truth to this criticism, but I will not spill much ink on it.

---

[1] The relationship between mind and life is not my topic here, but much of what I say has implications for philosophy of mind.

[2] I am not saying that she is unaware of this heterogeneity; it may simply have been for brevity. While the claim, by Fitzgerald and Callard (2015), that Malabou has had a 'limited engagement with peer-revied scientific publications' (12) seems harsh, there is some merit to the claim that

Instead, I will build on but go somewhat beyond Malabou's theory, not to debunk but strengthen it. I do this by assessing whether her views align with, and/or could be informed by, the organisational approach. In the next chapter, I stage a dialogue between Hegel and niche construction theory, which is undergoing a revision which seems to bring out its (original) dialectical character more fully. Here, I elaborate on some of the central facets of Hegel's understanding of how organisms relate to and change their environment. In the current chapter, however, the main task is to make sense of plasticity.

I structure the chapter as follows: I begin by examining notions of plasticity and canalisation in scientific literature, apart from Malabou's understanding. By providing a general discussion of canalisation and plasticity understood as *norms of reaction*, I try to show what is potentially missing in these conceptions. In short, I argue for the dialectical interplay between canalisation and plasticity, and later show that such a view can be found in Malabou's theory. I also touch upon the limitations of viewing plasticity as a result of other processes instead of a driver of evolution. I claim that plasticity does not depend on natural selection to have evolutionary significance. After these more general discussions, I move to Malabou's concept of plasticity, a process through which the organism familiarises itself with its environment and self-determines through this mediation. I tie this back to the notion that possibilities are not pre-given but produced through the activities of the organism. Similarly, habit formation is a change of disposition towards change that alter the possible interactions between organism and environment. It thus indicates 'a virtuality already inscribed within essence itself' (Malabou 2005: 74).

Next, I move to a central concept for Malabou, the notion of 'explosive plasticity', which she finds lacking in the scientific discourse on plasticity. It concerns radical raptures in the normal functioning of plasticity which institutes a new identity for the subject. I try to make sense of this radical negation through what we have already discussed concerning unpredictability and qualitative changes in evolution. Based on this, I argue that plasticity should not be understood as either a tendency toward destruction or creation but as an organisational principle, as the constant possibility that a given organisation will disintegrate and have to reorganise. This allows us to see the link between explosive plasticity and the virtuality explored in relation to habit formation.

I then expand on the notion that plasticity is an ecological concept and how it decentres any notion of a central controller, through Lambros Malafouris' (2013) notion: *metaplasticity*, which is a higher-order emergent plasticity produced through the interplay of plastic instances. It brings plasticity and organisation together and helps us understand the ruptures in identity that Malabou theorises through explosive plasticity. In the penultimate section, I relate plasticity to the notion of extended criticality, before concluding. My aim is to make the notion of explosive plasticity

---

she is mostly interested in unearthing scientific findings that 'will confirm, verify and/or deny the theoretical insights of cultural and social theory' (13)—instead of exploring the constitutive relation between the natural and social sciences. This limited engagement is also why I do not deal much with her book on epigenesis (Malabou 2016), without saying this work is irrelevant to my project.

more intelligible by looking at how organisms undergo constant transformations to keep their identity, and that explosive plasticity can be understood as ruptures that reconstruct the space of possibilities.

Canalisation and Plasticity

The literature usually presents canalisation—which is when a phenotype remains invariant under changing conditions—as the opposite of plasticity. It involves phenotypic traits that are insensitive, viz. not easily perturbed by environmental changes, while plasticity is considered the capacity to generate novel responses to such changes. Moreover, canalisation—a fundamental way for organisms to survive while undergoing change—is often used synonymously with robustness.[3] In traditional models, genetic canalisation is either a function of selection against the 'deleterious effects of mutation and recombination' (Gonzalez and Barbeito-Andrés 2021: 1068) or a by-product of environmental canalisation.

In these contexts, both plasticity and robustness are characterised by the decoupling of genetic and phenotypic variation. Canalisation stabilises the phenotype from environmental and genetic perturbations (Nijhout et al. 2017). Plasticity, on the other hand, decouples by producing multiple phenotypes from one genotype. But there are multiple reasons to question this simple opposition. For one, plasticity is a more general phenomenon than canalisation, making it misleading to consider them simple inversions. Secondly, the mechanistic basis of canalisation is lacking, as a quantitative model is not the same as a mechanistic explanation of 'actual physical interactions' (Hallgrimsson et al. 2019). Stated differently, generic populational approaches to genetics do not linearly translate to developmental models of specific interaction.

To address this issue, some have proposed a systems approach that does not simply identify the particular mechanisms involved in canalisation:

> The alternative to specific evolved mechanisms that modulate canalization is that the modulation of phenotypic variance occurs via emergent or embedded features of development rather than specific, dedicated mechanisms. In this case, canalization of a phenotypic trait is produced by the same genes and processes that underlie variation of that trait. (Hallgrimsson et al. 2019: 73)

Instead of dedicated mechanisms, canalisation is directed by the same sources that produce the variation of the trait, and it is enabled by network properties. The task of a systems approach is therefore not to identify specific mechanisms for each of these processes but to understand how plasticity and canalisation interact to maintain the organism within an interval of viability:

[3] Waddington (1953) represented it in his epigenetic landscapes as deep valleys, to illustrate phenotypes buffered against perturbation. This view considers development a stochastic process where multiple levels of organisation and complex interaction produce tremendous variation, most of which are deleterious, and some of which are made useful by the organism (Hallgrimsson et al. 2019).

> In a dynamically stable phenotype there are always factors that vary in order to maintain stability. So, the overall system may show stability or robustness in the face of perturbation, even while the underlying components are showing plasticity. The seeming paradox of both stability and plasticity in the regulatory system therefore stems in part from differences in the "level of analysis" selected by the researcher. (Nijhout et al. 2017: 181–182)

This reveals a need for 'multi-scale integration of imaging and molecular data, as well as the development of theoretical models to make sense of the large and complex datasets that such approaches generate' (Hallgrimsson et al. 2019: 76). In some cases, plasticity may enable canalisation and vice versa. The systems view thus underscores why I have used 'canalisation' differently in this study, as I have spoken only of the emergent canalisation involved in biological organisation, which differs from specific environmental or genetic canalisation—since it concerns the general ability of a living organism to respond to and constrain material processes. In this view,

> organization controls the dynamics of the organism, and prevents deleterious variations that would threaten its very existence. Accordingly, there is an important sense in which organization, by grounding stability, *counters, canalizes and uses* variation. (Mossio et al. 2016: 32)

Organisation grounds stability while propagating and preserving variation. From this perspective, it is misleading to speak of robustness as a property of genes; instead, we should understand it as a system's property. Moreover, there is an interaction between levels and forms of canalisation and plasticity—not simple opposition, but productive contradiction. Since canalisation at one level may contribute to plasticity at another, they are not mutually exclusive (McDonald et al. 2018).

For example, a selected adaptive trait concerns the passive aspect of plasticity, the capacity to be changed from without. This view of plasticity is not specific to biology but also applies to inorganic objects (Boonstra and Slagter 2019). It occurs when a trait is strongly canalised by genetic factors, as natural selection has fixed the trait. This view is implicit in Waddington's understanding, as the canalisation of one trait may lead to the accumulation of cryptic variation that is not subject to selection. It forms a wellspring of variation that may become useful if environments change. The cryptic variation may enhance the capacity to evolve (evolvability), and thus illustrates how less variation on one level could lead to more variation at another. Thus, different levels and their interaction must be kept in mind simultaneously. We cannot assume levels to interact linearly but should instead think in terms of unpredictable bio-resonance.

I argue that plasticity—like canalisation—is not about a specific mechanism, like a synapse changed by behaviour, but a systemic property that emerges from the dialectics of different plastic processes. The plasticity of one part of the system depends on its constitutive interaction with the whole organisation. As emphasised by Sultan: 'In a unified developmental approach, plasticity and canalization do not demand distinct causal mechanisms' (2015: 29). Instead, we should seek to understand how canalisation and plasticity, as 'patterns of variation', are intertwined. In this view, adaptation is not the function of selection in the past but an ongoing emergence through the tight interaction between the organism and the environment. Privileging the past over the present and future led biology to forget the historicity of biological systems (Montévil 2022). Plasticity sheds light on the pitfalls of such reasoning.

The Process of Plasticity

Before moving to Malabou's theory, I introduce how plasticity is discussed within biology. The prevalent version, developmental or phenotypic plasticity, concerns how organisms make use of the leeway left in their genetic repertoire to accommodate changes in the environment. In her seminal study, Mary Jane West-Eberhard (2003) defines plasticity as 'the ability of an organism to react to an internal or external environmental input with a change in form, state, movement or rate of activity'. In short, plasticity pertains to phenotypic changes that the organism may initiate in response to changes in the environment. West-Eberhard emphasises general responsiveness and flexibility: The process begins with environmental change to phenotypic accommodation and then leads to genetic accommodation. A changed phenotype is selected prior to genetic changes in allele frequencies that may stabilise the trait. This model underscores how the environment *in-forms* the organism as much as genetic factors, generating new phenotypes for selection (Sultan 2015). Here, environmental induction drives evolution, while genes are followers, as West-Eberhart says.

We should, however, ask whether this model truly challenges the MS explanation of evolutionary novelty, or just places more emphasis on the in-formative role of the environment. In the MS perspective, plastic changes are either not stable enough to be evolutionary, or simply a consequence of selection in the past. As discussed, the fractioned state of biology has created the impression that the causes of evolution are relatively independent. By way of scientific idealisations, the complex nexus of causes in evolution is overlooked and even misrepresented (Uller et al. 2020). Likewise, some conceptions of plasticity, like Schmalhausen's *norms of reaction*, run the risk of diminishing its evolutionary role by making it a consequence of selection in the past rather than a driver of change. Accordingly,

> if environments are construed by parents to systematically improve performance, or offspring respond appropriately to those environments, this is ultimately to be explained in terms of selection of genetic variation. Consequently, those who grant that 'genes may be followers rather than leaders in adaptive evolution' (West-Eberhard 2003) may nevertheless ascribe the ability of plasticity to contribute constructively to evolution in terms of past selection of genes […], a view that appears consistent with plasticity being a relatively minor 'add-on' […] to evolutionary theory. (Uller and Helanterä 2019: 364)

As they explain, the quasi-independence of causes effectively reduces the role of development, as developmental causes must 'survive selection and inheritance' to count as evolutionary. In other words, developmental plasticity can facilitate a novel phenotype, but this phenotype must be canalised by natural selection to be stably inherited. This means that plasticity is considered adaptive by default—since it must have been selected to exist—and that it only registers as evolutionary when changes in gene frequencies of a population have occurred because of a plastic response, not before. In short, while there are multiple norms which may be actualised under certain conditions, the norms are ultimately determined by genetic factors.

But this idea of plasticity as a simple add-on or fine tuner only works if organisms only exploit norms of reaction that are the product of selection in the past. Plasticity as a genotypic norm of reaction is a passive form of plasticity that occurs automatically if a certain trigger is in place. I subtle privileging of genetic factors, as if this is where causality and control ultimately resides, is maintained. Passive plasticity is simply 'variation minimised by environmental canalization' (Hallgrimsson et al. 2019: 69), without an active response from the organism. Against this, we should consider plasticity as an active evolutionary *process*, not just a result. This breaks with adaptationist explanations not only by seeing plasticity as an evolutionary cause but also by viewing it as spanning the whole life cycle of the organism, instead of a limited period of development.

We should therefore maintain a difference here between what is an automatic or physiological behavioural response and a behavioural response which demands an active choice on the part of the organism. The former has 'nothing to do with an organismal behavioral choice because it lacks all three key defining items: there is only one outcome, and it is not a choice, nor does it involve a drive of the organism as a whole' (Diogo 2017: 4). The latter, on the other hand, is an emergent property which involves the whole system. For this reason, it does not require consciousness or a central nervous system, according to Diogo. If we follow this distinction, we have to delimit behavioural choices to cases that requires an active choice and involves the whole living system.

In this view, as I argue below, plasticity enables the ongoing reproduction of organisation. This implies that passive (automatic) plasticity is affected by the activities and entanglements of the organism, and therefore part of and enabled by a larger process of self-determination. Even if the reaction is automatic, it is not unperturbed by the actions of the organism since it *became automatic* through a process which is not.

If development and inheritance were separate processes, the inherited material would not be changed by plastic modification. With changing environmental conditions, it is simply *the expression* of this material which changes, not the potential itself. Even if genes are followers in evolution, they are still considered quasi-independent from other processes. This notion limits what counts as evolutionary (between generations) and what counts as developmental (within generations). While plasticity, in response to environmental change, may drive evolutionary change, it is only genes that are stably inherited, and therefore any reaction norm is premised on the past inheritance of genes. Clearly, this does not challenge the core of the MS, as plasticity serves as a 'phenotypic scaffold', enabling new traits to be stabilised by genes, but having no evolutionary impact if it is not genetically canalised (Godfrey-Smith 2016). It simply favours genetic paths that would not otherwise be selected.

This diluted plasticity, depending on natural selection to 'make it' evolutionary, seems to have fed into some models of niche construction. The ultimate explanation, selection of genetic variants, is maintained, and the capacities of the current generation to modify its niche are premised on the prior selection of traits that enable this construction. Future generations may also inherit the changes only if they are selected. This diminishes the active role of the organism. We will see this more clearly in the next chapter; for now, I accentuate a simple point: Even if an alternative phenotype is elicited directly from an environmental change *without any activity from the organism*, the latter may still decide how the changed trait affects its fitness. It may for instance change where and how it lives, by behavioural change and niche construction. Speaking about a plastically induced phenotype change, Uller and Helanterä write:

> These characters, which affect the developmental and selective niches of the coat colour phenotype, may be adaptive but they were not originally selected to enable a match between coat colour and environment, which is the explanandum. Furthermore, in contrast to coat colour, these behaviours may not map straightforwardly onto any particular genetic variant. (Uller and Helanterä 2019: 365)

Selectionist explanations do not capture such changes, which may be adaptive irrespective of selection. Even if a mouse cannot actively decide its coat colour, it can change its niche to accommodate such changes. Crucially, while these changes may be adaptive, they were not *selected in for* this active matching. Also, these behavioural changes cannot be tied to a specific genetic variant (Uller and Helanterä 2019).[4]

Thus, neither development nor inheritance is premised on genotypic reaction norms alone, because these norms are not generated by genetic processes in isolation (see Sultan 2015). The notion that reaction norms are produced by genetic factors, casts the environment as a proximate factor. Even if phenotypes cannot be formed without genotypes, this is merely an extended and unilinear genotype model of norms of reaction.[5] As such, it illustrates the problem with some versions of systems biology: While they acknowledge the complexity of processes of genotypic expression, they nevertheless support the view that genes and selection drive evolution (Noble 2006). How genetic material is expressed may be altered, but not *what* it is possible to

[4] In his comment on this argument, Thomas Dickins claims that: 'If there is any sense in which genes provide a necessary and prior condition for developmental processes to unfold, then recurrence is readily fitted back into standard theoretical interpretations' (Dickins 2021: 198). For lack of space, I have not been able to discuss his views properly in this book. I should note, however, that he represents the rather hard-line view that the MS can accommodate all the changes proposed by the EES (Shan 2024). Commenting on Dickins and Dickins (2023) and the view that 'that there must be strong genetic control of extragenetic inheritance for evolution to produce adaptation' (Lala et al. 2024: en243, 310), the authors state that 'only a small fraction of the data are consistent with this' and furthermore that a 'degree of independence from genetic variation is necessary for extragenetic inheritance to play a role in short-term adaptation' (Lala et al. 2024: en243, 310, citing Anastasiadi et al. 2021, among others). As such, it seems to constitute an open empirical question. I do not pretend to do justice to Dickins position, which centres on the notion of information. I simply outline what a reply from the proponents of the EES would look like. I also believe that some of the claims he raises against dialectical causation are addressed laterally by the perspectives I outline in this book, especially since his treatment of dialectics is superficial, at best.

[5] Cf. Lala et al. (2024): 'The characterization of development as under the control of a genetic program, and of phenotypes as genetically specified reaction norms, can also be viewed as subtle forms of genetic determinism. In reality, organisms help to construct their own developmental trajectory through constant interactions with their internal and external environment, while reaction norms are products of developmental systems' (232).

express, which is given at birth and decided by the preceding selection of genes. Factors outside genes are of minor importance, even if they must be considered to understand the expression of genes.

A more radical version of plasticity than the one depicted above is found in the notion that plastic traits may spread in a population *without* any genetic change. As such, the plastic changes can be adaptive without being selected for; also, the selective stabilisation of the plastic response may be *enabled by the plastic response itself*, and not by a random mutation, as natural selection may act directly on the response irrespective of genetic change (Godfrey-Smith 2016). The focus on the longevity and genetic stabilisation of plastically induced traits obfuscates how short-lived environmental and epigenetic factors can have significant effects by altering the phenotypes that are available for selection. As Sultan (2017) writes:

> Because genotypes respond differently to these influences, developmental response systems are themselves subject to selection, but as 'entangled' evolutionary entities; the impact of selection on genotypes is attenuated by highly complex environmental interactions. (6)

Yet again, we see that a genotype does not contain the information to specify a phenotype in isolation. Sultan underscores that developmental information is partial and must be completed by the complex regulatory interaction between organism and environment. Developmental causes become evolutionary not just by genetic transmission but simply by *recurring*; in the process, the potential information contained in the genome is not merely released but also constructed.

We should also note that environmental challenges would elicit plastic responses in multiple individuals at once, allowing for more rapid changes in the population. This not only make new traits less likely to be lost, but would also make them more ubiquitous. Plastic responses are also more likely to produce adaptive phenotypes than mutations since they are 'tailored to the inducing environment' (Lala et al. 2024). In what follows, I aim to shed light on how Malabou's concept of plasticity, combined with the insights we have elucidated above, can inform these debates.

Malabou's Plasticity

> The meaning of the notion of plasticity is the same as its way of being. Plasticity is what it is, plastic. Indeed, the originary operation of receiving and giving form is not a rigid and fixed structure but an instance which can evolve, which means that it can give itself new forms. The temporal differentiation of plasticity makes possible the historical deployment of the substance-subject. (Malabou 2005: 186)

The above quotation outlines the basic historicity and reflexivity of plasticity; namely, how it is itself subject to change. The process of formation does not operate in a given manner throughout history but undergoes changes. This principle is important for what follows.

In a more recent work, Malabou claims that '[n]o robust theoretical discourse has yet constructed itself on the basis of contemporary epigenetics so as to conceptualize its contributions and successfully integrate its interpretative metaphors – be they readings or musical performances' (Malabou 2016: 152). I have tried to show that this claim is incorrect—that the discourse she is calling for is being developed and integrates the metaphors she mentions.[6] In this chapter, I display how this discourse aligns with Malabou's concept of plasticity.

As we will see, Malabou is not intervening directly in the debates depicted above but proposing her own interpretation. Her notion of plasticity is therefore somewhat idiosyncratic, but it is not unrelated to—or irrelevant for—a more scientific understanding. Her version of plasticity is paradoxical, both as an empirical phenomenon and as a concept, since it pertains both to the formation and construction of form, or creation *through* destruction. In its positive dimension, it is the dual capacity of giving and receiving form. But since Malabou emphasises how the organism, by resisting determination from without, actively shapes itself, plasticity is not only positive. It is, for instance, not just the strengthening and regeneration of synapses based on input from within or without, but also negative and destructive, as seen in the negativity of form inherent in every determinate content in the form of lesions or other ruptures. This capacity to resist is also the basis for creativity and is described as destructive or explosive plasticity, explored more later.

Plasticity is a balancing act between constancy and creation, where the organism is split in a way we have already seen: It must remain organisationally closed while exposing itself to the outside to survive. Malabou understands plasticity as the generation, reception, and destruction of forms (Malabou 2008). The dual character of compliance and resistance to form means that it moulds itself by opposing moulding from without. This notion challenges the adaptationist view of organisms by highlighting plasticity as the capacity to resist determination, to sublate the conditions that affect the organism. As such, it is also a kind of canalisation since it means that the organism is not completely polymorphous but constrained by its history.

Plasticity is placed in between 'the irreversible character of formation (determination) and of a remobilization of form (the capacity to form oneself otherwise, to displace, even to nullify determination: freedom)' (Malabou 2008: 17). Above, we depicted this as the capacity to sublate physical and chemical determinations and thereby self-determine. The subject is this infinite activity, a process of negativity 'whose identity is in this way made true, concrete, and mediated, and which actualizes itself in its internal self-differentiation' (Bourgeouis 1988: en3, 201, quoted by Malabou 2008: 11).

There is a close link between plasticity and self-determination. It has a passive and an active dimension: The first is the capacity to receive form, which is when the environment directly induces a new trait with any activity on the part of the organism. The latter is found in the capacity to give form through resistance to passive formation. This is a process of actualising possibilities that are produced through the process itself. The organism transforms changes coming from the outside

---

[6] Whether it is on 'the basis of contemporary epigenetics' is another question, but I have presented a broad reading of epigenetics that converges on the organicist's perspective.

into internal changes, producing its own essence retroactively. We see this in how Malabou challenges the notion of reading implied when we imagine development as simply reading out a text that is already produced:

> Far from basing itself purely and simply on the content of what is read, this new reading must in return express the content; and to do so, to form new propositions, all it can do is to *transform* the content of its reading, that is to *interpret* it. (Malabou 2005: 180)

This notion of reading indicates how individuality is formed through a negation of determination, 'by virtue of a resistance to form itself' (Malabou 2008: 71)—which could be another way of describing the way the organism posits its own externality and relates to itself through the other. As such, it converges on Hegel's notion of active and passive substance from *the Logic*: The process through which active substance posits its passive ground is also the process whereby this ground gains actuality, how it becomes able to act in the world (Malabou 2005). Similarly, active plasticity posits passive activity as its precondition, as we will see in relation to habit formation.

Reading as a mere unrolling of a given potential also obfuscates the point I made above, how inheritance is about *recurrence*, not only transmission of traits. Hence, it fails to consider how genetic information is only potential information that becomes actual through its interaction with environmental information (Danchin 2013). In this view, there is no static content, like an invariant genetic script, waiting to be read out. While the mechanism of genetic transmission is stable, its content is not. Instead, the interpretation is also a transformation of the content. This coincides with the view that the genetic information contained in DNA is incomplete. Again, we see how negativity is the form (regulatory process) which acts on the material (DNA) to produce a specific content (information), a determinate negation.

Plasticity suggests a general capacity to adapt, adaptability, which is about the ongoing plastic response of the organism, a response that is not codified or discrete but creative and continuous. We could perhaps consider this a kind of counter-canalisation. I will also argue that plasticity is an emergent organisational feature, in which the plasticity of the processes and constraints interact to produce a plastic totality. By contrast, adaptation *to a given form* (a static environment) is the eradication of individuality since it follows a logic of imitation (Malabou 2008).

To keep its identity the subject must expose itself to the risk of exploding, a risk that is ever-present. Malabou writes: 'All current identity maintains itself only at the cost of a struggle against its autodestruction: it is in this sense that identity is dialectical in nature' (Malabou 2008: 71). It is because of this contradiction inherent in every form—between constancy and creation—that transformation is possible. Identity is shaped through the dialectics of formation and explosion. It is what gives the organism the capacity not only to receive but to give form. Plasticity, then, is the premise for and process of the transmutation of causes that enable self-determination or canalisation—by imposing constraints on how the organism may change.

Prose and Poiesis

> Something we possess as a kind of ἕξις [habit] (for example, knowledge) is not strictly speaking present, if by present we mean something which is standing there and here, in front of us, ready at hand. Habit is in fact *virtuality*. Now the *virtual* is just that—what is never exactly 'here'. (Malabou 2005: 55)

In *The Future of Hegel* (2005), Malabou states that the notion of plasticity is both a structure and a means of making Hegel intelligible. It functions, as Hegel says apropos the concept, not 'like an empty receptacle, rather as a power that can fashion its own content' (Malabou 2005: 5). Plasticity is a power that implies both the construction and annihilation of form—suppleness and solidity. It is both 'those concrete shapes in which form is crystallized (sculpture) and to the annihilation of all form (the bomb)' (Malabou 2005: 9). Moreover, Hegel understands substance as plastic, in that it gives form while being shaped in return. This movement of self-determination is at the core of dialectics:

> The dialectical process is 'plastic' because, as it unfolds, it makes links between the opposing moments of total immobility (the 'fixed') and vacuity ('dissolution'), and then links both in the vitality of the whole, a whole which, reconciling these two extremes, is itself the union of *resistance* (*Widerstand*) and fluidity (*Flüssigkeit*). The process of plasticity is dialectical because the operations which constitute it, the seizure of form and the annihilation of all form, emergence and explosion, are contradictory. (Malabou 2005: 12)

On one hand, dialectics is plastic since fixity and dissolution are coupled within a larger whole, which is itself a unity of difference; on the other hand, plasticity is dialectical because it is constituted through a process where contradictory moments, the seizure and explosion of form, coincide. As such, both plasticity and dialectics are tied to temporality, denoting 'the *formation* of the future itself' (Malabou 2005: 12). The virtual dimension, evident in habit formation, is present in the process through which the organism interiorises its exterior to make it a principle of its own self-movement. It is like the bending we spoke of before—an interruption in the flow of cause and effect, whereby the subject sublates externally imposed changes into its own becoming. It is a change in the dispositions of the organism, a change in how it deals with change. Felix Ravaisson, whose theory of habit informs Malabou's take, writes:

> Habit does not simply introduce mutability into something that would otherwise continue without changing; it suggests change within a disposition, within its potentiality, within the internal character of that in which the change occurs, which does not change. (Ravaisson 1984: 10, quoted by Malabou 2005: 58)

Habit is the capacity to re-determine the conditions that enable organisational reproduction. It introduces a change in potentiality, in the possible relationship to these conditions. Ravaisson reiterates the notion that possibilities are produced through the history of the organism. Habit, as teleological repetition, is a process of shaping both the organism and its surroundings. It is a 'turning point, enabling the individual self to forego immediate reactions and responses to the surrounding world' (Ferrini

2020: 259). Organisms interrupt external causes, such that the effect is canalised to its own ends—a notion that we also found in Robert Rosen's (1991) concept that organisms are 'closed to efficient causation'. This is a process of converting existent conditions into action (actuality).

The dual nature of habit is rendered through Plato's *pharmakon*, which is both cure and poison (Žižek 2012; Malabou 2008). This duality is found in Hegel's theorisation: On the one hand, habit is about deadening and ritualisation; on the other, it is the precondition for any exercise of freedom, the historical basis upon which creative activities may appear. Hegel states that habits both involve the interest and consciousness of the subject matter at hand, while also being withdrawn from it. Subjects are so immersed in bodily activities that the activities become mechanical. But this automation paradoxically opens the possibilities for something new to occur. Take learning an instrument or language: The repetitive learning of rules which are made unconscious is the precondition for virtuoso improvisation. In this way, the organism makes something external into its own principle, canalising it to its own ends:

> If all external change is repeated, it turns into a tendency internal to the subject. The change itself is transformed into a disposition, and receptivity, formerly passive, becomes activity. Thus habit is revealed as a process through which man ends by willing or choosing what came to him from outside. Henceforth the will of the individual does not need to oppose the pressure of the external world; the will learns gradually to want what is. (Malabou 2005: 70–71)

In other words, the process of habituation fosters the ability to in-habit the world differently. It is a case of novelty arising through canalisation, of attunement. I should note that the duality of the pharmakon applies to constraints as well since they simultaneously limit and constitute possibilities (Montévil 2022). We could say that biological constraints are produced and maintained through plastic changes. Together, these processes introduce biases in the trajectory of an organism and thus enable some things to happen at the expense of others.

Malabou underscores how, in Hegel's view, nature is always second nature, and thus denaturalised since it is not an ahistorical essence outside us but a product of the process of habituation. When we move from nature to spirit, nature is not overcome but reduplicated (Malabou 2005). As such, it remains a problem for the organism. Through the idealisation of its surroundings, the organism completes the first step towards the development of a self. The process through which second nature is shaped, and individuality emerges, is a repetition of the first immediacy that the organism experiences, making it a posited immediacy (Malabou 2005). It is a process of simplification, which produces singularity in the way each organism relates to its milieu (as we shall see more clearly in the next chapter). Malabou relates this to sublation: to preserve and maintain its stability, habits also act to suppress and simplify. In the very act of maintenance, the immediacy of what is preserved is lost since it is preserved only by being taken out of its original context.

There is no pure repetition in historical systems; each iteration introduces irreversible differences in the living organisation. Livings systems display openness to novelty, to the immanent emergence of new forms. For example, every experience leaves neural traces, but these traces do not faithfully store experiences in the brain. Instead, they are increasingly divorced from the original experiences, plastically changed in each repetition. Hence, experiences change through the way they are integrated into the ongoing organisation. We might understand, then, why Malabou claims that the teleological or repetitive dimension of habit enables organisms to create new habits:

> Hegel's philosophy announces that the future, from now on, depends on the way the shapes and figures already present can be put back into play, on the way the extraordinary and unexpected can only arise out of the prose of the well-known and familiar. (Malabou 2005: 190)

Plasticity, like retroactivity, is the capacity to repeat the past, to make it plastic and thus reveal virtual possibilities that *could have been* within another totality of conditions. It is not that we cannot predict from within a range of events which event will occur; we cannot even know the range of events that are possible. As Ravaisson said above, habit formation is 'change within a disposition, within its potentiality, within the internal character of that in which the change occurs, which does not change' (1984: 10). Habit is not so much about changes in actual properties but instead about relational or dispositional changes which alter the response to changes.

This indicates how Hegel overcomes the dualism between nature as fully determined and man as free from nature's contingencies. Žižek writes that a 'natural organism has to regulate its exchange with its environment, the assimilation of the environment into itself, through habitual procedures which "reflect" into the organism, as its inner dispositions, its external interactions' (2012: 345). By internalising externally imposed changes through habit, then, living organisms differentiate themselves. Such self-differentiation is not displayed by inorganic beings but is *non-optional* for living beings—as the only means through which they survive. They remain unities through a process of differentiation and integration. Malabou writes:

> The organic being is characterized by its effort in maintaining its own unity through the synthesis of differences: the difference between the organism and its environment and the difference between the heterogeneous elements which make up the organism. Such an effort is nothing less than habit itself: from the beginning to the end of the organism's life, habit is busy applying its power from the inside, fulfilling the individual development of the organism's faculties. (Malabou 2005: 58)

Habit formation is a principle of self-movement which changes how the organism experiences its environment. It is thus a movement enabled by constraints which retroacts on these constraints and imposes its own, either by reinforcing or upending what it picks up. As a principle of self-organisation, habit challenges the assumed causal primacy of natural selection. From this perspective, selection does not explain the genesis of plastic forms by itself. Instead, plasticity partakes in the production of variation that the organism depends on to stay alive, regardless of whether natural selection has acted on it or not.

Explosive Plasticity

The virtuality inscribed in essence itself, it 'brings out one of the fundamental aspects of the Hegelian theory of substance: the recognition of the essential status of the a posteriori' (Malabou 2005: 74; cf. Ng 2009). As we have seen, the default state is 'progressively erased' but never overcome (Malabou 2008). For Malabou, like Žižek, this virtual essence is operative in nature itself. Virtuality is the dimension of what 'might not be', according to Malabou, a possibility which may never be actualised but still exerts influence. It is different from outright confirming or denying something because there is an ambivalence. For instance, a habit is never fully present, but perpetually in becoming. It has a virtual status as a counterfactual towards which the organism strives—and which thus shapes its dispositions—without being positively given. Even if the goal is never reached, it still has material effects. We shall see this more clearly in the next chapter.

Plasticity has the ever-present possibility of 'autodestruction', says Malabou. It points to the instability of every plastic form, the negativity of plasticity that is a resistance to form, or to negativity itself *as* form. This inherent explosivity is an ahistorical and formal possibility. Sparrow says that this does not imply that plasticity always tends toward destruction even without any external perturbations. He writes:

> This is not to say that plasticity tends naturally toward fixity or suspension, rather than creativity, but rather to assert that the material conditions of identity require that a number of disparate forces, foreign and domestic, must conspire for change to occur. Plasticity should therefore not be regarded as an impulse of any kind, creative or destructive, but instead as the generalized disposition of material bodies such as ours. (Sparrow 2015: 228)

Sparrow gestures towards a more ecological reading of plasticity, where one form of plasticity is not creative or destructive by itself but enabled by a network of constraints and processes. Explosivity could then be viewed as a constant (real) possibility of every organisation to break down, to become something other.

Explosive plasticity marks the emergence of unpredictable novelty from within the current organisation, of contingency in the radical sense described above, not as dependence on the totality of conditions but a change of this totality itself. Plasticity also concerns transitions from one level to another. Malabou speaks about this kind of emergence in relation to the interplay of the neuronal and the mental. It is enabled by the contradiction between them:

> There is no simple and limpid continuity from the one to the other, but rather transformation of one into the other out of their mutual conflict. We must suppose that mental formation draws its being or identity from the disappearance of the neuronal, born of a sort of blank space that is the highly contradictory meeting point of nature and history. Only an ontological explosion could permit the transition from one order to another from one organization to another, from one given to another. (Malabou 2008: 72)

Bergson is again the reference point, with his emphasis on the simultaneous creation and explosion of forms. Malabou quotes him saying that to make use of explosion, to harness their power, is an 'essential preoccupation of life'. This dovetails with the notion that organisms must canalise contingency to maintain themselves. Moreover, we emerge as autonomous (yet dependent) organisms through the sublation of the 'biological matrix', which does not mean that it is overcome. Organisms explode at each transition; they lose a bit of themselves to stay alive. Had they not been changed, had they not lost anything in their cycles of regeneration, organisms would be ahistorical and dead. The formative contradiction between formation and explosion is based on a more fundamental biological contradiction:

> In the central nervous system, as we have seen, the formative contradiction—formation/explosion—proceeds from a more original contradiction: that between the maintenance of the system "homeostasis" and the ability to change the system, or "self-generation". The nervous system, like any system, is self-regulated, self-organized, which means that it expends considerable energy in assuring its maintenance. Basically, in order to preserve itself from destruction, it must keep itself in the same state. Thus it continuously generates and specifies its own organization. (Malabou 2008: 74)

This quote underscores how plasticity should be understood organisationally. The decentring of the brain and its plasticity overlaps with the emphasis on self-generation and specification found in the organisational approach. Malabou also indicates the limits of homeostasis by saying that its contradiction with the ability of self-generation is the basis of the contradiction between formation and explosion. From what we developed above, it seems to imply *homeorhesis*, as the ongoing restoration of a biological trajectory (even if Malabou says 'state', not 'trajectory'). Finally, the contradiction Malabou speaks of seem to indicate the energetic openness and organisational closure of living systems.

The capacity to interrupt causal flows and redirect them, to become self-determining, demands ontological explosion. Simultaneously, it means that every living organisation has the constant possibility of being overturned. But plasticity is neither constructive nor destructive in itself. It is actualised and made possible through their dialectical relation, by the interaction of different plastic instances. This takes us back to the view that plasticity is itself plastic. Slight changes occur in each cycle of maintenance, while the system remains open to bifurcations that cannot be integrated into the current organisation.

In *The New Wounded*, Malabou lists several examples, unemployment, relational difficulties, and illness, which are not restricted to neuronal events:

> In all these cases what was striking was that once the metamorphosis took place, however explicable its causes (unemployment, relational difficulties, illness), its effects were absolutely unexpected, and it became incomprehensible, displacing its cause, breaking all etiological links. (Malabou 2012: 13)

The unpredictable character of such changes points to the limitations of anticipation and habituation. The process of familiarising itself with the world is never completed; the organism does not deal with an external historical development but a process it is constitutively entangled with. Malabou says that seeing what is coming (*voir venir*) is a process by which the organism is increasingly able to navigate in its environment in a fluid manner. Yet, this anticipation is restricted since the organisation is fragile. Explosive or destructive plasticity is the severing of links with the former self, radical ruptures when plasticity reaches breaking points constructed through the process of self-determination. At such junctures, the current plastic organisation disintegrates.

Modern neuroscience, on its part, reduces plasticity to flexibility, which is the capacity to regain one's initial shape. This is closer to homeostasis, restoring a specific state. Flexible materials can be stretched in every direction without breaking but they lack the resistance required for organismal creativity. Neglecting the historical aspect of plasticity, historical changes are considered reversible. Against this 'ideological avatar' of plasticity, Malabou emphasises the irreversibility of plastic changes. Unlike flexibility, plasticity does not display endless polymorphism, but 'imposes (a very strict) restriction on the capacity for deformation, re-formation, or explosion' (Malabou 2008: 15). As such, it involves the canalisation of indefinite possibilities. Flexibility, on the other hand, is passive and supple, because there is no tension between constancy and change. Every transition is smooth and reversible. This model is too harmonious to account for phenomena like the emergence of the psyche from the brain. Plasticity is a historical phenomenon with a *specific* form, which also contains canalisation since it means that possibilities that were at some point equiprobable are not anymore. Plasticity biases the trajectory of the organism since no identical iteration or reconstruction is possible for historical systems (Longo and Montévil 2014).

Neuroscience implicitly acknowledges this interplay of destruction and creation (Malabou 2012). For example, in the modulation of synaptic efficacy, the strengthening of certain synapses based on organismic activity comes at the expense of weakening others. This is a law of life, say Malabou: 'The fact that all creation can only occur at the price of a destructive counterpart is a fundamental law of life. It does not contradict life; it makes life possible' (Malabou 2012: 4). This casts plasticity as a balancing act between giving and receiving form which is crucial to the maintenance of a living organisation. But it does not exhaust the concept, as it misses the radical metamorphosis which emerges without mediation. There is no way to subjectivise the change within the current form since the change pertains to the identity of the subject itself.

Explosion is a *negative possibility*, a 'surprise resource' which is not integrated into the history of the subject but open to another future. We could think of this like above where the point is not to redeem negativity but to enact it since it opens the possibility of refunctionalisation. Yet, it goes beyond exaptations—which is when a trait, like feathers that were originally for temperature regulation, is repurposed to serve another function, such as flight. It transcends this notion because it concerns the organisation within which an exaptation can be identified and deemed functional in the first place. It is a negative possibility because it concerns the exhaustion of possibilities. Could we not interpret this as the exhaustion of *a space of possibilities*?

Malabou writes: 'What should we do with this threshold of non-presence that doubles the present, this negative halo that surrounds effectiveness with what could have been, since it keeps coming back?' (2012: 86). What *could have been* is subject to change, and this negative halo around the present cannot be eradicated. The similarities of this notion to what we have already developed are evident. In every decision, there is a negation of a generic potential. The irreversible canalisation of a certain pathway enables some while it makes other options inaccessible. This is not a positively given wellspring of possibilities but a negative halo, like the adjacent possible.

The reader might recall the example from above, about Facebook and Instagram. Malabou (2005) uses a similar example apropos virtuality. By her accounts, a generic definition of humans does not allow us to infer specific instances, like the philosopher, sculptor, or politician. These destinies are *virtually present in the genus* but cannot be predicted based on their generic quality. (In the next chapter, I tie this in with Hegel's view that the genus underdetermines the individual.) It is only through the process of repetition and practice that such determinations can become essential to human beings. Thus, she concludes, '[h]abit is the process whereby the contingent becomes essential' (Malabou 2005: 74). Habit contains the future of the genus since it is differentiated and actualised through habit formation.

Moreover, this passage from possibility to actuality 'manifests a virtuality already inscribed within essence itself' (Malabou 2005: 74). Virtuality is somewhere between possibility and actuality, a possibility which is *actualised as possible.* It has an influence on what is actualised without itself being actualised. We understand, then, why a virtual change is a change of dispositions, a change of possibilities. Malabou speaks about how plasticity's contradictory nature gives rise to explosions, 'a detonator which strikes the organization of the vital forces' (Malabou 2005: 187). This is a phase shift, from quantitative growth to a qualitative leap. She writes, 'plasticity, where all birth takes place, should be imagined fundamentally as an ontological combustion (*déflagration*) which liberates the twofold possibility of the appearance and the annihilation of presence' (Malabou 2005: 187). This underscores the openness to the future that Malabou considers central to Hegel's philosophy, and why plasticity operates at different levels of organisation:

> If saturation follows from a closure of the horizon, vacancy, for its part, opens up perspectives. This contradictory unity of saturation and vacancy is exactly what appears in the very form of the Hegelian System, which integrates while it dissociates, which unifies everything while letting what comes come. Plasticity designates the future understood as future within closure, the possibility of a structural transformation: a transformation of structure within structure, a mutation 'right at the level of the form'. (Malabou 2005: 192)

A mutation at the level of form is a mutation of self-relating negativity. Also, unity is a processual category, which must allow for alterity to remain intact. Furthermore, 'the possibility of a closed system to welcome new phenomena, all the while transforming itself, is what appears as plasticity' (Malabou 2005: 193). Plasticity is situated between the annihilation and emergence of form, and therefore contains the constant (virtual) possibility of its own destruction, a 'future within closure' fundamental to the maintenance of a living organisation.

No Central Executive

> [T]he centrality of the brain does not lie in its ability to constitute mentality by internalising and representing the world; rather, it lies in the ability of the brain to connect, to attend, to respond, to attune and relate to the world using its extraordinary plasticity and sensitivity. In other words, brain operations are inseparable from the rest of the body and its surrounding relevant environment. Importantly, there is no central executive. No single part of this dynamical system is responsible for central processing. What is often described or seen as central processing is in fact an attribution of agency. (Malafouris 2019: 6)

The above quotation from Malafouris complements Malabou's view on the role of the brain and indicates how its role in the emergence of the mind. He asserts that the plasticity of the brain is not decidable by looking only at the brain alone. Constrained by the larger system in which it operates, it cannot determine its own plasticity. Within the brain, assemblies of neurons lack the causal specificity to decide their own usage, constrained by the organisation they also enable. The phrase 'involved in'—which replaces 'determined by' in relation to genes—applies here as well. In other words, 'mind is not brain-based but brain-enabled. It is contextual. Mind and matter inhabit a single realm' (Schneider and Sagan 2005: 66). We cannot ascribe a central executive role to any of the parts within the organisation of the mind. While the brain enables the mind, it is enabled in return by the larger organism-environment nexus in which it is embedded.

Similarly, Malabou (2010) stresses the ecological or systemic notion of plasticity, as a self-organised regime that 'integrate the modifications that it experiences and to modify them in return' (61). In this view, the brain is conceived as an anticipatory structure, constituted by habit, which actively moulds and anticipates its surroundings (Boonstra and Slagter 2019). The brain enables the mind, which extends into the world, changes it, and alters the brain in return. There is no beginning or end to this process. It is not a linear extension of the capacities of the brain, as in some models of such extensions (Clark and Chalmers 1998), but a reciprocal constitution of outside and inside. Hence, the whole organism is 'an interactive space' where there is a constant tension between faithful transmission and radical transformation (Malabou 2016).[7]

If the power to resist determination is obfuscated, the organism is reduced to a passive object. The constitutive interaction between the organism and the environment breaks down. To maintain this dialectic, living beings must constantly deal with possibilities that are currently not present, but virtual. When the organism strives towards a goal, the possible effect, the goal, becomes the cause of the activity. It is not a cause in the mechanistic sense, but in a teleological sense, says Walsh (2015). Moreover, it 'captures the normative dimension of the relation between a goal and its

[7] Here, we slide into debates in philosophy of mind, but it is not my aim to settle such debates here, as it would require a much more systematic review than I can provide here. But see Sandnes Haukedal (forthcoming). Moreover, even if we change registers here somewhat, and speak about principles that would only apply to animals with nervous systems, or human being (since we speak about institutions later), I emphasise principles that are common to all living systems.

means' (Walsh 2015: 199). It explains why the organism acts as it does by reference to the goal it strives to achieve. This implies having a disposition, with no guarantee that its aim will be reached. It is not that the goal causes the disposition (as if we reverse time); rather, it constrains and enables it. This is how virtual possibilities shape actuality. That they are not actual does not mean they have no consequences: 'Goals may be nonactual states of affairs, but it is certainly not dabbling in the occult to invoke them' (Walsh 2015: 200).

Organisms are guided by the goals they seek towards. The brain must operate non-algorithmically because it cannot form a map of what *can* happen in advance. Malabou writes:

> [T]he analogy between the cybernetic domain and the cerebral domain rests on the idea that thinking amounts to calculating, and calculating to programming. The computer and the brain would in the end both be "thinking machines," that is, physico-mathematical structures endowed with the property of manipulating symbols. The discovery of the plasticity of brain functioning has rendered such a comparison moot. (Malabou 2008: 35)[8]

The brain does not work as a computer, a passive container with codified reactions to input and outputs. Further, organisms do not have the same control capacities as machines. There are no pregiven set points for the functioning of organisms, and no central controller; rather, control is distributed and self-organised (Stotz 2014). We should understand this in terms of extended processes of self-organisation, as developed in the chapter on Haraway. They form cognitive assemblies that extend in time and space and involve human and non-human actors and lessen the cognitive burden of the individual agent (Constant et al. 2022). This outsourcing also increases the organism's dependence on the environment.[9]

Consequently, the plastic brain mirrors its world, but this reflection is always a *mediation* that presents the world and its historical becoming in a biased manner. Hence, the reflection metaphor breaks down, as there is no absolute outside to reflect. 'Reflection', understood as simple mirroring without mediation, concerns external, not dialectical relations.[10] Because the organism has the capacity to decide how the world affects it by resisting its commands, the plastic brain does not work on its environment from the outside but is moulded in and through its entanglements with it. It is thus always beyond itself, constituted only through its relation to what it is not, with consequences for how we understand autonomy, as I touch upon later. While the brain is involved in all acts of cognition, this does not warrant the claim that cognition is located in or caused by the brain (Gabriel 2017).

By definition, a relational phenomenon must be open to the outside. As we move from naming the components involved to 'a fuller understanding of the process

[8] Malabou has since changed her position—arguing that machines, like living organisms, have the capacity to break down and reorganise themselves (Malabou 2019). This change does not affect the current analysis since the question is not about what computers can do. I doubt that the capacities I have outlined can be embodied by computers, but this is not our concern here (see next footnote).

[9] This also has implications for questions about the possibility of Artificial General Intelligence, to which I can only gesture here (see Kauffman and Roli 2021, 2023; Roli et al. 2022).

[10] But see Miller (2021) for a more dialectical understanding.

itself' (DiFrisco and Jaeger 2020: 8), a systemic perspective is demanded. The plastic brain is itself an environment, 'a metabolic place', both embodied—encompassing mind and body—and embedded, which means that it develops cognitive capacities through the entanglement with its environment (Malabou 2017). From this constitutive interaction, a higher-level plasticity of the organism-environment system emerges. This emergent *metaplasticity* defines the affordance landscape that the organism is faced with and involves the reuse (or exaptation) not only of brain areas but bodily and cultural practices. Such interaction and reuse change the significance of the niche (Gallagher 2017), as it involves 'the bidirectionality of influences between the genetic, behavioral, environmental, and socio-cultural levels of analysis' (Malafouris 2013: 40).[11]

Like Malafouris, Malabou (2008) speaks of the entanglement of different forms of plasticity *within* the brain, and also of the plasticity *between* the brain and other systems. Metaplasticity, both intra- and extra-neural, is a symptom of the incompleteness and lack of stability between different forms of plasticity, and a cause of further destabilisation. It distributes plasticity temporally and spatially: 'We create new things, embodied situated practices, and institutions which in turn make up our minds and ourselves' (Malafouris 2015: 2). It is therefore impossible to treat the informational content of the brain apart from action and the niche in which it operates. Metaplasticity brings out an ecological view of plasticity which is present in Malabou's work. It also converges on the broad conception of information and epigenesis that we have discussed.

Malabou's concept of plasticity always denotes plurality (Moder 2015). The brain is constrained by the life cycle it helps to maintain. As Malabou states in *What Should We Do with Our Brains?*, 'I am insisting upon the community between different kinds of systematic plastic organizations' (Malabou 2008: 6). Moreover, the 'neuronal functioning and social functioning *interdetermine* each other and mutually give each other form' (Malabou 2008: 9). Not only are there interacting forms of plasticity within the brain—undermining notions of central control—but there are interactions between these plasticities and plastic organisations outside the individual. Malabou states that this interaction 'sketches an organization that does *not at all* correspond to traditional representations of the brain as a machine without autonomy, without suppleness, without becoming' (Malabou 2008: 29–30, my emphasis). The historical dimension of plasticity challenges any notion of a program. Also, the notion that there is interdetermination between the social and the neuronal, implies a certain *underdetermination* of these instances, which opens them to each other, and makes the relations between them internal and enabling. I return to this topic in the next chapter.

Ian James says that 'there is plastic transformation because one instance or form of material existence in some way relates to another in the very plastic process of

[11] This implies a *heterarchical*, not hierarchical, organisation. In such an organisation, multiple constraints act on and canalise the same process, without being orchestrated by a central controller. In short, it means that control is distributed instead of centralised and that there is no permanent ranking of processes and constraints (Bich and Bechtel 2022).

transformation itself' (James 2016: 14). It is not only that the instances involved are plastic but that their interplay produces higher-order plasticity. Longo suggests something similar:

> The brain is not an input-output machine, but an always super-active organ, *constrained* by a changing context. Its continual activity is canalized by and works only in its preferred ecosystem: The skull of an animal as part of a sensing body in an ecosystem—and in history, as for humans. Its material flesh, the only one we can witness, is essential for this. (Longo 2020: 74, emphasis original)

The brain is canalised by the whole organism and ecosystem and canalises them in return. The processes are not in equilibrium. At times, the brain will influence the rest of the organism more than it is affected in return and vice versa. We are dealing with co-enabling constraints where the relative weight of each node is variable.[12] This mutual canalisation gives rise to an organisation where either part depends on the other, even if some agents may exert stronger influence at certain junctures.

We see, then, how the notion of enablement also applies here and sheds light on how we should understand metaplasticity. It also offers another perspective on the critique of neuroscience: If what is lacking in neuroscientific reasoning is the destructive aspect of plasticity—and this translates into a mechanical view of the brain (Rand 2011)—then the failure to include it is a failure to think in terms of organisation. From this view, destructive plasticity concerns a change in enablement relations.

Invariance and Chaos

> Contrary to the widespread view, Hegel does not deny contingency, nor does he deny that any specific thing can happen. But he does claim that, in the face of what occurs, it is useless to place necessity and contingency within *an order of occurrence* (*un ordre d'arrivée*). What occurs does not arise out of a pre-existing foundation, nor is the accident itself the foundation. Contingency and necessity support one another in such a way that spirit is free from their division and can simply let go its two-sided claim: it could have been otherwise, it could not have been otherwise. It would be futile to want to determine some ontological priority of essence over accident, or accident over essence, for their co-implication is primary. (Malabou 2005: 162–163, emphasis original)

The epigraph captures some of the topics discussed above, about the dual character of contingency as both grounded and ungrounded, as dependence on a totality of conditions and the contingency of this totality itself. Malabou (2005) calls this co-originarity of essence and accident the 'fundamental truth' about Hegel's thinking.

---

[12] As Ryan and Gallagher (2020) write: 'Instead of necessarily treating all aspects of the cognitive system as co-equals in the cognitive process, a proponent of ecological psychology may argue that there is reason for at least decomposing parts of the overall agent-environment system in ways where the balance of power, so to speak, itself can vary from circumstance to circumstance. Indeed, the appeal to enabling constraints […] may be a way of grounding this sort of response' (383). Here, reference is made to ecological psychology, a topic I deal with in forthcoming work.

Because of their complicity, no ontological ordering of necessity and contingency is possible. This 'negative result' concerns the limits of knowledge, which seems to correspond to the unfinished status of reality. These limits thus coincide with nature itself, which cannot actualise all its possibilities and therefore lacks sufficient reason, in the sense that we cannot deduce its contingent products (Padui 2010).

Again, we see the implications of the weakness of nature to actualise the concept—which makes it too variable for any conclusive symbolisation but also opens it to more complex structuration, as Furlotte (2016) argues. Hegel speaks of the process of setting oneself apart from the world, while simultaneously in-forming it, as a power of resistance. This transformation of what Hegel calls the natural 'soul' into a historical being *is* the dialectic itself, says Malabou (2008). In the next chapter, I elaborate on the Hegelian concept and tie it to mediational niche construction.

I have argued that the way to deal with a constantly fluctuating world is to make use of—instead of attempting to domesticate—randomness. This takes us back to the idea of bio-resonance and how 'different forms of randomness, at all levels of organization, may causally contribute to phenotypic changes and to biological stability by adaptivity and diversity' (Longo 2018: 91). Resonance is more than a metaphor, as any attempt to treat the brain in isolation ignores the historical development of the brain and thus reify it (Raja 2018). Against this view, biological systems are characterised by continuous variation. Hence, destruction and creation go hand in hand; choosing one path is at the expense of another. Comparing her concept of plasticity to Darwin's notion of variation, Malabou (2016) writes that: 'Characteristic of variability, plasticity designates the quasi-infinite possibility of changes of structure authorized by the living structure itself' (50). This echoes what we discussed earlier; how phenotypic variation is canalised by the biological organisation in place and must be compatible with it. Since it is not produced by natural selection, it implies additional sources of normativity in evolution.

It also indicates why Malabou (2012) speaks of plasticity as taking on *oriented form*, which is not 'a polymorphism without structure or rule' (145). Or, in the words we have applied, the changes it takes on are constrained or canalised by the viability of the organisation. Because of constraints imposed by the current form, it is not the case that any aspect can be changed at any moment, even if we cannot know which direction change will take. This is another way of saying that there is no ahistorical form of plasticity (Malabou 2011; Schuster 2018). While mechanistic physical systems mostly evince stable structures (symmetries) that break at certain instances (phase changes), biological systems must undergo continuous changes to maintain themselves. This indicates why Longo and Montévil (2013) define variation as a fundamental biological principle. They depict it as a continuous critical transition from one phase to another. It is a process in which the organism keeps itself in the range between order and chaos to allow critical transitions. To say that such transitions are continuous means that they do not happen at certain critical junctures, and thus rarely, as in classical physics—where criticality has to do with the singular limit where a phase transition occurs, and you go from one state to another, defined in advance. This limit is pregiven as a 'specific well-defined value of the control parameter' (Bailly and Longo 2011: 225). The most common example is the transition from liquid to ice. Such changes are, however, exceptions in physics, where conserving symmetry is the rule.

Against this, it has been proposed that biological systems do not display pointwise transitions, but continuous transitions—'ubiquitous symmetry changes' that are affected by global (systemic) effects (Bailly and Longo 2011; Longo and Montévil 2014, 2017). Criticality is also extended in the sense that it lasts throughout the life of the organism and follows from how biological systems impose global constraints on their components. Bailly and Longo (2011) mention two reasons for this: (1) that the 'structural stability of life' is not punctual but consists of many parameters because of the complexity of living systems; (2) that a fixed landscape of possibilities cannot be defined for such systems, as the possibility space constantly changes. In homeorhetic systems, there are no given set points for transitions; instead, such points are defined in an ongoing manner.

Biological systems undergo constant local changes to maintain their global organisation. The organism reconstructs itself continuously with variation produced through its dynamic coupling with its environment. Plasticity is the ongoing generation of different phenotypes from the same genotype, differences which can be inherited by the next generation. Organisms produce their own conditions of existence and pass them on to future generations. They survive by remaining within this interval of viability: 'From this point of view, the existence and the maintenance of living organisms would then be assimilable to the existence and to the maintaining of a situation (or zone) of extended criticality' (Bailly and Longo 2011: 230–231).[13]

The idea that organisms undergo constant symmetry breakings brings out why biological objects are not interchangeable (generic) but historical and specific. Two biological systems cannot be expected to behave in the same manner, even in the same context. This is also the case if they are genetically identical twins because they would make use of their genes differently. As early as the first cell differentiation of what becomes the foetus, differences are introduced between the systems. Each such mitosis (cell division) can be considered a critical transition (Longo and Montévil 2013). Also, genes from other organisms and symbiont besides the parents will affect the individuals, making the story even more complex (Pradeu 2011).

This, again, exemplifies how plasticity canalises the trajectory of the organism through the negation of virtual possibilities. Further, it underscores why history

[13] Here, I should also note that organisation is a form of *anti-entropy*, a notion that I will touch upon in the penultimate chapter. Anti-entropy differs from negative entropy because it not only cancels entropy but also produces entropy, which maintains variation that the organism may harness: 'The idea is that anti-entropy represents the key property of an organism, even a unicellular one, to be describable by severa levels of organization (also a eukaryotic cell possesses organelles, say), regulating integrating each other—they are parts that functionally integrate into a whole and the whole regulates them. This corresponds to the formation and maintenanc of a global coherence structure, in correspondence to its extended criticality: organization increases, along embryogenesis say, and is maintained, by contrasting the ongoing entropy production due to all irreversible processes. No extended criticality nor its key property of coherence would be possible without anti-entropy production, since always renewed organization expresses and allows the maintenance of the extended critical transition' (Longo and Montévil 2014: 20).

is central to understanding living systems. In every reproduction, there is slight variation, as emphasised by Darwin. This variation enables evolution:

> The space of the phenotypes, organisms, species, which are the relevant observables, changes constantly, even if minutely, at each reproduction (sometimes in a radical way, when the "hopeful monster" becomes viable). This suggests the impossibility of transferring the physico-mathematical concept of laws to the dynamics of biology and, *a fortiori*, to those of cognition. This is a "negative result" [...] which nevertheless opens the way to the positive (constructive, creative) role of history in the understanding of living systems. (Lanfredini and Longo 2016: 47–48)

This quotation suggests a way to understand explosive plasticity—not as something which only pertains to extreme cases of change but as a principle that is operative in every critical transition: 'In receiving form, it destroys its old form, and in giving form it destroys the form of the thing to which it gives a new form' (Rand 2011: fn35, 351). As such, plasticity has the dual character of negativity that I emphasised above. It points to how, in every construction, there is something explosive or destructive, either changing the form of something else or being changed from without. These explosions enable the change from one organisation to another but also undergird the functioning of the current organisation by opening it up to other levels of organisation (Malabou 2008).

Lanfredini and Longo (2016) state that the changing state space of living systems reveals the 'efficacy of history', the historical canalisation of possibilities. The path dependence of invariance in biology, like organisation or DNA, means that it is produced relationally and historically and is subject to constant change. There are no laws in biology because the space of what is possible constantly changes. Are the slight changes at each iteration not like the notion that there is always a loss involved in the changes that the organism undergoes?

To underscore the conceptual similarity, I return to the previous passage: 'Paradoxically, if we were flexible, in other words, if we didn't explode at each transition, if we didn't destroy ourselves a bit, we could not live. Identity resists its own occurrence to the very extent that it forms it' (Malabou 2008: 74). These explosions, Malabou writes, are 'discharges of energy', novel bursts that allow us to negate 'natural' determinations, without overcoming them fully. She does not only emphasise not only continuous changes within a space but a discharge of potentiality that may overturn any such space. In a sense, we might say that living systems undergo constant changes to remain open to more radical ones.[14]

From within the constant changes emerge singular points, whose specific values are not given in advance but produced through the process of symmetry breaking. I claim that these are instances where such changes are incompatible with the current 'organisation of vital forces' in place (Malabou 2005). Malabou writes: 'How can we imagine this beyond the limits of transformation except as the work of destructive plasticity, which sculpts by annihilating precisely at the point where *the repertory of viable forms* has reached exhaustion and has nothing else to propose?' (2012: 54, my

---

[14] It has been argued that self-organised criticality emerges from within extended criticality (Lovecchio et al. 2012), but I cannot explore this further.

emphasis). When the repertoire of viable forms is exhausted, hopeful monsters might become feasible (Lanfredini and Longo 2016). There is an ongoing re-production of the biological organisation, which is also an 'irreversible loss of possibility' (Meloni 2019). Thus, re-production is also a breaking of symmetries through the actualisation of possibilities.

The example that Malabou uses to explain these explosions is taken from the *Phenomenology of Spirit*: A foetus, which after a period of quantitative growth undergoes a qualitative leap, birth. Hegel describes this as 'dissolving bit by bit the structure of its previous world' wherein 'the vague foreboding of something unknown, are the heralds giving the signs that something other is in the offing' (Hegel 1977: 54, quoted by Malabou 2005: 187). Through a period of simultaneous degeneration and formation of form, a new world emerges as if from nowhere—i.e. as something that could not be predicted from within the confines of the old, even if there might be indications that something new is gestating.

Conclusion

James (2019) emphasises that we should not consider the distinction between the neuronal and mental level as a breach but instead think of it in terms of mutual groundlessness as an opening that enables their interplay:

> The ontological groundlessness of both the neuronal and the mental indicates that neither is a self-sufficient substance in itself. Both are articulated only in their relationality, and the relation of transition, transformation, or exchange from one to the other is made possible by the material spacing, differentiation, and singularization that ontological groundlessness makes possible in the first instance. (James 2019: 198)

The interdetermination that we spoke of above implies underdetermination or lack of stable ground, which we cognised through quantum physics earlier. The tensional continuity between the neuronal and the mental is a mark of their co-constitution or reciprocity. The mental is an emergent level of organisation that retroacts on the neuronal level. Again, we are dealing with loopy or reciprocal causality. We could understand this through John Dewey and his emphasis on how we must understand the continuity between levels of organisation neither as 'complete rupture', nor as 'mere repetition'. As such, a reduction is impossible and yet there is no absolute gap between them. The same relation is found between the genome, brain, body, and environment. According to Dewey, this continuity is illustrated by the development of the organism (see Renault 2013).

As we saw above, separation and connection are two sides of the same coin. Plasticity is the in-between (*entre-deux*), 'the dialectical tension that at once binds and opposes naturalness and intentionality' (Malabou 2008: 82). Hence, there is no neat distinction between the transcendental and the empirical. This point will be brought out more clearly in the next chapter. We are dealing with a position which is neither realist nor dualist, but emergentist. We might say that the organism tries to dominate chaos, the tendency towards entropy, by engaging in habit formation, but that this attempt is ultimately unsuccessful. Had it succeeded, it would be detrimental to its existence, as it would make it more vulnerable to changes in the environment because it would lack the functional randomness that generates variation at the biological level.

In the section on mechanism in the *Logic*, where Hegel speaks about mechanical objects lacking the power of *resistance to the concept*, which is the principle of self-determination. He thus seems to be saying that they are (1) generic, and thus always decided in relation to the theory or experiment (as seen in classical physics and its form of randomness) and (2) that they lack plasticity, which is found in the power to resist determination. With life, however, the organism gains actuality, qua the power to act in the world, a capacity that inanimate objects lack. In the next chapter, I will discuss life as the ground of cognition, and how we may relate Malabou's notion of plasticity to niche construction. Here, I argue that niche construction is not only about material changes to the environment but that the way we experience or mediate it is a form of niche construction as well.

So, after all of this: what is plasticity? First, it is a generic capacity to take and give form. It is, as seen in habit formation, a process of responding to changes from without and extending the organism into its environment. Moreover, I have argued that plasticity is never isolated but connected to other forms of plasticity, at different levels. As such, plasticity enables the process of organisation, of moulding, and operates within the organisation it enables. This explains why it is not the opposite of canalisation but involves the cancellation of certain possibilities at the expense of others, a historical trajectory. This sets it apart from flexibility. The precondition for the plastic power to resist is the groundlessness or contingency that we saw earlier. Like negativity, plasticity is a process of differentiation, or a capacity to initiate this process and an ongoing response to it. In the end, this might be the most important aspect of plasticity, the opening to be changed from without, which also affords the capacity to modify this outside. The next chapter expands on this process of reciprocal modification.

Bibliography

Anastasiadi, Dafni, Clare J. Venney, Louis Bernatchez, and Maren Wellenreuther. 2021. Epigenetic Inheritance and Reproductive Mode in Plants and Animals. *Trends in Ecology & Evolution* 36 (12): 1124–1140. https://doi.org/10.1016/j.tree.2021.08.006.

Bailly, Francis, and Giuseppe Longo. 2011. *Mathematics and the Natural Sciences: The Physical Singularity of Life* (Advances in Computer Science and Engineering: Texts). London, UK: Imperial College Press. https://doi.org/10.1142/p774.

Bich, L., and Bechtel, W. 2022. Organization Needs Organization: Understanding Integrated Control in Living Organisms. *Studies in History and Philosophy of Science* 93: 96–106. https://doi.org/10.1016/j.shpsa.2022.03.005.

Boonstra, Evert A. and Heleen A. Slagter. 2019. The Dialectics of Free Energy Minimization. *Frontiers in Systems Neuroscience* 13. https://www.frontiersin.org/articles/10.3389/fnsys.2019.00042.

Bourgeouis, Bernard. 1988. Translator's Introduction. In *Encyclopédie Des Sciences Philosophiques, Tome III: La Philosophie de l'esprit*, 7–93. Paris: Vrin.
Clark, Andy, and David Chalmers. 1998. The Extended Mind. *Analysis* 58 (1): 7–19. http://www.jstor.org/stable/3328150. [accessed 10 January 2018].
Constant, Axel, Andy Clark, Michael Kirchhoff, and Karl J. Friston. 2022. Extended Active Inference: Constructing Predictive Cognition beyond Skulls. *Mind & Language* 37 (3): 373–394. https://doi.org/10.1111/mila.12330.
Crockett, Clayton. 2010. Foreword. In *Plasticity at the Dusk of Writing: Dialectic, Destruction, Deconstruction*, xi–xxv. New York, NY: Columbia University Press.
Danchin, É. 2013. Avatars of Information: Towards an Inclusive Evolutionary Synthesis. *Trends in Ecology and Evolution.* https://doi.org/10.1016/j.tree.2013.02.010.
Dickins, Thomas. 2021. *The Modern Synthesis: Evolution and the Organization of Information.* Cham: Springer. https://doi.org/10.1007/978-3-030-86422-4.
Dickins, T. E., and B. J. Dickins. 2023. *Evolutionary Biology: Contemporary and Historical Reflections Upon Core Theory.* Berlin: Springer.
DiFrisco, James, and Johannes Jaeger. 2020. Genetic Causation in Complex Regulatory Systems: An Integrative Dynamic Perspective. *BioEssays* 42 (6). https://doi.org/10.1002/bies.201900226.
Diogo, Rui. 2017. *Evolution Driven by Organismal Behavior, Evolution Driven by Organismal Behavior: A Unifying View of Life, Function, Form, Mismatches, and Trends.* Cham: Springer. https://doi.org/10.1007/978-3-319-47581-3.
Farnsworth, Keith D. 2022. How an Information Perspective Helps Overcome the Challenge of Biology to Physics. *Biosystems* 217: 104683. https://doi.org/10.1016/j.biosystems.2022.104683.
Ferrini, C. 2020. Hegel's Philosophy of Natural and Human Spaces. In *The Palgrave Hegel Handbook* , ed. M. F. Bykova and K. R. Westphal, 241–266. Cham: Springer International Publishing. https://doi.org/10.1007/978-3-030-26597-7_13.
Fitzgerald, D., and F. Callard. 2015. Social Science and Neuroscience Beyond Interdisciplinarity: Experimental Entanglements. *Theory, Culture & Society* 32 (1): 3–32. https://doi.org/10.1177/0263276414537319.
Foster, John Bellamy. 2020. *The Return of Nature: Socialism and Ecology.* New York, NY: Monthly Review Press.
Foster, John Bellamy. 2022. The Return of the Dialectics of Nature: The Struggle for Freedom as Necessity. *Historical Materialism*: 1–26. https://doi.org/10.1163/1569206X-20222279.
Furlotte, W. 2016. Lack and the Spurious Infinite: Towards a New Reading of Hegel's Philosophy of Nature. In *Rethinking German Idealism*, ed. S. J. McGrath and J. Carew, 191–214. London, UK: Palgrave Macmillan. https://doi.org/10.1057/978-1-137-53514-6_9.
Gabriel, Markus. 2017. *I Am Not a Brain: Philosophy of Mind for the 21st Century.* Cambridge, UK: Polity.
Gallagher, Shaun. 2017. *Enactivist Interventions. Rethinking the Mind.* Oxford, UK: Oxford University Press.
Godfrey-Smith, Peter. 2016. *Other Minds: The Octopus and the Evolution of Intelligent Life.* London, UK: William Collins.
Gonzalez, Paula, and Jimena Barbeito-Andrés. 2021. Canalization: A Central but Controversial Concept in Evo-Devo. In *Evolutionary Developmental Biology: A Reference Guide*, ed. Laura Nuño de la Rosa and Gerd B. Müller, 1061–1072. Cham: Springer. https://doi.org/10.1007/978-3-319-33038-9_55-1.
Hallgrimsson, Benedikt, Rebecca Green, David Katz, Jennifer Fish, Francois Bernier, et al. 2019. The Developmental-Genetics of Canalization. *Seminars in Cell & Developmental Biology* 88: 67–89. https://doi.org/10.1016/j.semcdb.2018.05.019.
Hegel, Georg Wilhelm Friedrich. 1977. *Phenomenology of Spirit*, translated by A.V. Miller. Oxford, UK: Oxford University Press.
James, Ian. 2016. (Neuro)Plasticity, Epigenesis and the Void. *Parrhesia* 25: 1–19. https://doi.org/10.1177/1470412915619458.

James, Ian. 2019. *The Technique of Thought: Nancy, Laruelle, Malabou, and Stiegler after Naturalism.* Minneapolis, MN: University of Minnesota Press.

Kauffman, Stuart, and Andrea Roli. 2021. The World Is Not a Theorem. *Entropy.* https://doi.org/10.3390/e23111467.

Kauffman, Stuart A., and Andrea Roli. 2023. What Is Consciousness? Artificial Intelligence, Real Intelligence, Quantum Mind and Qualia. *Biological Journal of the Linnean Society* 139 (4): 530–538. https://doi.org/10.1093/biolinnean/blac092.

Lala, K. N., T. Uller, N. Feiner, M. Feldman, D. Andrews, et al. 2024. *Evolution Evolving: The Developmental Origins of Adaptation and Biodiversity.* Princeton, NJ: Princeton University Press.

Lanfredini, Roberta, and Giuseppe Longo. 2016. Epistemology of the Inert and Epistemology of the Living. *HUMANA.MENTE Journal of Philosophical Studies* 9 (31): 37–55. https://www.humanamente.eu/index.php/HM/article/view/43.

Longo, Giuseppe. 2018. How Future Depends on Past and Rare Events in Systems of Life. *Foundations of Science* 23: 443–474. https://doi.org/10.1007/s10699-017-9535-x.

Longo, Giuseppe. 2020. Information, Science and Democracy, for an Ethics of Scientific Knowledge Trust. In *Trust: A Philosophical Approach*, ed. Adriano Fabris, 63–79. Cham: Springer.

Longo, Giuseppe, and Maël Montévil. 2013. Extended Criticality, Phase Spaces and Enablement in Biology. *Chaos, Solitons and Fractals* 55. https://doi.org/10.1016/j.chaos.2013.03.008.

Longo, Giuseppe, and Maël Montévil. 2014. *Perspectives on Organisms: Biological Time, Symmetries and Singularities.* Berlin and Heidelberg: Springer. https://doi.org/10.1007/978-3-642-35938-5.

Longo, Giuseppe, and Maël Montévil. 2017. From Logic to Biology via Physics: A Survey. *Logical Methods in Computer Science* 13 (4). https://doi.org/10.23638/LMCS-13(4:21)2017.

Lovecchio, Elisa, Paolo Allegrini, Elvis Geneston, Bruce J West, and Paolo Grigolini. 2012. From Self-Organized to Extended Criticality. *Frontiers in Physiology* 3. https://www.frontiersin.org/journals/physiology/articles/10.3389/fphys.2012.00098.

Malabou, Catherine. 2005. *The Future of Hegel: Plasticity, Temporality and Dialectic, The Future of Hegel: Plasticity, Temporality and Dialectic.* London and New York: Routledge. https://doi.org/10.4324/9780203489338.

Malabou, Catherine. 2008. Addiction and Grace: Preface to Félix Ravaisson's of Habit. In *Of Habit*, vii–xx. London and New York: Continuum Publishing.

Malabou, Catherine. 2010. *Plasticity at the Dusk of Writing: Dialectic, Destruction, Deconstruction.* New York, NY: Columbia University Press.

Malabou, Catherine. 2011. *The Heidegger Change: On the Fantastic in Philosophy.* London, UK: Polity Press.

Malabou, Catherine. 2012. *Ontology of the Accident: An Essay on Destructive Plasticity.* Cambridge, UK: Polity Press.

Malabou, Catherine. 2016. *Before Tomorrow: Epigenesis and Rationality.* Cambridge, UK: Polity Press.

Malabou, Catherine. 2017. The Brain of History, or, The Mentality of the Anthropocene. *South Atlantic Quarterly* 116 (1): 39–53. https://doi.org/10.1215/00382876-3749304.

Malabou, Catherine. 2019. *Morphing Intelligence: From IQ Measurement to Artificial Brains.* New York, NY: Columbia University Press.

Malafouris, Lambros. 2013. *How Things Shape the Mind A Theory of Material Engagement.* Cambridge, MA: The MIT Press.

Malafouris, Lambros. 2015. Metaplasticity and the Primacy of Material Engagement. *Time and Mind* 8 (4): 351–371. https://doi.org/10.1080/1751696X.2015.1111564.

Malafouris, Lambros. 2019. Mind and Material Engagement. *Phenomenology and the Cognitive Sciences* 18 (1): 1–17. https://doi.org/10.1007/s11097-018-9606-7.

McDonald, Jeanne M. C. Shampa M. Ghosh, Samuel J. L. Gascoigne, and Alexander W. Shingleton. 2018. Plasticity Through Canalization: The Contrasting Effect of Temperature on Trait Size and

Growth in Drosophila. *Frontiers in Cell and Developmental Biology* 6. https://www.frontiersin.org/article/10.3389/fcell.2018.00156.
Meloni, Maurizio. 2019. *Impressionable Biologies: From the Archaeology of Plasticity to the Sociology of Epigenetics*. New York, NY: Routledge.
Miller, Elaine P. 2021. Hegel on Reflection and Reflective Judgement. *Hegel Bulletin* 42 (2): 201–226. https://doi.org/10.1017/hgl.2018.34.
Moder, Gregor. 2015. Catherine Malabou's Hegel: One or Several Plasticities? *Filozofija i Drustvo* 26 (4): 813–829. https://doi.org/10.2298/fid1504813m.
Montévil, Maël. 2022. Understanding Living Beings by Analogy with Computers or Understanding Computers as an Emanation of the Living. *Trópos. Rivista Di Ermeneutica E Critica Filosofica* 13 (2): 57–74.
Mossio, Matteo, Maël Montévil, and Giuseppe Longo. 2016. Theoretical Principles for Biology: Organization. *Progress in Biophysics and Molecular Biology* 122 (1): 24–35. https://doi.org/10.1016/j.pbiomolbio.2016.07.005.
Ng, K. 2009. Hegel's Logic of Actuality. *The Review of Metaphysics*, 63 (1): 139–172. http://www.jstor.org/stable/40387730.
Ng, Karen. 2013. Life, Self-Consciousness, Negativity: Understanding Hegel's Speculative Identity Thesis. In *The Freedom of Life: Hegelian Perspectives*, ed. Thomas Khurana, 33–67. Berlin: August Verlag.
Nijhout, H. Frederik, Farrah Sadre-Marandi, Janet Best, and Michael C. Reed. 2017. Systems Biology of Phenotypic Robustness and Plasticity. *Integrative and Comparative Biology* 57 (2): 171–184. https://doi.org/10.1093/icb/icx076.
Noble, D. 2006. *The Music of Life: Biology Beyond Genes*. Oxford, UK: Oxford University Press.
Padui, R. 2010. The Necessity of Contingency and the Powerlessness of Nature. *Idealistic Studies* 40 (3): 243–255. https://doi.org/10.5840/idstudies201040316.
Pradeu, T. 2011. A Mixed Self: The Role of Symbiosis in Development. *Biological Theory* 6 (1): 80–88. https://doi.org/10.1007/s13752-011-0011-5.
Raja, Vicente. 2018. A Theory of Resonance: Towards an Ecological Cognitive Architecture. *Minds and Machines* 28 (1): 29–51. https://doi.org/10.1007/s11023-017-9431-8.
Rand, Sebastian. 2011. Organism, Normativity, Plasticity: Canguilhem, Kant, Malabou. *Continental Philosophy Review* 44 (4): 341–357. https://doi.org/10.1007/s11007-011-9196-3.
Ravaisson, Felix. 1984. *De l'habitude, Corpus Des Oeuvres de Philosophie En Langue Française*. Paris: Fayard.
Renault, E. 2013. The Naturalistic Side of Hegel's Pragmatism. *Critical Horizons* 13 (2): 244–274. https://doi.org/10.1558/crit.v13i2.244.
Roli, Andrea, Johannes Jaeger, and Stuart A Kauffman. 2022. How Organisms Come to Know the World: Fundamental Limits on Artificial General Intelligence. *Frontiers in Ecology and Evolution* 9. https://www.frontiersin.org/journals/ecology-and-evolution/articles/10.3389/fevo.2021.806283.
Rosen, R. 1991. *Life Itself: A Comprehensive Inquiry Into the Nature, Origin, and Fabrication of Life*. New York, NY: Columbia University Press.
Ryan, Kevin J, and Shaun Gallagher. 2020. Between Ecological Psychology and Enactivism: Is There Resonance? *Frontiers in Psychology* 11. https://www.frontiersin.org/articles/10.3389/fpsyg.2020.01147.
Schneider, Eric D., and Dorion Sagan. 2005. *Into the Cool: Energy Flow, Thermodynamics, and Life*. Chicago, IL: University of Chicago Press.
Schuster, Joshua. 2018. After Deconstruction? The Challenge of Malabou's Plastic Biohistory. In *Thinking Catherine Malabou: Passionate Detatchments*, ed. Thomas Wormald and Isabell Dahms. Lanham, MD: Rowman & Littlefield Publishers.
Shan, Yafeng. 2024. The Extended Evolutionary Synthesis: An Integrated Historical and Philosophical Examination. *Philosophy Compass* 19 (6): e13002. https://doi.org/10.1111/phc3.13002.

Sparrow, T. 2015. *Plastic Bodies: Rebuilding Sensation After Phenomenology*. London, UK: Open Humanities Press.

Stotz, Karola. 2014. Extended Evolutionary Psychology: The Importance of Transgenerational Developmental Plasticity. *Frontiers in Psychology* 5. https://www.frontiersin.org/article/10.3389/fpsyg.2014.00908.

Sultan, S. E. 2015. *Organism and Environment: Ecological Development, Niche Construction, and Adaptation*. Oxford, UK: Oxford University Press.

Sultan, S. E. 2017. Developmental Plasticity: Re-conceiving the Genotype. *Interface Focus* 7 (5): 20170009. https://doi.org/10.1098/rsfs.2017.0009.

Uller, Tobias, and Heikki Helanterä. 2019. Niche Construction and Conceptual Change in Evolutionary Biology. *British Journal for the Philosophy of Science* 70 (2). https://doi.org/10.1093/bjps/axx050.

Uller, Tobias, Nathalie Feiner, Reinder Radersma, Illiam S. C. Jackson, and Alfredo Rago. 2020. Developmental Plasticity and Evolutionary Explanations. *Evolution & Development* 22 (1–2): 47–55. https://doi.org/10.1111/ede.12314.

Waddington, C. H. 1953. Genetic Assimilation of an Acquired Character. *Evolution* 7 (2). https://doi.org/10.2307/2405747.

Walsh, D. M. 2015. *Organisms, Agency, and Evolution*. Cambridge, UK: Cambridge University Press. https://doi.org/10.1017/CBO9781316402719.

West-Eberhard, Mary Jane. 2003. *Developmental Plasticity and Evolution, Developmental Plasticity and Evolution*. Oxford, UK: Oxford University Press. https://doi.org/10.1093/oso/9780195122343.001.0001.

Žižek, Slavoj, ed. 2012. *Mapping Ideology*. London, UK: Verso Books.

Chapter 7
The Organisation of the Environment

Abstract This chapter critiques traditional adaptationist views in biology by exploring the dialectical relationship between organisms and their environments. Drawing on Lewontin, Hegel, and contemporary theorists, it explores niche construction—how organisms actively construct their niches rather than passively adapting to pre-existing conditions. Expanding this concept somewhat, *experiential niche construction* is introduced to highlight how organisms mediate their surroundings through perception and behaviour, reshaping selective pressures in the process. Challenging reductionist models, the analysis emphasises the qualitative and reciprocal nature of organism-environment interactions. Critiquing reductionist and externalist models, the chapter advocates for a more integrated approach that recognises the qualitative and relational dimensions of organism-environment interactions. Hegel's philosophical framework is employed to conceptualise this relationship as one of mutual determination, where purposive activity and habit formation enable organisms to transform their surroundings while being shaped by them in turn. The discussion highlights the *underdetermination* of adaptations by the environment, illustrating how affordances and constraints emerge from organismic agency rather than external forces alone.

Introduction

> Cartesian biology objectifies organisms. They are seen as the passive consequences of internal and external forces, genes, and environment. Organisms are objects, the internal and external forces the subjects. What dialectical biology attempts to do is to break down the alienation of subject and object, to insist on the interpenetration of gene, organism, and environment. Thus, in place of the metaphor of adaptation of organisms to a preexistent environmental "niche," dialectical biology emphasizes the way in which organisms *define and alter their environment in the process of their life activities*. Organism and environment are both in a constant state of becoming, mutually determining each other. (Lewontin 1983: no pagination, my emphasis)

R. Sandnes Haukedal, *Agency and Organisation: The Dialectics of Nature and Life*,
Library of Ethics and Applied Philosophy 44,
https://doi.org/10.1007/978-981-95-1924-8_7

Before Lewontin and Levins, Jakob von Uexküll (2010) articulated the distinctive way that organisms relate to their surroundings by distinguishing between *Umgebung* and *Umwelt*. In the terms we have applied, the positing done by the organism creates the *Umwelt*, a world, whereas the presupposition for this positing is the *Umgebung*, the external surroundings. By positing a world, the organism does not deal with external conditions directly; rather, these conditions are defined in relation to its activity.[1] Yet, this should not lead us to regard the *Umwelt* as purely subjective. While there is a subjective interiority that has relative autonomy from external conditions, this interiority is already extended outwards, not formed in isolation. This takes us back to importance of thinking in terms of sympoiesis.

As I argue below, the *Umgebung* has a virtual and ambiguous character, since the way the organism deals with its environment is mediated by its activity and status. It actualises a mere subset of its possible interactions with the physical world, in an interplay of possibility and actuality. This is the logic of 'no (actual) niche without an organism'. The lines are blurred, yet do not collapse completely. In other words, the niche is 'indefinite in features prior to proliferation with variation and selection revealing what will co-constitute "task closure" for the organism' (Longo and Montévil 2014: 13). This reminds us of the higher-order closure we spoke about previously, in which the environment is integrated into the closure of constraints (Heras-Escribano and Saborido 2023).[2]

As we shall see, the physical surroundings of the organism *underdetermine* adaptations (Walsh 2022). A complete description of the ecosystem is impossible, as it is a co-evolving process where new functions emerge from changes in the organism-environment relation. The epigraph from Lewontin hints at this circumstance. Organism and environment are both incomplete, and their constant becoming opens them to each other. This underscores why explanatory externalism that accounts for 'properties of organic systems in terms of properties of the environments' (Godfrey-Smith 1996: 30) does not work. Either adaptation as such obfuscates how evolution works, or the problem is simply how we consider it. We may, for instance, combat adaptation*ism* by highlighting that it is not populations that adapt but individual organisms; or we could attack adaptation itself as a misleading metaphor with residues of natural theology (Depew 2017). It is not clear which of

---

[1] We could call this a difference between effective and fundamental exteriority (James 2019), or the fundamental and realised niche (Hutchinson 1957). The tension between these constitutes the recalcitrant problem of nature.

[2] The organisational theory of ecological functions was first proposed by Nunes-Neto et al. (2014). It does not, however, entail treating environments as organisms. In a recent paper, the authors state: '[W]e do not need to treat ecosystems as organisms to ascribe organizational functions to their components. On the contrary, there are important differences between ecosystems and organisms, as the former typically lacks the sort of agency and regulation that characterize the latter. Moreover, ecological systems do not show the same degree of stability and cohesion observed in organisms, or at least in many cases of organismality. Accordingly, there is no requirement that one is committed to an interpretation of ecosystems as superorganisms to apply the organizational theory. What is necessary to ascribe function to components of a system based on this theory is just organizational closure, conceived as closure of constraints' (El-Hani et al. 2024: 299).

these positions presents the clearest challenge to the MS, nor are they necessarily incompatible.

The challenge to populational thinking represented by the focus on development makes genetic factors the consequences and not causes of evolution. Here, adaptation is involved but it happens in developmental systems that are defined together with their environments. The more radical position—that talk of adaptations as such is misleading—goes to the core of what evolution is about. I have touched upon both options without deciding between them. In what follows, the question of whether adaptation should be discarded is posed more clearly through an extended notion of niche construction. The question is not only about who adapts and how long it may take but whether we can usefully speak of adaptation in the first place, whether seeming adaptation obfuscates construction. By expanding what counts as construction, there might not be much explanatory work left for adaptation, at least not without a reconceptualisation that underscores the internal relation between adaptation and construction (Walsh 2022).

The extension of evolutionary inheritance, through the notion of ecological inheritance, implies niche construction—especially epistemic niche construction and how we inherit norms that shape how we behave in and experience the environment (Sterelny 2003). As we shall see in the concluding chapter, this also has implications for how we understand scientific research.[3] In this chapter, I emphasise the mediational aspects of such construction, and how the organism exploits its plasticity to change the impact of natural selection. I also revisit and expand some of Malabou's concepts to see how she may inform and be informed by these developments. I will, however, spend more time on Hegel's philosophy and relate it to dialectical biologists like Lewontin. The fundamental principle here is that there is no organism without an environment and vice versa.

The outline: I begin with some general remarks on what is missing in models of organicism and niche construction and connect it to Hegel and the previous discussion of habits. I then introduce experiential niche construction, and how it answers to a perceived restriction in previous models of niche construction: They focussed too much on external changes and thus missed more subtle experiential and relational changes that do not necessarily involve material changes to the environment. This indicates why natural selection cannot be reduced to environmental factors alone but is also shaped by the activities of the organism. After these sections, I move on to discuss the notion that the environment underdetermines affordances and adaptations.

[3] A wider debate, which I cannot touch upon here, is whether we can meaningfully distinguish cognitive niche construction from biological niche construction. This takes us to issues of the impact of research, which is not only an expression of human evolution but also affects other species. I think this follows from a materialist understanding of scientific practice. As Rouse (2024) argues, speaking about scientific niche construction as simply conceptual, epistemic, or cognitive niche construction—and thus only analogous or metaphorical extensions of biological niche construction—forgets that these kinds of practices are expressions of human biological niche construction. These kinds are always dynamically entangled (see Trappes and Leonelli 2025; Rouse 2024). I think the notion of mediational niche construction, which I deal with in this chapter, speaks to the same concern; moreover, Hegel's view highlights how conceptualisation is an inherent part of organismic activities.

Since the environment cannot be defined except through its relation to an organism, it seems that the notion of an external environment is a mere abstraction. Adopting Walsh's notion of underdetermination, I argue that the physical environment should be considered as a 'reference environment', an abstraction which does not need to be stable, constant, or autonomous. I then return to experiential niche construction, to get clearer about what we are dealing with and why it is important. I also tie it back to the notion of habit and anticipation. Next, I try to summarise what the notion of experiential niche construction means for natural selection. This ends part one of the chapter.

In the second part, and to get more concrete about the philosophical implications of this reconceptualisation, I describe Hegel's take on how the subject relates to and shapes its world. Positing is like the quantitative surroundings into a qualitative world of affordances which have normative value to the organism (Crippen 2020). Building on Karen Ng's work, I relate this to what Hegel means by the *concept*, and how it may be interpreted as embodied in the purposive activities of organisms. I also discuss the importance of understanding life as the ground of cognition, as proposed by Ng. I then tie what I said above concerning actuality and possibility in previous chapters to Hegel's notion of purposiveness and the ontological status of the niche. I argue that we find the same kind of underdetermination that Walsh speaks of in Hegel's conception of how the organism relates to its genus, and that this implies enablement or reciprocity, which are also implicit in the notion of causal spread that we find in the notion of experiential niche construction. Finally, I bring the two parts together and argue that they underscore the metaplasticity of mind and world. Based on this, we should be able to perceive why the extension of niche construction is a return to a dialectical understanding of the organism-environment interaction.

Part I: Niche In-habituation

> There is no organism without an environment, but there is no environment without an organism. There is a physical world outside of organisms and that world undergoes certain transformations that are autonomous. Volcanoes erupt, the earth processes on its axis of rotation. But the physical world is not an environment, only the circumstances from which environments can be made. The reader might try describing the environment of an organism that he or she has never seen. There is a noncountable infinity of ways in which the bits and pieces of the world might conceivably be put together to make environments but only a small number of those have actually existed, one for each organism. (Lewontin and Levins 2007: 32)

The epigraph articulates the central notion that there is no niche without an organism. It also reminds us of the discussion of actuality and possibility, as there is no way to abstractly prestate the possibilities of the environment. The Hegelian backdrop implied sheds light on limitations in current conceptions of niche construction and organicism. According to Emmeche (2004), organicism failed to consider the experience of organisms. He highlights how this idea hinges on the analytic divide between primary and secondary qualities. Organicism has focussed on the 'objective' or measurable dimensions of higher-level properties and entities.[4] Thus, it forgets how the objective qualities are not experienced in the same manner by all organisms but depend on how they actualise (and construct) the potential of the environment.[5] Against this, organicism has been committed to the view that emergence pertains only to 'material properties of highly self-organized matter' (Emmeche 2004: 207).[6] In short, they focus on the external, not the internal aspects of emergence. Only the material surroundings, not the *world* of the organism, are considered. As we shall see, this view seems to have influenced NCT.

A quick recap: NCT is concerned with how organisms actively shape their environment and thus change the selection pressures that they are under. It takes two different forms: perturbation and relocation. The former is about the physical changes that an organism makes, and the latter pertains to how it may change the factors it is under by changing its habitat. In this view, niche construction occurs.

> when an organism modifies the feature–factor relationship between itself and its environment by actively changing one or more of the factors in its environment, either by physically perturbing factors at its current location in space and time, or by relocating to a different space-time address, thereby exposing itself to different factors. (Odling-Smee et al. 2003: 41)

Aaby and Ramsey (2019) have argued that these two forms are inconclusive. Odling-Smee, Laland, and Feldman imply but fail to grasp the relational change between the organism and the environment. There are two main issues with this definition: First, it 'excludes many ways in which organisms can actively modify their relationship with their environment' (Aaby and Ramsey 2019: 4). It overlooks how internal changes to the organism's constitution also change how selection pressures act on it. Second, the categories are somewhat arbitrary. They fail to encompass the full scope of changes in the relationship between organism and niche and are therefore blind to other processes that could be considered niche construction. Aaby and Ramsey therefore propose a return to the initial formulation of the theory—the dialectical version formulated by Lewontin and Levins.

---

[4] It performs a functional description from the perspective of the observer, not the agent, which was the case also for systems theory and autopoiesis. Di Paolo et al. (2018) call the first approach Kantian, whereas their enactive approach is Hegelian (see Gambarotto and Mossio 2022).

[5] The view I outline also answers the problem of how to reconcile the view that we may have multiple perceptions with the notion that they all refer to the same world. It moreover achieves this without introducing a notion of representation—which creates the problem of how representations match the outside world. In short, the richness of the stimuli offers multiple ways of perceiving the same environment without relying on such notions (Nonaka 2020). I develop this more fully in Sandnes Haukedal (2025).

[6] He also mentions authors such as Kauffman, Oyama, and Depew as committed to the quantitative view. The emphasis on qualitative experience highlights shortcomings of organicism that were lost to many of its proponents. While they acknowledge qualitatively new properties, the properties are not considered from the perspective of the organisms and their experience. Since the time of this publication, however, Kauffman has become increasingly aware of the semantic aspect of information, as seen in Kauffman and Roli (2021).

I cannot elaborate on the changes Aaby and Ramsey propose for NCT, only thematise what they call *constitutive* niche construction—the term applied for the mode that I describe below.[7] These are cases where there is an active change in an organismic feature which changes the relation it has to environmental factors. They define the position discussed below as concerned with cases where there is 'a causal relationship between a change in an organism's form or capacities (its features) and the factors of the environment that it experiences, without there being a change to the environment itself' (Aaby and Ramsey 2019: 9). These constitutive changes are also causal since they modify the interaction between an organism and its environment. I aim to cast light on the causality involved and the status of the external environment it implies in what follows.

Against the narrow emphasis on quantitative properties of the environment (factors), Hegel (2010) claims that 'objective' processes are immediately ruptured when the organism relates to the world. This is not to say that it is not affected by or affects physical and chemical processes, just that it does not interact with them *as such* but sublates them. The external environment does not rigidly specify the experience that the organism makes of it. Nor is the organism self-sufficient, says Hegel, as witnessed in the act of consuming natural things to further one's existence (Hegel 1991).

As soon as you claim something is objective, you have subsumed chemical and physical processes through the purposive activity that pertains to the whole organism, and not just a specific trait (Sultan 2015). Concomitantly, organisms are not only acting in the world as differentiated Hegelian wholes but are also *acted upon* as differentiated wholes. For instance, in temperature-dependent sex determination, specific molecular effectors are targeted by the whole system of enablement relations (Bizzarri et al. 2020). This view reverberates in the *Logic* when Hegel says: 'In so far as the object confronts the living being in the first instance as an indifferent externality, it can act upon it mechanically; but in doing so it is not acting on a living being; where it enters into relationship with a living being it does not act on it as a cause, but *excites* it' (Hegel 2010: 685). Unlike mechanical objects, a living being is not acted upon by the outside world but is *excited* by it since the relation is one of reciprocity, not mechanical causality. I return to this quotation at the end of the chapter.

Evolution concerns the whole organism and its extended self-maintaining system. The *Umgebung* is virtual and collapses as soon as the organism tries to delimit it. The ambient quantities are only accessed through their relations to organisms, so they are never truly intrinsic. This perpetual becoming of both the *Umgebung* and *Umwelt* through each other displays a fundamental dialectical insight about how both include each other in their constitution. The relation is not linear: The organism does not simply uncover what is latently there in its surroundings but constructs it through

[7] We should note that their categories: relational (about changes in relations between organism and environment and between organisms themselves), constitutive (changes in constitution or phenotype), and external (their name for perturbational niche construction) are not mutually exclusive but overlap.

its activity, both in a material and epistemic sense. Since it has already changed and been changed by this relation, it makes no sense to search for a beginning or end. There is a dialectical interplay between what is called the external and the experienced environment, as the former is not given but constructed through the activities of the organism. This takes us back to habit formation:

> The animal allows the surroundings in which it lives to subsist, which transforms those surroundings already into a 'world': something unified, a space remade to suit. Accordingly the 'habit' of the animal already functions as a way of 'inhabiting' the world, and thus involves a particular relation to temporality. (Malabou 2005: 63–64)

The contradiction is that construction does not only pertain to material construction but activity in a broader sense, thus becoming more like its apparent opposite, adaptation. What is presented below is what Godfrey-Smith (1996) has called a *constructivist* explanation of adaptation. We are reminded of the *plastic reading* introduced at the beginning of Hegel's logic—a reading which actively rewrites what is interpreted. Through this process, the organism is able 'to see (what is) coming' (*voir venir*), says Malabou. The future is both visible and invisible:

> The future is not the absolutely invisible, a subject of pure transcendence objecting to any anticipation at all, to any knowledge, to any speech. Nor is the future the absolutely visible, an object clearly and absolutely foreseen. It frustrates any anticipation by its precipitation, its power to surprise. (Malabou 2005: 184)

Seeing what is coming is also not seeing, not being able to predict what will happen. It points to a future that 'is neither present to the gaze nor hidden from it' (Malabou 2005: 184). Habit makes the future more predictable but never erases its contingent becoming. It is a process of constructing a habit(at), an externalisation and alienation that enable the reverse process of internalisation or recollection as it makes the environment familiar.[8] The organism interprets underdetermined ecological affordances based on its dispositions and history of interactions with the environment. Affordances are produced through the coupling between the action possibilities of the environments and the capacities of the subject to make use of such possibilities (Walsh 2015).

As with the screwdriver, we cannot state in advance all such possibilities. Only by combining the physical forces of the environment with the action of the agent do we arrive at ecological information that the organism can make use of. From this perspective, plasticity is the power to configure the world through 'the sensible-intelligible interactions and contact of material bodies with their surrounding environment' (James 2019: 193). It is about the reciprocal giving of form between material and mind, between the neuronal and the mental; it hereby points to the fact that no living organisation is self-sufficient; finally, it is their mutual openness that enables them to transform themselves and each other (James 2022).

---

[8] When Hegel speaks of 'alienation' in this context, he simply means self-exteriorisation. We should not understand it in the properly Marxist sense, which Hegel has (somewhat misleadingly) been interpreted as promoting (Renault 2020).

The Construction of Selection

> That which is mediated is itself an essential moment of what mediates it, and each moment is the totality of what is mediated. (Hegel 2010: 624)

Underscoring my previous claim, Heras-Escribano and de Jesus (2018) state that NCT has been overly concentrated on 'the objective dynamics of organism-environment dynamics' (256), and that this approach must be complemented by a perspective that seeks to understand how these interactions become meaningful to the organisms engaged in them. They understand this in terms of sign interpretation and consider a sign—anything the organism perceives as meaningful—*a concrete actualisation of affordances*. Below, I use experiential niche construction to highlight the salience of distinguishing between potential and actual information.

Just as the MS maintains that natural selection always has an abundance of variation to work on and thus does not have to wait, the view presented here claims that the organism cannot wait for natural selection to decide what is adaptive. It would take too long to allow it to survive in the first place. The organism uses whatever might further its existence, exploiting the leeway left by genetic factors to actively increase its chances of survival—through behavioural or physiological changes that are not tied to competition but increased complexity (Sultan 2015). Each organism actualises and reproduces the same potential information in a unique manner, irrespective of source. As Wagner and Danchin (2010) write: 'a given fact can provide different information to different parties' (206).

But the question remains as to whether and how these changes in mediation affect selection. Here, the distinction between the selective niche and the developmental niche is useful. Lynn Chiu (2019) points to the work by Stotz (2017), which defines the developmental niche as 'a multi-dimensional space of environmentally induced and developmentally regulated, heritable resources that scaffold development' (Stotz 2017: 2), i.e. the part of the niche which the organism interacts with during ontogenesis. The selective niche, on the other hand, concerns the selective sources acting on the variations produced through the organism-environment interplay. In other words, developmental niche construction constrains natural selection by forming the variants it may act on.

I am not going to split hairs over this distinction; what is important for us are the cases where the difference between these forms of niche construction breaks down—where the developmental response may be said to 'create a differently experienced environment and thus a different selective environment' (Chiu 2019: 308).[9] This exemplifies the entanglement of different forms of causality, on different levels of organisation. Despite differences, selective and developmental niches overlap:

---

[9] Chiu notes that Stotz does not consider changes to the developmental cues that alter the phenotypes of the organism as part of selective niche construction. Stotz sees the selective niche as only the 'intrinsic properties of the external world' (Chiu 2019: 308) and thereby overlooks the entanglement of the developmental and selective niche.

> The external factors that determine selection can often be the same as those that influence individual development through phenotypic plasticity. Thus, the snapshot of phenotype–fitness relationships we often use to measure natural selection empirically may be a subset of much broader and highly dynamic eco evo devo processes. It follows that these relationships could be viewed as evidence for coordinated variation between dynamic intrinsic developmental factors, which affect the outward phenotype, and current ecological factors. (Skúlason et al. 2019: 1802)

This questions the notion that we can isolate natural selection from the organism it acts on. Instead, a more complex model is required. We find this in Lewontin's model of niche construction, which undermines the idea that natural selection is a form of environmental selection and thus external to the organism. Instead, he considers these processes interdependent, in the sense that one cannot take for granted that the environment is experienced in the same way: 'organisms determine which elements of the external world are put together to make their environments and what relations are among the elements that are relevant to them' (Lewontin 2000: 51). Reinterpreting the organism a subject in evolution demands that we take its experience seriously.

In this light, we understand why Chiu argues for a *causal spread* of the agents of natural selection. Here, natural selection is not the same as environmental selection. Natural selection is also caused by 'internal' agents:

> Theories of niche construction that leave out the experiential variety tend to maintain an externalist characterization of natural selection and argue that niche construction feeds into the environmental causes of natural selection. Natural selection, however, does not adapt a population to its environment when different organisms of a population construct and experience different environments. Instead, in these scenarios, the causes of selection are spread across varying organisms and their varying constructed environments. I argue that experiential niche construction helps maintain the spread of selective causes across organism and environment interactions. It thus creates the conditions for a kind of natural selection that is not "externalist." (Chiu 2019: 300)

This makes the reification of the environment as a mere driver of natural selection untenable. It hereby challenges the MS view—in which environmental sources of selection are privileged. In this perspective, even if the capacity of organisms to alter their environment is acknowledged, it is a result, not a cause. At the same time, it is only the parts of the environment that contribute to competition and predation that are relevant for selectionism. The environment cannot induce behavioural changes directly, only via natural selection, which is seen as an outside force with given selective pressures. Consequently, saying that these pressures are alterable might still leave the environment as a process acting on the organism from the outside. Even if it gives the organism some power, this power is ultimately caused by natural selection.

To be precise: While the organism may alter the environment in standard models of niche construction, the modification belongs to the environment. As Chiu emphasises, the altered 'properties are intrinsic to the environment despite having an organismal origin' (2019: 315). Meditational niche construction, however, is not intrinsic, or physically detectable in the environment, and thus not accessible to conspecifics in the same way. Also, changing location would not cancel these differences. Chiu claims that such changes would modify the *patterns of selection* without altering the selective environment. This means that niche construction and natural selection are not separate processes. Chiu does, however, not pick one or the other option but considers them complementary:

> When a decoupling strategy is appropriate, natural selection can be treated as environmental selection and niche construction as the causal contributor to the sources of selection. However, when a commingling strategy is applicable, natural selection supervenes on the varying niche constructing activities and outcomes of organisms. Niche construction is constitutive of natural selection. Mediational niche construction can help retain variation in constructed environments, and thus provides the strongest support for commingling strategies. (2019: 316)

In the latter scenario, natural selection emerges from within, not apart from niche construction. Altering the physical environment is not necessary; it is enough to change the relationship between the organism and the environment, and the former's experience of the latter. I believe this takes us closer what Kauffman means by *niche creation*, the adjacent possible space we have discussed which is enabled by the current system. Through exploring this space, we construct a new *problem domain*, with specific norms and motivations (Kauffman 2019; Werner 2024). The ability to produce such a domain relies on and reinforces autonomy; and crucially, it should not be considered be individual act.[10]

To say that selection supervenes upon the niche construction activities of organisms is to say that it depends on them, that natural selection changes as niche construction changes. As such, the causes of selection are 'distributed across organism-environment relations' (Chiu 2019: 316). (This already indicates the mediation or experience is not an individual act.) Chiu concludes that by embracing the notion of niche construction as constitutive of natural selection we may offer a clearer opposition to the MS, as the usual retort against NCT: 'we knew this already, but it is of minor importance', does not stand against the more entangled conception of causation offered in the extended version of NCT. Also, the sentence about retaining variation points to the idea that niche construction not only alters selection pressures

[10] Although I cannot develop this further here, I believe the notion of 'shared enaction' could be useful to cast light on this phenomenon, and give a fuller explanation of what is meant by experiential niche construction, which is still somewhat underdeveloped. Kauffman emphasises how organisms create niches for each other; moreover, we should explore the social dimension of experience. In an article about the emergence of institutions, Werner (2024) uses the example of a plain of grass in a park, which affords the possibility of playing football but does not specify this activity in any rigid manner. By engaging in football, using threes as goal posts, this environment is repurposed collectively, in an open-ended manner. Trees are repurposed for the game but other options would be used. He writes: The important fact is that this lawn does not provide any proper infrastructure, either for playing football or doing anything else that is clearly defined. The domain of action is therefore as undetermined as it is possible to be in real-life cases (where it is always determined to some extent). First, this implies that the game of football itself, and thereby some of the actions performed by the players, establishes a defined space' (Werner 2024: 16). A new problem domain, with behavioural norms, is produced through such choices. I think this example comes close to what is meant by mediational niche construction since not material change is required. It also comes close to 'niche creation', as there is no mechanistic causality involved; the change is instantaneous in a specific sense, even if the activity of playing and the norms are enacted over time.

but may isolate the organism from selection altogether. This highlights the importance of thinking in terms of enablement: Natural selection is not an outside force acting on individual organisms. This would reproduce the simplistic notion of two opposed processes, instead of perceiving it as an internal and complex relation.

Environment = Abstraction?

> What is true of the environment of such simple creatures as bacteria is also true of more complex lifeforms. More complex agents are likewise selectively open and differentially sensitive to an environment of potentialities and possibilities based on norms that originate in their own autonomy, in the sustaining of their own identity as individuals. More generally, the significance the environment has for agents is virtual. This significance finds its actualisation—it is enacted—through the agent's active engagement with the environment. (Kiverstein and Rietveld 2018: 151)

It is not wrong to say that organisms constrain natural selection, but we must keep in mind that constraining is also enabling. Walsh (2022) stresses that the environment is not a discrete and autonomous force that determines what counts as an adaptation. Instead, adaptation is an evolutionary response to ecological affordances. In this view, the environment fails to rigidly specify the affordances that organisms perceive and act on. Hence, the environment, qua external and autonomous force, cannot individuate and explain adaptations. This shift requires a more circumspect way of studying the environment, as unique to each individual (Abrams 2009):

> The simple point is that the conditions to which adaptive evolution molds form are not autonomous from organismal form itself. They are constituted of the organism's adaptation to, construction of, and transduction of causal influences distributed throughout the organism/environment system [...]. To be sure, there is an environment, and it is external to organisms, but the intrinsic properties of the external environment taken in isolation cannot adequately account for the adaptedness of organisms. (Walsh 2022: 74)

Since the environment underdetermines adaptations, 'the concept of being an adaptation is not the concept of being suited to the conditions of one's external autonomous environment' (Walsh 2022: 75). The environment underdetermines adaptation because it cannot decide how the organism is going to experience its inputs. The problem for generic models of adaptation as a property of populations, then, is how to translate them into the concrete adaptive capacities of individuals. The solution proffered by people like Fisher is simply to assume that the members of a population are already adapted since they would not be present in the population if they were not. The problem is that this does nothing to explain *why* organisms are adapted; it only provides a semblance of an explanation.[11] As Lewontin (1974) writes:

---

[11] It seems to commit *the ecological fallacy*, where individuals and their traits are deduced from statistical data at a group level. Here, it would be deducing the individual organism from the populational level. To combat this fallacy, the deductions must be confirmed or disproven by studying individual behaviour (see Trappes et al. 2022).

> To concentrate only on genetic change, without attempting to relate it to the kinds of physiological, morphogenetic, and behavioral evolution that are manifest in the fossil record and the diversity of extant organisms and communities, is to forget entirely what it is we are trying to explain in the first place. (23)

If you want to explain the specific adaptivity of a population, it is not enough to look at what is common to them. You must specify how they differ and why. Walsh (2022) calls this the Anna Karenina problem: '*All adaptive populations are alike; every adapted population is adapted in its own way*'. This problem cannot be solved by generic models but requires concrete examination. An upshot of the affordance view, however, is that every material change to the environment also involves an experiential change, but that this relation is not transitive since there are changes in the experiences without material changes (Walsh 2015). In other words, experience is the process through which the organism internalises the constraints that it is also responsible for shaping. Constraints arise from and shape organismic activities, which is to say that they are historical and relational. Compare to Tahar (2022):

> They [constraints] arise from the evolutionary process as it is shaped by the activities and practices of living beings, but they also channel this process, and even generate its future possibilities, *through internalisation, i.e.,* the way in which living beings experience these norms and actualise them through practices in the specificity of their unique situation' (19, emphasis original).

Constraints gain normative causal power through the agency of the organism. The process of internalisation is not psychological. It only demands that organisms experience and regulate their actions in accordance with the affordances made available through their activities. It might have disadvantageous consequences, like blurring distinctions between physical and experienced environment, and not allowing for any non-experienced physical changes. From our perspective, this seems inadmissible, as it creates the impression that external nature is completely internalised or overcome by the organism, leaving us in subjective idealism, where self and world are indistinguishable. It would mean that the problem of nature is done away with, and all we have left is an idealised notion of nature, where everything is mediated. Hence, it reduces nature to lifeless matter.

The sceptical retort to the very notion of experiential niche construction is that organismic responses to selection pressures should not be confused with changes in the environment itself. If conflated, the constructive capacity of organisms is overstated (Godfrey-Smith 1996, 2009). While Baedke et al. (2021) do not share this general scepticism, they argue that it is still 'unclear how shifts in organisms' experiences can change selection pressures acting on them or establish individualized niches' (48). They say that this lack of clarity stems from the view that organism and environment are viewed as inextricable (a view they attribute to Walsh 2015; Sultan 2015; Levins and Lewontin 1985, among others). For analytical purposes, it seems we need the notion of the external environment, but we should not take it to mean that it is absolutely outside the organism. There is no clear distinction between the organism and the environment even if they are autonomous. The negativity or underdetermination of the external environment opens the space for the organism to act on and make its environment familiar (Heron 2021).

Again, authors such as Sultan (2015) acknowledge that there is an external environment apart from the experienced one. But the relationship is not grasped by a quantitative approach. She proposes that one may study the difference between the ambient light conditions (photon flux density per square metre) and compare this to the total photons captured by a plant. Comparing plants, then, would allow one to study the differences in access to light. The same could be done for temperature. So, while this perspective troubles simple models, it does not make research impossible. In Sultan's view, this is a call for realism, for overhauling the artificiality in much research done in lab contexts. It demands that we view the external environment as an abstraction, a necessary heuristic but 'not an indispensable feature of evolutionary metaphysics' (Walsh 2022: 78).

The physical environment can be kept in view as a 'reference environment', an abstraction which does not need to be stable, constant, or autonomous. It simply provides a coarse-grained approximation of the physical environment, a proxy for affordances which make models more tractable (Walsh 2022). This move does not eradicate all difficulties, however, as the notion that the physical environment is a mere abstraction could lead to the view, tied to Uexküll, which says that the whole environment is 'nothing but the experienced environment' (Baedke et al. 2021). In this view, there is an 'internal creation of a perceptual cue, which is then attributed to the environment' (Froese 2022: 9). Thus, we would simply replace externalism with internalism instead of trying to understand how inside and outside relate. How to overcome this dilemma?

The Materiality of Experience

> An important element in epigenetic factors in fact derives from the environment, the outside, and, as we shall see with brain epigenesis, learning, the milieu, habit, in a word, experience. (Malabou 2016: 82)

Above, Malabou underscores the importance of the environment in the construction of the organism, and ties this directly to experience. This seems to imply a broad conception of epigenesis, between organism and environment, not just within the organism. It involves organismic agency, e.g. 'the system's capacity to transduce, configure, and respond to the conditions it encounters' (Sultan et al. 2022: 5). One way of understanding it is ontological co-constitution: organism and environment are commingled and 'form a single interacting system that cannot be meaningfully disentangled' (Baedke et al. 2021: 2). From this, the authors identify two problems with the concept of experiential niche: (1) the lack of epistemic boundaries and (2) the lack of integration between experiential and physical kinds of reciprocal causation. Importantly, the point is not to establish a particular boundary but establishing 'a way of demarcating biological individuals from their surroundings by clarifying the causal structure of each interacting component' (Baedke et al. 2021: fn5, 13). I do not presume to conclude the debate here, merely to indicate the stakes at hand, and how they can be informed by a dialectical approach.

But I believe that the distinction between closure of constraints and closure of *processes* points in the direction of a solution; the external environment would be those processes which are subject to closure of processes but not closure of constraints. I did not highlight this distinction previously but it is useful for our present purposes: while closure of constraints demands that a given constrain is also enabled by the system it constrains, many processes would either enable or constrain but not vice versa. If so, they would be part of closure of processes, but not organisational closure.[12] Again, this distinguishes biological systems from dissipative systems; namely, that only the former refers to the conditions of existence of the components involved.

In this way, we would have a dynamic way of distinguishing what should count as inside and outside by judging which processes both enable *and* constrain each other at any given time. Those that are reciprocally entangled in such a way would count as part of the organisational closure, and hence be *inside* in this specific sense (which does not entail they are physically located inside the organism). Those that only fulfil one of these roles at any given moment would only be part of the operational closure.

This allows us to define closure in a way that does not erect a rigid border between organism and environment but considers the border plastic. Organisational closure may extend outside the physical borders of the organism; moreover, an ecosystem may also be subject to closure of constraints (cf. footnote 2). Such an understanding also underscores the continuity between questions in philosophy of biology and philosophy of mind which I have hinted at.

Similarly, the eco-evo-devo perspective proposed by Sultan, Gilbert, and Chiu makes niche construction not about two opposed causal chains operating in opposite directions but into a more complex process of co-constitution. There is no niche *as such*, apart from its relation to organismic activity. This shift from dynamic interaction to dynamic construction entails that not only is a niche modified materially and then (subsequently) the changed niche harbours new pressures, but that the pressures

[12] As Leonardo Bich states: 'It [closure of constraints] focuses on the distinctive capability of living systems to contribute to their own conditions of existence. It establishes a generative dependence between components through a closed topology of transformation processes in which internally produced components constitute a subset of the conditions of existence for those processes. However, this circularity at the level of constraints should not be confused with any cycle of activities at the level of processes. Cycles are captured by a different circularity that is also known as *closure of processes* or *operational closure*, which stands for the recursion between the operations of the components of a system: a closed network of operations in which all the actions of the components have an effect inside the system. [...] The circularity realized by closure of constraints encompasses not only the activities of the components but their conditions of existence, provided by their participation in the organized system they continuously realize. While the notion of cycle, or operational closure, says nothing about the origin of the components, organizational closure points to their internal generation as well as to the properties they need to satisfy in order to contribute to self-production, that is, to be able to participate in processes of production—transformation and degradation—of other components' (2025: 21).

themselves are processes that are constantly modified, either in material or mediational terms. There are not two distinct processes running in opposite directions but a complex nexus of processes of niche construction and selection, where selection also pertains to how *the organism selects how natural selection affects them* (Uller and Helanterä 2019).

The evo-devo perspective explains evolution through inherited changes in development. But it is not enough to say that development produces the variation on which selection acts and thereby constrains it:

> Evolution and development are [...] integrated not only in the production of new variants upon which natural selection can act; they are also integrated in the formation of new niches and the integration of the organism into the habitat it has helped to create. (Chiu and Gilbert 2020: 475)

This underscores why separating the selective and developmental niche is difficult in practice. It also underlines why evo-devo requires the prefix 'eco'. Without it, the ecological dimension of co-construction between species is missed, as is the notion that environmental changes may induce phenotypic changes at different levels. As integrated (holobiont) systems, organisms self-maintain through 'instances of reciprocal scaffolding of developmental processes and mutual construction of developmental, ecological, and evolutionary niches' (Chiu and Gilbert 2015: 191). This indicates why perturbation and mediation are two sides of the same coin. While either may initiate the process of construction, their relation becomes recursive, which makes it (almost) impossible to isolate one from the other.

We touched upon the interspecies dimension in relation to the extended phenotype, as an add-on to organisms formed beforehand, an extension of a host phenotype. But this model only concerns the environmental effects of genes (Trappes et al. 2022). Chiu and Gilbert (2020) summarise my point:

> It may be tempting to think of symbiont microbiota as building blocks that supply the animal with extended phenotypes and expanded niches. On this view, microorganisms harbor specific, functional traits that are recruited and added to the host organism. However, the transition from carnivory to herbivory is not just the gradual addition and removal of adaptive traits. It is also a dramatic shift in the significance and relevancy of the environments the microorganisms and host organisms find themselves in, mediated by their plasticity and abilities to construct their environments. (461)

Inter-species relationships change the organisms involved. These evolutionary transitions, explosions, or symmetry breakings are also changes in the significance of the environment since it alters the organism-environment relation. Such entanglements are central to understanding plasticity and habit and take us from the opposition between automatism and novelty to 'the continuity between *biological autonomy* and *sense-making*' (Ramírez-Vizcaya and Froese 2019: 4). This points both to life as the ground of cognition but also to a metaplastic organisation of habits.

The fact that niches cannot be isolated from organisms is another version of the problem of pre-defining the possibilities of a state space. The possibilities of a niche are produced in and through the mediation and material construction of the niche by specific organisms. This suggests that the organism's experience of its niche has causal import. By way of habit, we may call it the precondition of adaptation. Malabou underscores the interplay of activity and passivity involved in this process:

> The living being is at once identical to and different from its non-living origins and surroundings. Because sameness and alterity are related, habit becomes a condition of adaptation in a double sense, being a form of *contemplation*—absorbing the environment, passively lending itself to what is given—and a kind of *exercise*, informing and transforming the surroundings, appropriating the given conditions for its organic functions. (Malabou 2005: 60)

Experience not only shapes the organism through the reception of form but also involves resistance, which shapes this reception itself. Malabou highlights the duality involved in the process: on one side, the organism must expose itself to external factors such as light stimulus. This is the contemplative part, a kind of submission (Malabou 2005). The other dimension is the exercise of habit on the stimulus, a capacity which is acquired through the act of submission. Through a process of training, the organism familiarises itself with its environment and thus negates and mediates its content. This provides it with increasing leverage: 'What was initially only submission becomes, with repetition, the power to initiate movement. The blunting of sensation has as its correlative an increase in judgement and discrimination' (Malabou 2005: 63). This increase in the interpretive capacity makes the organism 'a being who invents', who self-actualises through its other. Thus, form is not externally imposed on the content but part of its construction (Žižek 2012).

The mechanistic repetition of habits is the premise for anticipation and creation. Linking habit to construction, we move beyond the duality of passivity and activity. A quantitative and passive approach misses how the organism integrates and mediates the information obtained from the environment, while its opposite fails to account for the constraints from which qualities emerge. From this perspective, we understand why focusing on the most evident cases of niche construction is misleading. Taking the most dramatic examples, such as beaver dams and coral reefs, as representative of niche construction as such, has led researchers to overlook more subtle changes in the organism-environment relationship, changes that do not necessarily involve changes to the external environment (Sultan 2015).[13] By overlooking such changes, they promulgated an externalist definition, which also gives the impression that conscious intent is needed for niche construction to occur.

From Intrinsic to Relational

> Natural selection is not a consequence of how well the organism solves a set of fixed problems posed by the environment; on the contrary, the environment and the organism actively codetermine each other. (Levins and Lewontin 1985: 89)

[13] This is not to say that such changes in the experience do not affect the niche in absolute terms. Sultan (2015) emphasises that the mediation of the external niche eventually feeds back to change external conditions in absolute terms.

While the term 'experiential niche construction' was coined by Sultan (2015), Lewontin already highlighted the way organisms make use of their reservoir of possibilities to cope with their environment, thus filtering away some while making other factors relevant. Chiu (2019) remarks that since many failed to make sense of this notion, they have discarded it:

> They thus opt to identify constructed environments as the intrinsic properties of the external world, albeit those relevant to the organism. An evolutionary theory of niche construction concerns the evolutionary significance of constructed, intrinsically defined environments. (299)

Others have warned of the danger of bloating, in which 'nearly every activity of an organism [...] causes a change in the relational properties of organism and environment' (Biasetti 2020: 289). This charge is somewhat peculiar, as the argument that something is ubiquitous (and therefore trivial) could just as well be levelled against the other sources of evolution, like natural selection and drift (Aaby and Ramsey 2019). Chiu combats this charge by stressing that not all situations are the same. She thereby reveals what is missing in such critiques: By taking a one-size-fits-all approach which treats niche construction and natural selection as distinct processes, it is no wonder that one would not accept something like experiential niche construction. The sceptics find this notion unacceptable because they believe that Chiu and others' views reduce all intrinsic properties of the environment to relational properties and that this is the case in all situations. Yet, they seem not to find a reduction of relational properties to intrinsic ones equally troubling.

But as we have seen, Chiu operates with different modes. In the decoupled mode, niche construction contributes to the environmental sources of natural selection. The two processes are thus dissociated and only externally related. In the other mode, by contrast, commingling, 'niche construction *constitutes* the conditions of natural selection' (Chiu 2019: 300). In the latter case, natural selection is neither external nor autonomous to niche construction. Here, a quantitative, material, change of the actual niche is not needed for a change in the organism-environment interplay. A qualitative change can be abrupt and acausal. It enables new actuals but does not cause them (at least not in a mechanical sense, see below). This underscores that the mistake made in some models of niche construction and selection is perceiving both as (efficient) causal processes, instead of in terms of enablement.

While experience produces causal changes, as the organism changes its behaviour or modifies its inner milieu over time, the initial change in the relationship that makes some options visible and possible for the organism is not causal in a linear sense. The change of what is possible is instantaneous and knows no simple locus. There is no exchange of energy, no impact between objects, just changes in the enablement structure. Yet, as these changes guides the actions of the organism, they are not without material effects, be it for the organism or the environment. There are downstream physical effects, but these are not at the same time interval as constitutive and experiential changes.

To further comprehend this dialectical model of niche construction, I return to Hegel to see how his analysis of the concept as embodied in the activity of organisms might shed light on our discussion. This is particularly evident in his notion of teleology—how organisms make the environment intelligible through purposive activity.

Part II: The Translation of the Concept

> Hegel repeatedly insists on this feature of the concept: although it is a logical form, the concept must not be considered like an empty receptacle, rather as a power that can fashion its own content. (Malabou 2005: 5)

Objective conditions are the preconditions of our activity. To paraphrase Marx, we are confined within constraints we cannot decide, but we are able to posit them as our own. Therefore, as Longo and Montévil (2014) highlight, we cannot understand organismic form based on physical constraints in isolation—only through the way organisms regulate and integrate constraints into their own functioning. As noted, the experiential dimension of this process was first emphasised by Lewontin (2000). In the following sections, I bring these views into dialogue with Hegel's notion of the concept as embodied in the teleological action of the organism. Malabou (2005) calls it a schematisation of reality, and the process whereby the concept unifies itself with external reality and thus becomes an *idea*. The process of actualising the concept gives it 'an objective existence in which internally purposive form is manifest' (Ng 2020).

I aim to show how Hegel anticipates the notion that we not only alter our environments materially but also cognitively, and how this does not imply that it is something only happening inside the organism, but that cognitive constructions are always tied to and shaped by objective reality. There is nothing in this understanding that denies that material changes are also involved. These two modes are dependent yet autonomous from each other, and their interaction is complex and unpredictable.

We have discussed how the organism is a negative unity, wherein the parts are distinct but inextricable, in that they only exist as distinct *through* each other. They are never positively given as stable things, but as part of an ongoing and negative process. This explains why Hegel also calls it an infinite unity, an 'infinite self-stimulating and self-sustaining process' (Hegel 2004; Ferrini 2011). It is a unity only through the other and is thus estranged in the sense that it is extended outwards and lack stable boundaries. Hegel thus emphasises the reciprocal interdependencies between organism and environment.[14] He writes that the notion of self-sufficient shapes is a 'suppression of what differentiating is in itself, namely, not to be in-itself

[14] I should note that Hegel does not regard 'reciprocity' as the highest point in his *Logic*. It is a stage on the way to Absolute Knowing, which is only achieved at the level of human beings. But his does not take anything away from the argument I am making here since reciprocity is a crucial step on this path (see Radnik 2024).

and to have no stable existence' (Hegel 2018: 108). Living systems diverge from inorganic nature, which are self-identical since they do not rely on anything else for their existence. The processes that enable life divide into groupings which are ceaselessly overturned and re-assembled. It is, as Malabou emphasises, both taking and giving of form, as much a sublation as a differentiation into groupings, which is also a negation of former groupings. Hegel writes:

> The whole cycle constitutes life. It is neither what is first expressed, namely, the immediate continuity and unmixed character of its essence, nor is it the stably existing shape and what is "the discrete" existing for itself, nor is it the pure process of all of this, nor again is it the simple gathering together of these moments. Rather, it is the whole developing itself, then dissolving its development, and, in this movement, being the simple self-sustaining whole. (Hegel 2018: 106)

A whole that develops itself through its own negation *is* life, in Hegel's account. This view betrays the close relationship between organisation and life. As we can see, Hegel enumerates several of the steps involved, from the immediate continuity to the gathering of these moments and concludes that life is not *in* any of these steps by themselves but in the movement of the whole. Likewise, life is not in specific processes or constraints but in their reciprocal determination. This explains why not all aspects of the physical environment are relevant to the living organism. It actualises a small subset of all the possible processes that may affect it, based on its current state and the ongoing process of producing itself as an infinite unity. Hence, Hegel anticipates Uexküll's distinction between specific environment and general surroundings, as well as 'recent conceptions of habitat and ecological niche' (Westphal 2020: 226; Ferrini 2010). Based on its organisation and functional needs, only subsections of the physical environment are relevant to the organism.

In Karen Ng's view, purposiveness provides an anchor for understanding what Hegel means by the *concept*.[15] Arguably, his whole logic is invested in rethinking what we mean by conceptuality, to trace it back to the activity of living organisms and how they self-determine themselves by being driven outside themselves. This points to a fundamental similarity between Hegel and pragmatism, as 'theoretical reason is rooted in practical reason, and [...] practical reason is rooted in our bodily interactions with our natural (and social) surroundings' (Westphal 2015: 173). As we shall see, the concept speaks to the mediation of externality by a process of concretisation through judgements. Also, the truth of substance is expressed through the concept: First as subjective essence, external to what it conceptualises. Then, by externalising itself, the concept gains objectivity. The formal concept makes itself its subject matter, says Hegel. By externalising itself it 'becomes a creator of nature' (Hegel 2010: 523). Simultaneously, any conceptualisation of nature deals with concepts already actualised as *ideas*. The concept 'is free inasmuch as in this real world, in its objectivity, it recognizes its subjectivity, and in this subjectivity recognizes

[15] I use 'concept' instead of 'notion' in correspondence with recent translations. Also, I do not capitalise the 'C' in 'concept', as it creates the impression that it is a technical notion, having to do with universal categories, which I believe it is not. Instead, it is tied to the practical activity of organisms and thus is not static.

that objective world' (Hegel 2010: 527). There is no pure world on one side and organism on the other since they are always already shaped by their relation. Its disorganisation—or lack of interiorised conceptuality—both opens it to organismic determination *and* imposes limits on this determination (Furlotte 2018).

Translating the concept, then, means trying to apply it to nature but finding that any such application fails because logical categories are too rigid to capture nature in its contingency. Yet, our categories nonetheless apply to nature since the very logical categories we apply emerge from an interaction with nature. They apply in a concrete, not an abstract way. Translation is therefore moving from the abstract to the concrete and back again. The weakness that resists conceptualisation is mirrored by the subject that emerges from nature. As such, we relate to our own weakness through nature.

Hegel arrives at the idea as the actualisation of the concept through Kant's notion of *inner purposiveness*. Ng notes how 'Hegel claims that to grasp living organization and inner purposiveness is to grasp 'the *concrete* Idea', and that purpose (*Zweck*) is 'the active Concept''. This means that the power of the concept cannot be understood except through 'the organization and activity of life' (Ng 2021). Both the concept and idea are entangled with purposive activity. The question is not whether purposiveness is a mere subjective projection or present in the objects themselves, as it was for Kant, but whether it is a true principle—which means that it applies both to the subject and the object. In other words, 'what matters is that the concept of purposiveness reflects the truth about the organization of nature' (Ng 2021: 459).[16]

Purposive behaviour is how the concept becomes for-itself, concretised as what Hegel calls 'free existence', through which it gains objective reality:

> The self-determining activity of the Concept that expresses what is actual is thus neither an autonomous rational force that determines reality without resistance and opposition from what is really possible and the things themselves, nor however is it a merely formal capacity, empty by itself without a given manifold of content, that is unable to determine things beyond a very minimal necessity and is incessantly haunted by the threat of empirical chaos and external contingency. (Ng 2009: 168–9)

In other words, concepts are not merely possible and external to the things they order, but *actual*—capable of acting in the world. Actual purpose is 'a concept that determines external objectivity' (Bordignon 2020: 70). It enables the organism to experience its environment as determinate and thereby opens the space for questioning nature's intelligibility. It has the same content as mechanism, but the content

[16] Similarly, Illetterati (2022) writes: 'Thought is rather something that, even though immanent to him, still transcends the dimension of the subject. Similarly, in developing *a non-objectivist conception of reality*, Hegel articulates an idea of reality in which it is not simply the other of thought, something that exists in opposition to and in front of thought, but is rather a reality permeated by thought, which is the condition of possibility for its intelligibility. What the entire Hegelian systematic project aims to demonstrate is that thought is what constitutes reality itself, in other words, something that is, of course, in the subject (in the one who thinks), but which is not, however, simply subjective; in this sense, reality, far from being what takes shape *in front* of subject's gaze, is rather the terrain within which both the world and the subject exist, that is, it is the space within which gain sense both the world's mode of being, in which the subject acts, and the very subject's mode of being' (144).

is no longer accidental or external, as in the case of mechanical collisions. It is intrinsic to the maintenance of organisms, sublated into their functioning.

The Objective Reality of Self-Determination

> [L]iving and spiritual activity can interrupt the flow of causes and effects even while remaining dependent on the totality of conditions that provide a context of action. (Ng 2020: 159)

The essential element of teleology is form. The form-activity of the organism negates determinate content to produce new content (determinate negation), in a process of differentiation. It is also a principle of self-determination that 'is still affected by externality as such and has an objective world over against it to which it refers' (Hegel 2010: 656). It is not the case, then, that the concept determines the external world at will, which would leave us in subjective idealism. As Frank Ruda (2022) explains, this lack of conceptuality is also nature's power over the concept, its refusal to be conceptually deduced. Therefore, we should place equal emphasis on the positing activity and the presuppositions of this positing. Material reality imposes constraints on how it can be posited, as it exhibits its own kind of rationality through its historical becoming. Hence, Hegel's approach is not a priori but questions the possibility of such an approach, as concepts are not made in a vacuum but the product of grappling with external reality.

Here, the role of contingency in Hegel's thinking is evident. He emphasises the concreteness of how the universal is embodied by organisms, as they are subject to conditions which lie outside their control. Because of the role of external factors, and their persisting contingency, the 'concept-governed structure' which nature exhibits 'can never be realized in an unadulterated, ideal way' (Stern and Weiss 2024: 28). But imperfect realisation is not the same as saying that there are no inherent principles of organisation in nature; the concept is not autonomous in nature, but also not absent, as Stern and Weiss explain. There is *reason in the world*, however imperfect. We deal with this further in the discussion of how the genus underdetermines the individual organism.[17]

The negativity of form and the necessity of content, says Zambrana (2015), establishes the historicity of intelligibility. 'Necessity of content' denotes how form needs specific content to work on to become concrete. Therefore, content is necessary:

> The advance of the logic depends on more than the movement generated by negativity; it depends on the content through which negativity finds traction as a *relation* of qualitative or quantitative opposition (in the logic of being) or retrospective determination of cause and effect (in the logic of essence). (Zambrana 2015: 127)

[17] We should note that the 'feebleness' of nature also opens up a space for multiple interpretation and opinion. Had nature perfectly instantiated the concept, had there be no contingency in nature, then interpretive freedom would be impossible.

Form needs content to work on; otherwise, there would be no way to get the process going. The contradiction between them is productive. It is similar to how science provides the content that enables philosophy to make nature intelligible. Since nature is always changing, there is no ahistorical standard against which we may judge its intelligibility. As an ongoing process, we cannot expect our models to be or remain adequate. Abstract universals are too rigid to grasp nature in its contingent becoming.

In Hegel's view, purposiveness is, first, the subjective drive to posit itself, not a force or a substance, but a *transition*—like habit formation—directed towards external reality. Moreover, since it is directed outwards, it exemplifies how self-determination is 'external to itself', actualised in reality. This is a process of positing:

> the movement of purpose can [...] be expressed as being directed at sublating its *presupposition*, that is, the immediacy of the object, and at *positing* it as determined by the concept. This negative relating to the object is equally a negative attitude towards itself, a sublating of the subjectivity of purpose. (Hegel 2010: 658)

Whereas chemical substances lose their identity through their relation to other substances, the organism maintains itself through difference. In other words, 'in the empirical relation to each other the individuality of any chemical substance *does not maintain its difference*' (Ferrini 2007: 10). Mechanical connections, on the other hand, do not mix at all, as they are externally related—not mixtures but aggregates. Purposive action differentiates the indifferent whole of mechanical or chemical nature while maintaining itself through this differentiation. It thus actualises what is only possible at the previous stages by organising them into its functioning.

Through this process, externality is reshaped by the concept, making the latter 'an objective structure endowed with the impulse to its own realisation' (Bordignon 2020: 71). It is both the principle guiding the process and its result. The internal telos, characteristic of living organisms, returns to itself through the other, mediating both the subject and the object. This kind of negation of the organism from itself, outward into indifferent externality, is a self-repulsion (*Gegenstoss*) through which the organism maintains itself, by 'sublating the subjectivity of purpose', making it objective. As an infinite unity, it extends into (but is not able to dominate) its environment to reproduce its organisation.

Furthermore, the concept is internally divided into judgements. It is not *one* thing, but a process of differentiation and return to itself. The concept only makes sense as a concrete totality of judgements.[18] Through conceptual judgements, which are evaluative or teleological, organisms differentiate themselves and their environment simultaneously. This is why the determination of the environment is also self-determination, which means that both become determinate or actual (Zambrana 2015). In other words, the subject posits itself by positing its niche. In this process, the activity of judging 'plays an irreducible role [...] as an act of self-determination

[18] This dovetails with an ecological understanding of concepts: 'Concepts and categories do not represent the world in the mind, as is generally assumed, but are a participating part of the mind-world whole. Therefore, they only occur as part of a web of meaning provided both by other concepts and by interrelated life activities. This means that concepts and categories exist only in concrete complex situations' (Gabora and Eleanor 2008: 95).

and self-constitution, an activity that is immediately manifest in the activity of life' (Ng 2020: 20). This logical concept of life is the basis upon which self-conscious organisms emerge. It concerns corporeity, relation to externality and to the species, which '*enable and constrain* the activities of self-conscious cognition' (Ng 2021: 1165, my emphasis).

As such, it provides an account of judgement and subjectivity which is not psychological but pertains to all living organisms. Hence, we need not assume consciousness or intentionality to explain such judgements, which is to say that we are dealing with a naturalistic notion (cf. Mossio and Bich 2017). As Boonstra and Slagter (2019) write: 'Purposive activity stems not from the projection of intentions, goals, or plans onto the organism. Instead, the organism's internal purposiveness is grounded in the immanent necessity of self-maintenance: internal purposiveness is self-preservation' (4). It is not a cognitive notion but follows directly from the fact that the organism must develop itself continuously by externalising its inner purposiveness (Jaeger 2024). Organisms maintains themselves only by means of this circular process, whereby they internalise (or re-collects) the constraints they are part of producing (Hegel 2010). Since ascription of function depends on organisation, we arrive at a naturalistic account of normativity.[19]

From Limit to Restriction

> For actuality (Wirklichkeit) is that which has an effect (was wirkt) and preserves itself in its otherness, whereas what is immediate is receptive for negation. (Hegel 1991: 116)

Normative agency is one of the principles of the organisational approach. It is also a part of the non-naturalistic naturalism I have examined, which tries to capture nature not as dead but as living and imbued with intrinsic normativity since it is moulded by the purposive activity of organisms (Illetterati 2023). This is nature not as something outside us but as a continuous becoming. Without life, cognition would be empty and lack determinations; moreover, life explains the drive and negativity which animates the dialectic (Ng 2021). Life is a productive ground, says Ng—engaged in a dialectical relation with what it grounds. Thus, we move from a relation of opposition to a productive contradiction: It is not that we have a destructive force that threatens life, but that destruction is internal to life itself. Likewise, the environment is not in external opposition to the organism, as this would imply that a harmonic relation between them is possible in principle. Such equilibrium is only possible for lifeless objects. Without contradiction, there is neither organism nor niche.

---

[19] There is a normative dimension involved in the activity of the organism and the organisational definition of functions which is tied to this activity, which I could not highlight here. See Corti (2022) for a discussion of the normativity involved in the organisational approach through a Hegelian lens. See Bich (2024) for a discussion of the role of regulation to explain organisational teleology, which I believe point to a solution to Corti's critique.

To explain this, let us rephrase the notion of 'reference environment' in Hegelian terms: As abstract objectivity—quantitative surroundings—the environment is a nullity for the organism. It is experienced simply as a generic limit (*Grenze*) to its own activity. This is the environment as external or abstract. [But there is no pure abstraction, as the environment is never pristine. Relate to abstract and actualised possibility?] The limit impels the organism to go beyond itself and to make the generic limit into a specific other; when the organism goes beyond itself, it turns the limit into a *restriction* (*Schranke*), that it wants to assimilate. Through the process of moving outside itself, it experiences itself as limited. 'In order for the limit (*Grenze*) that is in every something to be a restriction (*Schranke*), the something must at the same time transcend it in itself' (Hegel 2010: 104). The organism overcomes the limit only to find itself restricted by the outside. The resultant feeling of deficiency produces an *activity of deficiency*, as the organism time and again tries to overcome its restriction but cannot, as it cannot survive except through the other. Hence, the lack cannot be eliminated.

As Michelini et al. (2018) write: 'What is at stake here is not the maintenance of some static entity reaffirmed in unchanged form; rather it is a ceaseless process, an activity that constitutes the immanent contradiction of the living being' (9). The restriction is thus productive or enabling, and we see why the environment is crucial to the establishment of the organism as a processual unity.[20] The above authors argue that this reveals how autonomy has an interactive dimension, and how this interaction does not 'simply derive from internal organisation'.

Khurana (2013) argues that the organism can only act on its environment to the degree that the environment is already implicit in its self-relation in the sensation of lack. This, again, underscores why the autopoietic approach is too internalist for our purposes, and why the limit between a system and its configuration space is plastic.[21] By going from the generic limit to the specific restriction, the organism produces a world which is permeated by its own activity—by what seems subjective, the concept. This world is 'no longer an object of investigation, a merely objective world without the subjectivity of the concept, but [...] an objective world whose inner ground and actual subsistence is rather the concept' (Hegel 2010: 734). Tarrying with the objective world, the organism actualises the concept, it makes the niche it occupies into its own product by an infinite self-relation through its other.

The concept that humans seek to actualise includes previous concepts as part of their genesis. The purposive activity of animals enables the emergence of the conscious activity of humans, as there would be no consciousness without life (Ng 2020).[22] By overlooking the active character of concepts—how they '*enter into connections* and thereby sublate their *immediate* determination' (Hegel 2010: 607,

[20] The third step involved has to do with reproduction, which is when the organism seeks to unite itself with other organisms, to create a higher unity or genus. I cannot explore this here.

[21] Here, we could understand the configuration space as the closure of processes, and the system as a closure of constraints.

[22] Like humans, they do not deal with the environment as empiricism believes (i.e. as abstract universality); instead, they deal with what is concrete and self-reflexively meaningful to their survival, and thus the concept is at play.

emphasis original)—, we miss how organisms constantly overturn their immediate determination to become cause and effect of themselves. As for the question of the external versus the mediated environment, Hegel is highlighting how the organism's 'constructive instinct' (*Bildungstrieb*) extends the organism into the environment. In ways similar to affordances, the organism makes external causes into 'external potencies'. As such, it 'is subject to a different mode of being determined: one in which it is already implied' (Khurana 2013: 178). Singularity emerges when the organism relates to the generic potential of the environment and makes it a specific niche. As abstract, the environment has yet to be negated and concretised by the subject. This positing is an act of *simplification*, of exploring and actualising one part of the space of possibilities at the expense of other parts (Malabou 2005). The organism relates to the environment and changes which parts of its surroundings are relevant to it; it does not merely react to quantitative changes from a generic environment but *responds* to qualitative differences that are related to its history and current status. This determination of the environment is also a determination of itself as an individual. Hegel calls this a double transition, 'not only the transition of one determinateness into the other, but equally the transition of this other into the first, its going back into it' (Hegel 2010: 279).

The actualisation of the idea is a 'joining with itself', as Hegel underscores that the move from logic to nature is not simply expanding logic to a sphere outside itself but as an actualisation of the activity of form that operates in nature itself (Ng 2020). The move from logic to nature is not a transition because the idea is already reality. It is not simply expanding logic to a sphere outside itself but grasping how the idea can only be actualised outside itself. It reaches completion only by escaping its own notional structure. We may understand Hegel's logical system as similar to an organic system, as both open and closed—as able to maintain its organisational closure only by opening itself to its outside. We have to understand this very relationship in the terms that logic has disclosed, namely in terms of reciprocity (Nuzzo 2024).

As Žižek et al. (2022) says, this is not about domination but instead 'means that we are 'totally exposed to the meaningless contingency of natural necessity' (50). It is when the incompleteness of the absolute idea (incomplete insofar as it is only subjective) 'literally manifests as and corresponds to the incompleteness of nature' (Žižek et al. 2022: 155). Thus, the free release connotes how nature is autonomous, not possible to grasp completely in logical terms. It is an acknowledgement of how the impotence of nature is reflected in our logical categories; the free release of the idea into nature is *not* a logical transition because our categories are incapable of dominating nature just as nature fails to determine itself (Zambrana 2015).

The idea is only realised in its opposite, in the contingency of nature, which does not abide by logical categories even if they are not completely absent either. It takes us back to the topic discussed above, Hegel also speaks of how the species or genus that an individual belongs to *underdetermines its individuality*. This dovetails with the notion that the environment underdetermines what counts as adaptation. Like the genus, the environment is a context for action; also, the alienation from the genus and the environment is enabled by the genus and environment themselves, due to their lack of causal specificity. The individual cannot be derived from the genus, even if you cannot grasp the individual except as an exemplar of the genus. Hence, the individual is constructed through the contradiction with its genus; it is a power to resist the determination of the genus (Ng 2020).

Likewise, you cannot derive an organism from physical surroundings, as the organism can resist these determinations, and make them accord with its purposive activity. Before the capacity to resist such determination, the genus and environment's power is one of violence, of determining a thing from without. A physical object is dominated by its context in a way that a living being is not. In both these cases, the individual retroacts on its condition of possibility and hereby changes what they were conditions *for* by actualising them in an unprecedented way. It is not simply releasing a given possibility but constructing it. In a word, it is a concrete universal because it has the power to instantiate the universal in a multitude of ways. But this instantiation is not made at will, since contingency is never eradicated.

The imperfect realisation of the genus in the individual organism is one of the examples that Hegel gives of the impotence of nature. It reveals the gap and persisting problem: the non-identity between our concepts and reality. Thus, Hegel is not stating that nature is left behind by spirit or that concepts create nature ex nihilo, but the opposite: He is stating that the idea is never fully actualised in nature, as nature is too contingent or weak for that to occur. This makes it impossible to reduce nature to conceptual determinations (Schülein 2021).

Mediating Mediation

> The threshold is crossed when the magic leap into the "autonomy" of the neural self-relating occurs, that is, when the neural activity starts to "glide around as if out of the control of solid earth" (335)—in Hegelese, as if it were to *posit retroactively its own presuppositions;* and it is this short circuit which generates the effect of "immediacy" proper to qualia: in it, the complex dynamic network of neural mediations is "sublated/*aufgehoben*" in the simple immediacy of direct perception. The "raw" character of our immediate experience is thus the result of a complex effort of mediation; its inertia is sustained by its very opposite, the lightness of the "free thought" freely gliding in the air. (Žižek 2006: 212)

In the above quotation, Žižek touches upon the immediacy of mediation or the mediation of immediacy; how, in Hegel's thinking, any immediacy is always already mediated by the organism. Yet, the immediacy of life, which enables self-consciousness, resists mediation. It is not that every immediacy is completely mediated but that in every mediation there is a remainder of immediacy. Mediation or positing is the ongoing process of digesting this surplus (Zambrana 2015).

To take this back to the organisational view, we could formulate a similar principle by reference to entropy. While I cannot develop this fully here, I points us to a central dimension of the organisational approach which I could not thematise above: namely, that another world for organisation (or biological complexity) is *anti-entropy*: Organisation is the attempt at controlling disorder or entropy; it is an attempt at binding *free energy*—the energy which is available for physical work—which creates a surplus of entropy.[23] As the organism attempts to bind available energy, it also produces further disorder. In each cycle of biological reproduction, the persistent symmetry breakings produce variability or entropy. Longo and Montévil write that.

> biological reproduction, as morphogenesis, is intrinsically associated to variability and, thus, it produces entropy also by lack of (perfect) symmetries. By this, it induces its proper irreversibility, beyond (and in addition to) thermodynamic irreversibility. (Longo and Montévil 2014: 218)

Disorder is hereby made historical, without saying it is hereby fully mediated or no longer disorder. Yet, we might say that it is no longer pure disorder; it is mediated by the failed attempt at being controlled, made historical. In other words, the amount of available energy (*exergy*) in the environment is not given; instead, it is a product of the interaction of the organism and the environment (Schneider and Sagan 2005):

> Thus, while producing new order (anti-entropy), life, as iteration of a never identical and an always *slightly disordered* morphogenetic process, generates also entropy (disorder), by the (somewhat disordered) reproductive process itself. (Longo and Montévil 2014: 219)

This offers another way to understand the mediation of immediacy, as an imperfect reproduction and construction of living organisations—in an extended sense also involves the environment. In this view, it is not that immediacy or contingency disappears, but that it is affected by the activities of the organism, in unpredictable and irreversible ways (see Sandnes Haukedal and Mylonas 2024 for a fuller discussion of this topic).

There are, according to Ng, different forms of immediacy in Hegel's thinking, and the immediacy of the form of life is constrained by corporeity, externality, and the process of the species. As such, objective reality is 'not the immediacy of the sheer givenness, but [...] always appears as shaped by the specific constitution of one's life-form' (Ng 2020: 199).[24] If life is the ground of cognition, and life is characterised by organisation, then organisation is the ground—the mediated immediacy—upon which we can make evolution intelligible. I return to this point below.

About how the organism deals with its externality, Hegel writes: 'In its self-feeling the living being has the *certainty* of the intrinsic *nullity* of the *otherness* confronting it. Its impulse is the need to sublate this otherness and to give itself the truth of

[23] As Bailly and Longo (2009) write: anti-entropy is correlated to the formation of *multilevel, integrated and regulated organization*, and *not only* to the appearing of order corresponding to a lowering of physical entropy (92, emphasis original). This is why we should not speak of negative entropy, which is the inverse of entropy and simply cancels it. Instead, the notion of anti-entropy is more dialectical. It also points to how the attempt at controlling entropy produces its own entropy. In this way, order produces further disorder, which it can make harness later. Anti-entropy hereby introduces a notion of proper biological irreversibility.

[24] Again, I think this dovetails with the notion of entropy, as indicated by Montévil (2020): 'Couplings are far more proteiform in biology than in the standard framework of thermodynamic. In artifacts and industrial processes, let us recall that the thermodynamic couplings correspond to the processes' purpose to generate usable work. In biology, couplings' plasticity corresponds to the variability of biological functions that is intrinsic to the historical changes of biological objects' (no pagination).

this certainty' (Hegel 2010: 684, emphases original). Here, certainty is a subjective category, concerning how we believe things are and express this belief. It relies on an external relation between the judgement and the object. In this case, the certainty that the environment is a nullity is confirmed through the process of sublation, which is when the intrinsic nullity is made into a relational *something* (determinate negation). The intrinsic nullity of that which confronts the organism should be taken in the sense above: no niche apart from an organism. It makes no sense to speak of intrinsic properties (presuppositions) of an environment irrespective of the organism that occupies it (positing). Even if we may isolate an external environment from the experienced one, in its practical dealings the reference environment is not external to the organism but shaped materially through this relation.

Also, as argued earlier, possibilities do not exist prior to actuality, as if being a latent in-itself that simply unfolds. Thus, the virtual environment is also historical, in that its possibilities are canalised by its prior relations to organisms, which have biased its future possibilities (or affordances for action). It does not mean that they are non-intrinsic, purely relational, but that it is only by being posited by a living organism that such presuppositions gain actuality, viz. the ability to act.

The negativity or contradiction between the in-itself (external) and for-itself (experienced) environment provides the necessary ambiguity, the disparity, says Ng, that 'drives consciousness towards self-consciousness' (2020: 105). This disparity is apparent between the subject and the object. Through its desire and movement outwards into the world, the organism establishes its self-identity. The disparity observed in nature is also consciousness' inner difference, since it is entangled with what it is not. Again, this is an infinite self-relation through the other.[25] This also relates back to the weakness of nature, 'the endless diversity of her formations' (Hegel 2010: 536), which cannot be deduced and opens the space of intelligibility and intervention.

Through recollection, the organism appropriates its own posited externality. Ferrini (2020) says that this self-grounding activity of life 'establishes its own presuppositions in order to be what it is' (255). Experience and self-consciousness are shaped through their internal relation to externality. We here face the contradiction between the environment as independent and the environment as co-constituted with the organism from a different perspective. As seen, Hegel emphasises the feeling of deficiency peculiar to animals, related to their self-feeling. It impels the organism towards its other, through which it preserves itself. This dependence is not contingent but necessary for its self-reproduction.

As Bordignon (2020) puts it, 'the organism is identical with itself through its own process of self-differentiation' (82). This is not an immediate transposition but a process of mediation. One example is digestion. In this process, the external environment is both negated and integrated. Hegel writes:

[25] While this is a transcendental argument, about the possible conditions of experience, it does not only apply to possible experience, but to actuality. It is therefore also a phenomenological argument, and these two arguments reinforce each other, as Ng (2020) demonstrates. The latter gives content to the former, which shows the necessity of a living relation to the environment.

> This process begins with *need*, that is, the twofold moment of self-determination of the living being by which the latter posits itself as negated and thereby refers itself to an *other* than it, to the indifferent objectivity, but in this self-loss it is equally not lost, preserves itself in it and remains the identity of the self-equal concept. The living being is thereby the impulse to posit *as its own* this *world which is other than it*, to posit itself as equal to it, to sublate the world and objectify *itself*. Its self-determination has therefore the form of objective externality, and since it is at the same time self-identical, it is the absolute *contradiction*. (Hegel 2010: 684)

That '[i]ts self-determination has therefore the form of objective externality' can be read in (at least) two ways: either as a subjective projection without material reality; or, as a materialist thesis of how we become external to ourselves by reshaping our externality. When we say that the organism is identical to the environment, it is always in a mode of alienation, of externalisation and self-loss, where it gets its own message in return in an altered version. The living being is an individual due to its capacity to contain this absolute contradiction. Also, notice the term 'indifferent objectivity', which the organism differentiates through positing some possibilities instead of others. It produces positively given reality through a negation of virtual possibilities.

The absolute contradiction, that organisms determine themself as objective externality while remaining self-identical, means that they 'are in themselves the *negativity* of themselves' (Hegel 2010: 684). They are externalised as the actualisation of the concept. In other words, there is no clear distinction between a living system and its configuration space (Jaeger 2019). Nonetheless, we might still differentiate between external and experienced environments—inseparable yet distinguishable—because there is always a remainder of negativity in both. But there is no way to conclusively disentangle them. Organisms may choose to ignore certain factors or actively modify them. The relation between these environments is, however, internal, as they mutually constrain each other.

Through these activities, the organism *organises its environment* (Ferrini 2010). It organises its interaction with the environment, and this concomitant organisation of self and other is the activity that constitutes life (Renault 2013). The organism perceives its material surroundings in relation to its current state and purposes; it partakes in its own externality. This underscores why the environment is internally related to the organism, and why neither can be understood without the other. Our qualitative experience is not determined by or outside quantitative surroundings but enabled by them.

Organisms gain a foothold on objective reality because they are part of it. Even a mechanistic description depends on teleology (as indicated earlier)[26] since the way we make nature intelligible in the first place is through a primary purposive relation to the world. This living relation to the outside enables self-consciousness. It involves another kind of causality—one which is not *violent* as external (mechanical) relations are but aims to accord with them. Experience, then, is constituted through the relation between life activities and self-consciousness (Ng 2020). Perception is made possible

[26] 'But the *purposive connection* has proved to be the truth of *mechanism*' (Hegel 2010: 652). In this sense, Hegel proposes a teleological kind of essentialism, of concrete universals (Pinkard 2024).

through embodiment, which allows the organism to posit the environment as both distinct and identical to itself:

> Only by the embodiment of inner determinations does the subject get to the stage of sensing them; for before they can be sensed it is necessary that they be posited both as distinct from the subject and as identical with it; but this happens only by the externalization, by the embodiment of the inner determinations of the senser. (Hegel 2007: 78)

The overlaps with enactive models of the mind are evident, but I cannot pursue them further here. I can only mention that Varela (1999) speaks of enaction as turning an unceasing flow into a meaningful world (see Zaslawski 2018). This takes us back to canalisation and worlding. The subject deals with an immediate externality which is already posited or canalised through its living relation to it. Experience is enabled and constrained by this relation. The subject (for-itself) passes over into objectivity (in-itself) but finds that the immediate objectivity is already mediated by its own activity:

> This identity [of purpose and objectivity] is on the one hand the simple concept, and the equally *immediate* objectivity, but, on the other hand, it is just as essentially *mediation*, and it is that simple immediacy only through this mediation sublating itself as mediation. Thus the concept is essentially this: to be distinguished, as an identity existing for itself, from its *implicitly existent* objectivity, and thereby to obtain externality, but in this external totality to be the totality's self-determining identity. (Hegel 2010: 669)

In other words, the organism relates to something that it implicitly is, an immediacy that it has always already mediated, as it would not be able to grasp it unless it was somewhat familiar in advance. But Hegel is not saying that immediacy or contingency is completely swallowed by mediation, nor does he think all immediacies are the same. Instead, they place different constraints on how the subject may actualise itself. As Ng (2020) underscores, the point for Hegel is not to say that no immediacy exists, but to find the appropriate immediacy, which she claims he finds in the activity of life. Life, as teleological ground, thus replaces *being* as an inaccessible and inchoate ground that can only ever be approximated. Life is the posited ground produced by the activities of the organism.

Conclusion

> [P]lasticity characterizes a regime of systematic self-organization that is based on the ability of an organism to integrate the modifications that it experiences and to modify them in return. (Malabou 2010: 61)

The epigraph summarises the relation between plasticity and mediational niche construction, where habit is the bridge between possibility and actuality—a collapse of the virtual surroundings into a niche through a process of simplification. Malabou ties the simplification sketched above to the labour of the concept as an abbreviation of the virtual multitude experienced by the organism. Importantly, this should not be understood as a transcendental manoeuvre, since the form is not external to its content:

> The sharpening of the 'point', which is inherent in the process of dialectical simplification, does not emerge from the activity of the 'I think' as if this were something that would take its simple form and 'attach' it onto the content [...]. Rather, the sharpening to a 'point' also, like the others, results from a process of release or letting-go immanent in the objective determinations themselves. (Malabou 2005: 158–159)

This simplification is essential to speculative teleology. There is an ontological opening for the process of simplification within nature itself. Plasticity is 'the form of our world', says Malabou (2008: 38). We should take form in the sense above, a process of negativity, and see form as internal to the content it in-form: 'We find here the poetical and aesthetic force that is the fundamental, organizing attribute of plasticity: its power to configure the world' (Malabou 2008: 39).

Through this process of organisation, teleology reverses its course, says Malabou, 'in that the forms already actualized discharge their potential energy and consequently liberate future possibilities of actualization' (Malabou 2005: 166). The overlaps with what we found in Žižek and Kauffman are evident: The organism actualises the virtual possibilities that were in an ontological limbo between possibility and actuality. Žižek (2012) emphasises that virtuality pertains to every becoming which is not simply an unfolding of a pre-given totality of possibilities. In other words, to say that possibilities have a historical becoming is to say that they are virtual or unprestatable. It is a possibility that has actual effects without ever being actualised, like attractors that guide action but are never reached. As such, it reveals the absolute relation between actuality and possibility. This is the negative halo that shapes habit formation. In biology, this is called a counterfactual, a virtual point of reference for the activity of the organism—something that *would have been* had the conditions changed, which we can only access in retrospect.

At last, we can return to the quotation that I mentioned initially to show how Hegel suggests something like enablement in the section on life in his *Logic*:

> The indifference of the objective world to determinateness and hence to purpose is what constitutes its external aptitude to conform to the subject; whatever other specifications there might be in it, its mechanical determinability, the lack of the freedom of the immanent concept, constitute its impotence in preserving itself against the living being.—In so far as the object confronts the living being at first as something external and indifferent, it can affect it mechanically, but without in this way affecting it as a living thing; and in so far as it does relate to it as a living thing, it does not affect it as a cause but it rather *excites* it. Because the living being is an impulse, externality impinges upon it and penetrates it only to the extent that in principle it is already *in it*; hence the effect on the subject consists only in that the latter *finds* that the externality at its disposal *accords with it*. (Hegel 2010: 685)

There is much to unpack here: first, the impotence of the physical constraints of the environment in preserving themselves against the living organism, viz. how the weakness of nature and of the concept are two sides of the same coin. This weakness is embodied in the plasticity of the environment. Secondly, the environment may affect the organism mechanically, but insofar as it does, it does not affect it *as living*. It is affected in this way when the environment does not cause but *excites* the living organism to the degree that it is already related to its activities. The effect of the environment thus cannot be decided in isolation from the organism that is *always already* part of it. The subject interrupts the exteriority it confronts and makes it a means through which it reproduces itself. We see, then, why the task is not to understand the external world as indifferent to the subject but rather to understand how the subject is included in this objective reality, i.e. how it is both subjective and objective. This, again, sets Hegel apart from Kant who only spoke about the possible conditions of experience not actual conditions—except as things-in-themselves, outside possible experience. In the former's theory, we find a contradictory notion of a subject that is always outside itself, actualising the concept as an objective idea.

Hegel speaks of causality as reciprocity because he holds that we cannot determine anything as a cause without the effect. The cause 'refers back to the effect' since they are reciprocally entangled. Thus, there is symmetry between the cause and effect, but this does not mean that all causes are the same or have the same causal power, just that there is no way to identify something as a cause without the result it caused. Reciprocity is implicit in any notion of causality. This view dovetails with the notion that we cannot establish anything is necessary or causal except through the retrospective logic of positing (Zambrana 2015). In this sense, even mechanical causation implies reciprocity as their condition of intelligibility.

Reciprocal causation is also crucial for the model of NCT presented above. As Walsh (2022) stresses, even if you may synchronically (snapshot) isolate a cause, you cannot explain its causal dynamics except diachronically. If the effect of the environment on the organism depends on the effect of the organism on the environment, you cannot treat them independently. The conditions to which organisms adapt 'are constituted of the organism's adaptation to, construction of, and transduction of causal influences distributed throughout the organism/environment system' (Walsh 2022: 74). If not, we miss how organisms sublate external determinations, how they 'exist as posited by the activities of the subject' (Khurana 2013: 187).

This is, in my view, like saying that organisms respond to affordances by enacting judgements and not to the external environment as such. From the externalisation and internalisation of conceptuality emerges a space of intelligibility. Through the activities and interaction of current and previous organisms, an affordance landscape, to which the organism responds, is produced. Like concepts, affordances are not objective properties of the world, nor are they subjective (Chemero 2009). They constitute relational possibilities for action shaped through the entanglement of the world and agent (Kauffman and Roli 2021).

The organism embodies the concept in its purposive actions, but the environment it organises is already idealised through its prior activity (and the activity of other organisms). Affordances are ontologically actual *as virtual*. They enable but do not cause organismic action. We see, then, how concepts work in practice, as the notion of enablement makes intelligible the 'differential causation' we have spoken of, where the set of external conditions constrain and enable certain actions but do not determine them rigidly. Organisms enact the world through the concept it embodies and hereby change the enablement relations through which they maintain themselves.[27]

In this light, niche construction seems like the ultimate example of the paradoxical externality of nature, the externality of something that is intimately related to the organism. As such, it is also an expression of freedom, as 'being at one with oneself in the other', as Hegel says at several junctures. It is embodied in the way the organism 'constitutes, sustains and reproduces its unity by way of assimilating its environment and by reproducing itself (its species) in relation to an other' (Khurana 2013: 22). It also highlights plasticity's externality to itself—how it is never one but a negative unity of different parts. Here, nature is not outside the subject, nor identical to it. Through the experience of the organism, nature is made into a world, a landscape of affordances. We cannot possess it as a thing we may quantify, just as we cannot know the full impact of our dealing with it since it is riddled with contingency. This explains why Hegel conceived of the problem of nature as the ultimate enigma. It sets limits to philosophy and science because its lack of rigid principles is not mere appearance.

Philosophy must take up the findings of science and consider them as an ongoing totality. Hence, we may understand why there is no neat distinction between idealism and materialism in Hegel's thinking. Ideas do not reside within the head of an organism but are embodied in their practical engagement with the world. As such, ideas are only actualised when they go beyond the merely logical and 'is confronted with the constraints and concrete structure of reality' (Illetterati 2021: 47). The point is not that thought now applies externally to other realms; rather, it is a matter of grasping how thought is always in relation to conditions and constraints outside itself, and to see how it sublates such constraints without ever overcoming them. Illetterati summarises it succinctly: 'In becoming other than itself the absolute idea recognizes this otherness at once as a result of its own action and as something through which the idea itself is constituted: something in which it comes to know itself' (Illetterati 2021: 48). This is not a mere extension of ideas outwards but as much an acknowledgement of how ideas are produced through their constitutive interaction with externality.

In this view, animals also seek to understand the world purposefully, to produce new problems spaces and norms; even if they cannot recognise this themselves, they engage in shared enaction, and hereby mediate their environment (Werner 2024). The difference between them and humans is that we relate self-consciously to the purposes that animals embody—to the living organisation which is shaped through habitual and purposive behaviour (Hegel 2015; Pinkard 2012). How do we tie this back to what we said about mediational niche construction? For one, the distinction between action and perception breaks down. We could say that conceptual content derives from conceptual activity. By acting in the world—by actualising itself through the overcoming of its perceived limit—organisms make the world into an enabling and concrete 'restriction' for its actions. And this can be viewed as a kind of *shared*

[27] Hegel thereby defends a peculiar kind of conceptual realism, in which concepts exist in the world but in an imperfect manner (Stern and Weiss 2024).

enaction, not limited to one organism in isolation (Werner and Kiełkowicz-Werner 2022).

This underscores why we should not view the different modes of niche construction as separate but entangled. Also, since these adaptive capacities to deal with the outside world depend on the construction of a world of affordances, a constructivist reading of adaptation is required. If we do not adopt this model, it is challenging to explain how the organism may orient itself in the world and make judgements. It does attach pre-formed categories to the world externally. Rather, through the labour of the concept—through habit formation—it differentiates itself and the environment simultaneously and thus not only reacts to external cues but responds to and integrates a world that it is internally related to and is therefore imbued with normative value.

We see, then, the background for Lewontin's dialectical perspective. This perspective, as I have outlined through Hegel, makes evident why we cannot see the organism and the environment as separate processes, and why mediation and perturbation are two sides of the same coin. I also contend that grounding philosophical concepts in living processes makes them more intelligible. I hope to have rendered why experiential niche construction should not be dismissed, and how it brings out the dialectical nature of NCT more clearly, even if I could not do so comprehensively. Furthermore, while mediation is itself a kind of construction, we must keep in mind the importance of thinking in terms of negativity to avoid the notion that we can overcome the problem of nature. It might seem an anthropocentric problem, but as subjectivity cannot be restricted to humans, I believe the same principle applies to all living organisms that maintain themselves through their other. Had they overcome the problem they would be lifeless. The notion that organisms embody the concept might seem equally problematic, but I do not think it is controversial to say that they make judgements based on their current state and the environment in which they find themselves. To make judgements and act purposively *is* to embody the concept.[28] We thus understand why Malabou says that the concept is a 'power that can fashion its own content' (Malabou 2005: 5).

Finally, the double transition between organism and environment might prove itself 'of great importance for scientific method' (Hegel 2010: 279) since it elucidates their constitutive interaction. We spoke about the *middle-out approach* and how the scientist must enter where she can 'get a foothold' by available means (Harrison 1921; Noble 2006). I propose that the 'middle' Noble is seeking could be found in life as the ground of cognition. The individual has a ground that is not beyond it but, as Zambrana (2015) says, 'rather refers to the articulation of concrete conditions' (82). There is no ground outside or without the mediating activity of the organism and this articulation *is* the concept. The scientist gains a foothold at the level of life because it is the condition of intelligibility for her work. It lies between physical/

[28] We can, of course, understand this in a dual sense. An organism embodies the concept in the way that it makes judgements about the world but also in the sense that it is a concrete universal, a kind, which is how we come to explain them scientifically (see Koch 2023). These two dimensions are, however, not separate, as the organism also understand the world through how it perceives other organisms as kinds.

chemical and societal processes; it is enabled by these processes and constrains them in return. Life is not only the condition of science, but it should also be a central area of study. We are taken back to the notion that the organism is a point of departure, which reveals 'organising relations at all levels, higher and lower, course and fine, of the living structure' (Needham 1942: 656, quoted in Haraway 2004: 139). I explore this idea further in the conclusion.

Bibliography

Aaby, Bendik Hellem, and Grant Ramsey. 2019. Three Kinds of Niche Construction. *The British Journal for the Philosophy of Science* 73 (2): 351–372. https://doi.org/10.1093/bjps/axz054.

Abrams, Marshall. 2009. What Determines Biological Fitness? The Problem of the Reference Environment. *Synthese* 166 (1): 21–40. https://doi.org/10.1007/s11229-007-9255-9.

Baedke, Jan, Alejandro Fábregas-Tejeda, and Guido I. Prieto. 2021. Unknotting Reciprocal Causation between Organism and Environment. *Biology & Philosophy* 36 (5): 48. https://doi.org/10.1007/s10539-021-09815-0.

Bailly, Francis, and Giuseppe Longo. 2009. Biological Organisation and Anti-Entropy. *Journal of Biological Systems* 17 (01): 63–96. https://doi.org/10.1142/S0218339009002715.

Biasetti, Pierfrancesco. 2020. Dialectical Thinking and Science: The Case of Richard Lewontin, Dialectical Biologist. In *Natural Born Monads: On the Metaphysics of Organisms and Human Individuals*, ed. Andrea Altobrando and Pierfrancesco Biasetti, 265–292. Berlin and Boston: De Gruyter. https://doi.org/10.1515/9783110604665.

Bich, Leonardo. 2024. Organisational Teleology 2.0: Grounding Biological Purposiveness in Regulatory Control. *Ratio* 37: 1–14. https://doi.org/10.1111/rati.12405.

Bich, Leonardo. 2025. Biological Organization. In *Elements in the Philosophy of Biology*. Cambridge, UK: Cambridge University Press. https://doi.org/10.1017/9781009393959.

Bizzarri, Mariano, Oleg Naimark, José Nieto-Villar, Valeria Fedeli, and Alessandro Giuliani. 2020. Complexity in Biological Organization: Deconstruction (and Subsequent Restating) of Key Concepts. *Entropy* 22 (8). https://doi.org/10.3390/E22080885.

Boonstra, E. A., and Slagter, H. A. 2019. The Dialectics of Free Energy Minimization. *Frontiers in Systems Neuroscience*, 13. https://www.frontiersin.org/articles/, https://doi.org/10.3389/fnsys.2019.00042.

Bordignon, Michela. 2020. Teleology, Backward Causation and Contradiction. Hegel's Dialectical Account of Organic Nature. In *Natural Born Monads: On the Metaphysics of Organisms and Human Individuals*, ed. Andrea Altobrando and Pierfrancesco Biasetti, 65–88. Berlin and Boston: De Gruyter. https://doi.org/10.1515/9783110604665-004.

Chemero, Anthony. 2009. *Radical Embodied Cognitive Science*. Cambridge, MA: MIT Press.

Chiu, Lynn. 2019. Decoupling, Commingling, and the Evolutionary Significance of Experiential Niche Construction. In *Evolutionary Causation: Biological and Philosophical Reflections*, ed. Tobias Uller and Kevin N. Laland, 299–322. Cambridge, MA: The MIT Press. https://doi.org/10.7551/mitpress/11693.003.0015.

Chiu, Lynn, and Scott F. Gilbert. 2015. The Birth of the Holobiont: Multi-Species Birthing Through Mutual Scaffolding and Niche Construction. *Biosemiotics* 8 (2): 191–210. https://doi.org/10.1007/s12304-015-9232-5.

Chiu, Lynn, and Scott F. Gilbert. 2020. Niche Construction and the Transition to Herbivory: Phenotype Switching and the Organization of New Nutritional Modes. In *Phenotypic Switching: Implications in Biology and Medicine*, ed. Herbert Levine, Mohit Kumar jolly, Prakash Kulkarni, and Vidyanand Nanjundiah, 459–482. Cambridge, MA: Academic Press. https://doi.org/10.1016/B978-0-12-817996-3.00015-3.

Corti, Luca. 2022. The "Is" and the "Ought" of the Animal Organism: Hegel's Account of Biological Normativity. *History and Philosophy of the Life Sciences* 44 (2): 1–22. https://doi.org/10.1007/s40656-022-00498-8.

Crippen, Matthew. 2020. Enactive Pragmatism and Ecological Psychology. *Frontiers in Psychology* 11: 538644. https://doi.org/10.3389/fpsyg.2020.538644.

Depew, David J. 2017. Natural Selection, Adaptation, and the Recovery of Development. In *Challenging the Modern Synthesis: Adaptation, Development, and Inheritance*, ed. Philippe Huneman and Denis M. Walsh, 37–67. Oxford, UK: Oxford University Press. https://doi.org/10.1093/oso/9780199377176.003.0001.

Di Paolo, Ezequiel A., Elena Clare Cuffari, and Hanne De Jaegher. 2018. *Linguistic Bodies: The Continuity between Life and Language*. Cambridge, MA: MIT Press. https://doi.org/10.7551/mitpress/11244.001.0001.

El-Hani, Charbel N., Felipe Rebelo Gomes de Lima, and Nei de Freitas Nunes-Neto. 2024. From the Organizational Theory of Ecological Functions to a New Notion of Sustainability. In *Organization in Biology*, ed. Matteo Mossio, 285–328. Cham: Springer. https://doi.org/10.1007/978-3-031-38968-9_13.

Emmeche, Claus. 2004. Organicism and Qualitative Aspects of Self-Organization. *Revue Internationale de Philosophie* 58 (228): 205–217.

Ferrini, Cinzia. 2007. Hegel's Confrontation with the Sciences in 'Observing Reason: Notes for a Discussion.' *Bulletin of the Hegel Society of Great Britain* 28 (1–2): 1–22. https://doi.org/10.1017/S0263523200000616.

Ferrini, Cinzia. 2010. From the Physical World to the Habitat: Biocentrism in Hegel's Interrelation of Animal Subjectivity with Its Environment. In *Hegels Naturphilosophie in Der Dritten Moderne. Bestimmungen, Probleme Und Perspektiven* (Ernst-Haeckel-Haus-Studien. Monographien Zur Geschichte Der Biowissenschaften Und Medizin 13), ed. by O. Breidbach and W. Neuser, 119–135. Berlin: Verlag für Wissenschaft und Bildung.

Ferrini, Cinzia. 2011. The Transition to Organics: Hegel's Idea of Life. In *A Companion to Hegel*, ed. Stephen Houlgate and Michael Baur, 203–224. New York, NY: Blackwell. https://doi.org/10.1002/9781444397161.ch9.

Ferrini, Cinzia. 2020. Hegel's Philosophy of Natural and Human Spaces. In *The Palgrave Hegel Handbook*, ed. Marina F. Bykova and Kenneth R. Westphal, 241–266. Cham: Springer. https://doi.org/10.1007/978-3-030-26597-7_13.

Froese, Tom. 2022. Scientific Observation Is Socio-Materially Augmented Perception: Toward a Participatory Realism. *Philosophies* 7: 37. https://doi.org/10.3390/philosophies7020037.

Furlotte, W. 2018. *The Problem of Nature in Hegel's Final System*. Edinburgh: Edinburgh University Press. https://doi.org/10.3366/edinburgh/9781474435536.001.0001.

Gabora, Liane, Rosch Eleanor, and Diederik Aerts. 2008. Toward an Ecological Theory of Concepts. *Ecological Psychology* 20 (1): 84–116. https://doi.org/10.1080/10407410701766676.

Gambarotto, Andrea, and Matteo Mossio. 2022. Enactivism and the Hegelian Stance on Intrinsic Purposiveness. *Phenomenology and the Cognitive Sciences*. https://doi.org/10.1007/s11097-022-09823-9.

Godfrey-Smith, Peter. 1996. *Complexity and the Function of Mind in Nature*. Cambridge, UK: Cambridge University Press.

Godfrey-Smith, P. 2009. *Darwinian Populations and Natural Selection* (Darwinian Populations and Natural Selection). Oxford, UK: Oxford University Press. https://doi.org/10.1093/acprof:osobl/9780199552047.001.0001.

Haraway, D. J. 2004. *Crystals, Fabrics, and Fields: Metaphors that Shape Embryos*. Berkeley, CA: North Atlantic Books.

Harrison, R. G. 1921. On Relations of Symmetry in Transplanted Limbs. *Journal of Experimental Zoology* 32 (1): 1–136. https://doi.org/10.1002/jez.1400320102.

Hegel, Georg Wilhelm Friedrich. 2004. *Philosophy of Nature: Part Two of the Encyclopaedia of the Philosophical Sciences*, With Foreword by J. N. Findlay, translated by A. V. Miller. Oxford, UK: Oxford University Press.

Hegel, Georg Wilhelm Friedrich. 2007. *Hegel's Philosophy of Mind Translated from the 1830 Edition, Together with the Zusätze*, ed. Michael Inwood, translated W. Wallace and A. V. Miller. Oxford, UK: Clarendon Press.

Hegel, Georg Wilhelm Friedrich. 2010. In *The Science of Logic*, ed. George Di Giovanni. Cambridge, UK: Cambridge University Press.

Hegel, Georg Wilhelm Friedrich. 2015. In *Encyclopedia of the Philosophical Sciences in Basic Outline. Part I: Science of Logic* (Cambridge Hegel Translations), ed. Klaus Brinkmann and Daniel O. Dahlstrom. Cambridge, UK: Cambridge University Press.

Hegel, Georg Wilhelm Friedrich. 2018. *The Phenomenology of Spirit*. Cambridge, UK: Cambridge University Press.

Hegel, Georg Wilhelm Friedrich. 1991. *Elements of the Philosophy of Right, Cambridge Texts in the History of Political Thought*. Cambridge, UK: Cambridge University Press.

Heras-Escribano, Manuel, and Paulo de Jesus. 2018. Biosemiotics, the Extended Synthesis, and Ecological Information: Making Sense of the Organism-Environment Relation at the Cognitive Level. *Biosemiotics* 11 (2): 245–262.

Heras-Escribano, Manuel, and Cristian Saborido. 2023. Affordances and Organizational Functions. *Biology and Philosophy* 38 (1). https://doi.org/10.1007/s10539-023-09891-4.

Heron, Kai. 2021. Dialectical Materialisms, Metabolic Rifts and the Climate Crisis: A Lacanian/Hegelian Perspective. *Science & Society* 85 (4): 501–526. https://doi.org/10.1521/siso.2021.85.4.501.

Hutchinson, G. Evelyn. 1957. Concluding Remarks. *Cold Spring Harbor Symposia on Quantitative Biology* 22: 415–427.

Illetterati, Luca. 2021. Philosophy as the Science of Freedom. In *Hegel's Encyclopedic System* (Routledge Studies in Nineteenth-Century Philosophy), ed. Sebastian Stein and Joshua Wretzel. New York and London: Routledge.

Illetterati, Luca. 2022. The Action of the Absolute Idea. On the Relationship between Thought and Reality in Hegel's Philosophy. *Revista Estudos Hegelianos* 19 (34): 125–165.

Illetterati, Luca. 2023. Beyond a Naturalistic Conception of Nature: Nature and Life in Hegel's Early Writings. In *Nature and Naturalism in Classical German Philosophy*, ed. Luca Corti and Johannes-Georg Schülein, 187–208. New York and London: Routledge.

Jaeger, J. 2019. Dynamic Structures in Evo-devo: From Morphogenetic Fields to Evolving Organisms. In *Perspectives on Evolutionary and Developmental Biology*, ed. G. Fusco, Padova: Padova University Press. https://doi.org/10.31219/osf.io/mprc8.

Jaeger, J. 2024. The Fourth Perspective: Evolution and Organismal Agency. In *Organization in Biology*, ed. M. Mossio, 159–186. Cham: Springer International Publishing. https://doi.org/10.1007/978-3-031-38968-9_8.

James, I. 2019. *The Technique of Thought : Nancy, Laruelle, Malabou, and Stiegler After Naturalism*. University of Minnesota Press.

James, Ian. 2022. Introduction. In *Plasticity: The Promise of Explosion*, ed. Tyler Williams. Edinburgh: Edinburgh University Press.

Kauffman, Stuart. 2019. *A World beyond Physics: The Emergence and Evolution of Life*. Oxford University Press.

Kauffman, Stuart, and Andrea Roli. 2021. The World Is Not a Theorem. *Entropy*. https://doi.org/10.3390/e23111467.

Kiverstein, Julian D., and Erik Rietveld. 2018. Reconceiving Representation-Hungry Cognition: An Ecological-Enactive Proposal. *Adaptive Behavior* 26 (4): 147–163. https://doi.org/10.1177/1059712318772778.

Khurana, T., ed. 2013. *The Freedom of Life: Hegelian Perspectives*. Berlin: August Verlag.

Koch, Karen. 2023. Mechanism, External Purposiveness, and Object Individuation: From Mechanism to Teleology in Hegel's Science of Logic. *Hegel Bulletin* 44 (1): 148–170. https://doi.org/10.1017/hgl.2022.44.

Levins, R., and R. Lewontin. 1985. *The Dialectical Biologist*. Cambridge, MA: Harvard University Press.

Lewontin, Richard. 1974. *The Genetic Basis of Evolutionary Change* (Biological Series). New York, NY: Columbia University Press.
Lewontin, Richard. 1983. The Corpse in the Elevator. *The New York Review of Books*. https://www.nybooks.com/articles/1983/01/20/the-corpse-in-the-elevator/.
Lewontin, Richard. 2000. *The Triple Helix: Gene, Organism, and Environment*. Cambridge, MA: Harvard University Press.
Lewontin, Richard, and Richard Levins. 2007. *Biology under the Influence: Dialectical Essays on Ecology, Agriculture, and Health*. New York, NY: Monthly Review Press.
Longo, Giuseppe, and Maël Montévil. 2014. *Perspectives on Organisms: Biological Time, Symmetries and Singularities*. Berlin and Heidelberg: Springer. https://doi.org/10.1007/978-3-642-35938-5.
Malabou, Catherine. 2005. *The Future of Hegel: Plasticity, Temporality and Dialectic, The Future of Hegel: Plasticity, Temporality and Dialectic*. London and New York: Routledge. https://doi.org/10.4324/9780203489338.
Malabou, Catherine. 2008. In *What Should We Do with Our Brain?* ed. Elaine Altman, Mary Boger Boger, Eric Holtzman, Lisa Maya Knauer, Eli Messinger, et al. New York, NY: Fordham University Press.
Malabou, Catherine. 2010. *Plasticity at the Dusk of Writing: Dialectic, Destruction, Deconstruction*. New York, NY: Columbia University Press.
Malabou, C. 2016. *Before Tomorrow: Epigenesis and Rationality*. Cambridge, UK: Polity Press.
Michelini, Francesca, Matthias Wunsch, and Dirk Stederoth. 2018. Philosophy of Nature and Organism's Autonomy: On Hegel, Plessner and Jonas' Theories of Living Beings. *History and Philosophy of the Life Sciences* 40 (3): 1–27.
Montévil, M. 2020. Historicity at the heart of biology. *Theory in Biosciences*. https://doi.org/10.1007/s12064-020-00320-8.
Mossio, M., and L. Bich. 2017. What makes biological organisation teleological? *Synthese* 194 (4): 1089–1114. https://doi.org/10.1007/s11229-014-0594-z.
Needham, J. 1942. *Biochemistry and Morphogenesis*. London: Cambridge University Press.
Ng, K. 2009. Hegel's Logic of Actuality. *The Review of Metaphysics* 63 (1): 139–172.
Ng, Karen. 2020. *Hegel's Concept of Life: Self-Consciousness, Freedom, Logic*. New York, NY: Oxford University Press. https://doi.org/10.1093/oso/9780190947613.001.0001.
Ng, Karen. 2021. What Is a Logical Concept of Life? Reply to Critics. *Hegel Bulletin* 42 (3): 454–471. https://doi.org/10.1017/hgl.2021.22.
Noble, D. 2006. *The Music of Life: Biology Beyond Genes*. Oxford, UK: Oxford University Press.
Nonaka, Tetsushi. 2020. Locating the Inexhaustible: Material, Medium, and Ambient Information. *Frontiers in Psychology* 11. https://www.frontiersin.org/journals/psychology/articles/10.3389/fpsyg.2020.00447.
Nunes-Neto, Nei, Alvaro Moreno, and Charbel N. El-Hani. 2014. Function in Ecology: An Organizational Approach. *Biology and Philosophy* 29 (1): 123–141.
Nuzzo, A. (2024). The Logic of Nature: Nature as the "Idea in the Form of Otherness." In *Hegel's Philosophy of Nature: A Critical Guide*, ed. M. F. Bykova, 97–116. Cambridge: Cambridge University Press. https://doi.org/10.1017/9781009023160.007.
Odling-Smee, F. John, Kevin N. Laland, and Marcus W. Feldman. 2003. *Niche Construction: The Neglected Process in Evolution*. Princeton, NJ: Princeton University Press.
Pinkard, Terry. 2012. *Hegel's Naturalism Mind, Nature, and the Final Ends of Life*. Oxford, UK: Oxford University Press.
Pinkard, Terry. 2024. The Prospects for an Idealist Natural Philosophy: Logic and Nature. In *Hegel's Philosophy of Nature: A Critical Guide* (Cambridge Critical Guides), ed. Marina F. Bykova, 259–272. Cambridge: Cambridge University Press. https://doi.org/10.1017/9781009023160.018.
Radnik, B. 2024. *Freedom, in Context: Time, History, and Necessity in Hegel*. London, UK: Bloomsbury Academic.

Ramírez-Vizcaya, S., and T. Froese. 2019. The Enactive Approach to Habits: New Concepts for the Cognitive Science of Bad Habits and Addiction. *Frontiers in Psychology* 10. https://doi.org/10.3389/fpsyg.2019.00301.

Renault, E. 2013. The Naturalistic Side of Hegel's Pragmatism. *Critical Horizons* 13 (2): 244–274. https://doi.org/10.1558/crit.v13i2.244.

Renault, Emmanuel. 2020. Hegel, Marx, and Presentism. In *From Marx to Hegel and Back: Capitalism, Critique, and Utopia*, ed. Victoria Fareld and Hannes Kuch, 55–72. London and New York: Bloomsbury Academic. https://doi.org/10.5040/9781350082700.ch-003.

Rouse, Joseph. 2024. Niches and Niche Construction in Biology and Scientific Practice. *European Journal for Philosophy of Science* 14. https://doi.org/10.1007/s13194-024-00604-6.

Sandnes Haukedal, R. 2025. The Concrete Universal of Embodied Concept. *Verifiche* 53 (The Concrete Universal: Relevance, Meanings and Perspectives): 97–122.

Sandnes Haukedal, Rasmus, and Nikolaos Mylonas. 2024. Entropy and Negativity: The Ecological Implications of Dialectics. *Technophany, A Journal for Philosophy and Technology* 2 (1): 1–25. https://doi.org/10.54195/technophany.15370.

Schneider, Eric D., and Dorion Sagan. 2005. *Into the Cool: Energy Flow, Thermodynamics, and Life*. Chicago, IL: University of Chicago Press.

Schülein, Johannes-Georg. 2021. Nature's Otherness: On the Status of Nature in Hegel's Encyclopedic System. In *Hegel's Encyclopedic System*, ed. Sebastian Stein and Joshua Wretzel, 133–154. New York, NY: Routledge. https://doi.org/10.4324/9780429022555-10.

Skúlason, Skúli, Kevin J. Parsons, Richard Svanbäck, Katja Räsänen, Moira M. Ferguson, et al. 2019. A Way Forward with Eco Evo Devo: An Extended Theory of Resource Polymorphism with Postglacial Fishes as Model Systems. *Biological Reviews* 94 (5): 1786–1808. https://doi.org/10.1111/brv.12534.

Sterelny, Kim. 2003. *Thought in a Hostile World: The Evolution of Human Cognition*. Oxford, UK: Blackwell.

Stern, Robert, and Leonard Weiss. 2024. The Feebleness of the Concept in Nature: A Challenge to Conceptual Realism? In *Hegel's Philosophy of Nature: A Critical Guide* (Cambridge Critical Guides), ed. Marina F. Bykova, 15–35. Cambridge: Cambridge University Press. https://doi.org/10.1017/9781009023160.003.

Stotz, Karola. 2017. Why Developmental Niche Construction Is Not Selective Niche Construction: And Why It Matters. *Interface Focus* 7 (5): 20160157. https://doi.org/10.1098/rsfs.2016.0157.

Sultan, S. E. 2015. *Organism and Environment: Ecological Development, Niche Construction, and Adaptation*. Oxford University Press.

Sultan, S. E., A. P. Moczek, and D. Walsh. 2022. Bridging the Explanatory Gaps: What can we Learn from a Biological Agency Perspective? *BioEssays* 44 (1): 2100185. https://doi.org/10.1002/bies.202100185.

Tahar, Mathilde. 2022. Biological Constraints as Norms in Evolution. *History and Philosophy of the Life Sciences* 44 (9). https://doi.org/10.1007/s40656-022-00483-1.

Trappes, Rose, and Sabina Leonelli. 2025. Conceptualising Research Environments Using Biological Niche Concepts. *European Journal for Philosophy of Science* 15 (1): 11. https://doi.org/10.1007/s13194-025-00640-w.

Trappes, Rose, Behzad Nematipour, Marie I. Kaiser, Ulrich Krohs, Koen J. van Benthem, et al. 2022. How Individualized Niches Arise: Defining Mechanisms of Niche Construction (Niche Choice, and Niche Conformance). *BioScience* 72 (6): 538–548. https://doi.org/10.1093/biosci/biac023.

Uller, Tobias, and Heikki Helanterä. 2019. Niche Construction and Conceptual Change in Evolutionary Biology. *British Journal for the Philosophy of Science* 70 (2). https://doi.org/10.1093/bjps/axx050.

Varela, Francisco. 1999. The Specious Present: A Neurophenomenology of Time Consciousness. In *Naturalizing Phenomenology*, ed. Jean Petitot, Franscisco J. Varela, Barnard Pacoud, and Jean-Michel Roy, 266–314. Stanford, CA: Stanford University Press.

von Uexküll, Jacob. 2010. *A Foray into the Worlds of Animals and Humans: With a Theory of Meaning* (Posthumanities). Minneapolis, MN: University of Minnesota Press.

Wagner, R. H., and É. Danchin. 2010. A Taxonomy of Biological Information. *Oikos* 119 (2): 203–209. https://doi.org/10.1111/j.1600-0706.2009.17315.x.

Walsh, Denis M. 2015. *Organisms, Agency, and Evolution.* Cambridge, UK: Cambridge University Press. https://doi.org/10.1017/CBO9781316402719.

Walsh, Denis M. 2022. Environment as Abstraction. *Biological Theory* 17 (1): 68–79. https://doi.org/10.1007/s13752-020-00367-2.

Werner, Konrad. 2024. Enacted Institutions, Participatory Sense-Making and Social Norms. *Synthese* 203 (5): 152. https://doi.org/10.1007/s11229-024-04569-0.

Werner, Konrad, and Magdalena Kiełkowicz-Werner. 2022. From Shared Enaction to Intrinsic Value. How Enactivism Contributes to Environmental Ethics. *Topoi* 41 (2): 409–423. https://doi.org/10.1007/s11245-021-09750-5.

Westphal, Kenneth R. 2015. Hegel's Pragmatic Critique and Reconstruction of Kant's System of Principles in the 1807 Phenomenology of Spirit. *Hegel Bulletin* 36 (2): 159–183.

Westphal, Kenneth R. 2020. Causality, Natural Systems, and Hegel's Organicism. In *The Palgrave Hegel Handbook*, ed. Marina F. Bykova and Kenneth R. Westphal, 219–239. Cham: Springer. https://doi.org/10.1007/978-3-030-26597-7_12.

Zaslawski, N. 2018. Neurodialectics: A proposal for Philosophy of Cognitive and Social Sciences. *Constructivist Foundations* 14 (1): 42–55.

Zambrana, R. 2015. *Hegel's Theory of Intelligibility*. Chicago, IL: University of Chicago Press.

Žižek, Slavoj. 2006. *The Parallax View*. Cambridge, MA: MIT Press.

Žižek, Slavoj. 2012. *Less than Nothing: Hegel and the Shadow of Dialectical Materialism.* London, UK: Verso.

Žižek, Slavoj, Frank Ruda, and Agon Hamza. 2022. *Reading Hegel*. Cambridge, UK: Polity.

Chapter 8
Conclusion: Whither Science?

Abstract This conclusion presents a radical critique of the Modern Synthesis (MS) in evolutionary biology, advocating for its replacement with a more dynamic and dialectical framework through the Extended Evolutionary Synthesis (EES). It challenges the notion of science as an impartial observer, arguing instead that scientific theories actively shape—and are shaped by—the material world they seek to explain. The author proposes a 'plastic' approach to science, one that embraces historicity, contingency, and pluralism, rejecting rigid paradigms in favour of open-ended, interdisciplinary inquiry. Drawing on Hegelian dialectics and organisational theories, the text calls for a fundamental shift from mechanistic causality to *enablement*, where organisms and environments co-constitute evolutionary processes through reciprocal relations. Beyond biology, the argument extends to a broader reimagining of scientific practice—one that integrates insights from the humanities and acknowledges the ideological and material constraints shaping research. Ultimately, the conclusion envisions a self-reflexive, context-sensitive science capable of grappling with life's complexity without resorting to reductive or deterministic explanations. Such a science, I contend, must remain perpetually unfinished, fostering a 'culture of theorisation' rather than enforcing doctrinal unity.

Introduction

> Of course, with intentional agents, categories in theories can acquire a causal role in the generation of behavior, and if the behavior involves the production of material systems, such categories or decisions using them can result in the generation or creation of physical, biological, psychological, social, and cultural order. But in this way, theories become parts of the physical world as well as lenses through which it is viewed. The interests and needs of human agents can become materialized in similar fashion, becoming instantiated through hardware and software technology, our choice of research projects, and of how they are to be pursued, producing […] 'changes both in the lens and in the picture it presents.' In this way, the picture I urge combines elements of a constructivism in a broader-based realism. In this picture it may be extremely hard—not to mention, in most cases, pointless—to tease the aspects of construction and realism apart. Nonetheless, it is plausible to assert that theories will become more causally efficacious in that world to the extent that theoretical categories

R. Sandnes Haukedal, *Agency and Organisation: The Dialectics of Nature and Life*, Library of Ethics and Applied Philosophy 44,
https://doi.org/10.1007/978-981-95-1924-8_8

map accurately onto natural categories in the world—or onto cost-benefit approximations to them. (Wimsatt 1994: fn43, 241)

It is time to conclude. This conclusion is more of a road map for further studies than an ending. As readers have probably noticed, we have been dealing with many unsettled debates. I have tried not to close them, but to repeat them, a task that has only just begun. Continuing in this track, I will not make grand statements about the new synthesis that we have searched for, but provide some thoughts on the way forward, and sketch what we should look for in a new scientific research programme, and even whether such a programme is desirable. (I therefore gesture at many perspectives which that I cannot develop but which I think offer fruitful avenues for research.) Based on this, I argue that the critique presented in this work does not only pertain to biology but points to a radical critique of modern science as such.

We have touched upon many different theories and perspectives and sought to render common threads among them. I will not attempt to gather all these threads in this chapter but some of them are suggested in the long epigraph above. Here, Wimsatt expresses several of my concerns, most importantly the notion that science and philosophy are not innocent practices but part of the world they model. He also points to the entanglement of construction and realism that we discussed through niche construction, and Hegel's understanding of the concept. Finally, Wimsatt emphasises the way the effect of science will be different based on how science approaches the world, and how the structure of a theory constrains its relation to external reality.

Of course, theories do not bring reality into being. But a scientific framework like the Newtonian theory (for instance) ushered in a worldview with practical changes that were unimaginable before. This indicates that it did map onto 'natural categories of the world', even if it did so partially and at the expense of other ways of shaping material reality. A new kind of science can equally become causally efficacious to the degree it maps onto reality. But this relationship is never simple since science is not outside what it discloses. It must map onto something it is part of. Thus, we must accept the non-innocence of science, and how it is (reciprocally) constrained by the material that it seeks to grasp. We have to grasp the reciprocity of theory and practice.

The MS has been efficient to the degree that it has managed to make biological systems intelligible, but its limitations have become increasingly evident, as many of its promises have failed to materialise. One reason might be that it is more a theory of stasis than of change. It is a theory of evolution as changes in populations, but since it lacks a theory of organisms—except as the result of evolutionary processes at other levels—it does not explain change so much as explain it away, reducing it to a passive medium between variation and selection.

How should we tackle this circumstance? If there is no theory of evolutionary change to jettison, it weakens the legitimacy of the retort against the EES: that it lacks empirical underpinning and theoretical coherence—because so does the MS! Would it not be enough to come up with more promising metaphors, ones that are more likely to become theories? Based on such concerns, I question whether the MS was ever the theory it was presented as.

As for the EES, the task might not be to include something new within old metaphors but to propose metaphors that might foster a theory in the first place. This means that it should not be restricted to biology, even if it emerges from this discipline. Also, this does not preclude it from aiming at fostering a specific research tradition, but it would prevent it from stagnation and hardening. It needs to be 'precise enough to be rejected' (Soto and Brilmyer 2020: 5), with theoretical constraints guiding research, able to encompass and give rise to multiple research agendas.— Hence, it should not only be a theory but a 'culture of theorization [...] the drive to have elaborated and consistent discourses that integrate with other relevant fields—overall the care for the rationality of discourses' (Chaix et al. 2021). This dovetails with Malabou's notion that sciences that are not plastic, and that do not incorporate insights of other sciences and discourses, will stagnate. It also hints at an organicist conception of scientific practice itself, that I explore below.

This depends on how we understand the scientific status of the MS. Insofar as the MS was not a coherent theory but a philosophy of nature, I think Rouse (2024) is correct to say that the EES should perhaps consider itself in similar terms—self-consciously, in contrast to the MS. Indeed, one of the major criticisms of the EES is its ambiguity and lack of cohesion (Ochoa 2024). But I add that we should understand this in the Hegelian sense, which is to say a philosophy of nature which is not modelled after physics, and which is therefore not external to its object. Hence, it does entail a strict naturalistic philosophy which undermines the autonomy of philosophy (Godfrey-Smith 2001).

Dialectical in kind, it should seek to merge theory and practice. Whereas the latter relates to the teleological attitude of making use of nature to maintain ourselves, and the theoretical approach seeks to grasp nature in its lawfulness, Hegel's method is conceptual. Hegel is against the view that considers nature as something static and unchanging, which a theoretical or metaphysical view; and he is against the view that sees nature completely fleeting and inexistent—'that individual things are nothing in themselves' (Hegel 2004: 9).[1]

Contrary to both attitudes, the point is to account for both the particularity and universality of nature, to *comprehend* (*Begreifen*) not only its necessity but its contingency—to see the universal in the individual without eradicating its self-determination by reducing it to generic principles. Hence, while physics prepares the material for philosophy, its method is inadequate and must be transcended by philosophy. In Nuzzo's words, 'the articulation of the "philosophy of nature" is ultimately the "self-determination of the concept" of nature itself' (Nuzzo 2024: 112). The very notion of philosophy of nature, as distinct from other parts of Hegel's philosophy, is immanent to the development of the concept, and therefore not arbitrary.[2]

---

[1] As Hegel remarks: 'Of a metaphysics prevalent today which maintains that we cannot know things because they are absolutely shut to us, it might be said that not even the animals are so stupid as these metaphysicians; for they go after things, seize and consume them' (Hegel 2004: 9).

[2] I rely heavily on Nuzzo's argument here, as she masterfully shows how philosophy of nature ends at the point where its contradiction cannot be resolved within its own confines. This is also tied to the

I should also note that a philosophy of nature is not precluded from constituting a part of (one or multiple) research program(s).[3] Moreover, whether a philosophy of nature can be useful to science is 'only known after the fact (Godfrey-Smith 2001: 289).

From this backdrop, in what follows, I argue for a broad conception of the interplay between science and society and propose that we should not merely aim to overthrow the MS but articulate metaphors that may stimulate a new theory of nature. As Gallagher asks: 'Does science itself, as one such social institution, and as one set of cultural practices, remain the same within this different kind of naturalism?' (Gallagher 2018: 117). I believe we are on the verge of a revolution not only in biology but in science itself; but I cannot do proper justice to this prospect here—only hint at what this new science might be and how the authors discussed in this volume could contribute.

Instead of trying to summarise their relative contributions, I think it is more useful to get back to the issues we began with. I therefore return to the question of what success might look like for an expanded evolutionary synthesis and the scientific ideal it should establish. I argue that the new synthesis must be plastic, open to both minor revisions and radical changes, and neither flexible nor rigid. If as it hardens, it is not a synthesis in the sense of an ongoing and unfinished process, but a synthesis in name only, a forced marriage like the MS. For this reason, I entertain the option that we should not seek a synthesis at all.

We must acknowledge the unruly nature of both nature and culture, how we are neither blank slates nor slaves to our biology. If our biology does not determine but constrains what social systems we may construct, and these systems, in turn, constrain our biology, and the very life of the planet, we see why constructionism and realism are not opposed externally but internally entangled.

Repeating Organicism

> After the Scientific Revolution, Natura no longer complains that her garments of modesty are being torn by the wrongful thrusts of man. From an active teacher and parent, she has become a mindless, submissive body. Not only did this new image function as a sanction, but the new conceptual framework of the Scientific Revolution—mechanism—carried with it norms quite different from the norms of organicism. The new mechanical order [...] and its

---

powerlessness of nature, namely that it cannot overcome the contradiction between contingency and necessity; this is why the transition to philosophy of spirit is required, since this is where freedom, as being with oneself in one's other, can be achieved. But this is not to say that nature ends here—as nature is operative both before and after philosophy of nature even if is only grasped *as nature* therein (Nuzzo 2024). Moreover, spirit is not outside nature but implicit in it. Again, we have to keep in mind that Hegel's philosophy is an organic whole in which form and content coincide.

[3] Of course, it can be argued that all research programs have an underlying philosophy of nature, but that this fact is obfuscated, denied, and/or not thematised by the practitioners of this research program.

> associated values of power and control [...] would mandate the death of nature. (Merchant 1989: 190)

If we follow Merchant, the overthrow of organicism meant the death of nature in the sense that an impression was created that nature was a servile servant for our actions and that we had overcome it—that we had left the problem of nature behind. This, as we have seen, is inadmissible from a dialectical perspective, where no stage is ever fully overcome. The falsity of this image has steadily become evident, even if there have been many attempts at hiding it. This vision of nature translates to a view of evolution works and implies a certain kind of science within a specific mode of production, says Merchant (1989). These are dead zones for biologists and philosophers alike, as they fail to grasp the historicity of the phenomena they examine. The result is distorting reification. Dialectical materialism—understood as an ecological and relational approach to the universe—represents an alternative to this view. It does not limit itself to socio-historical dialectics but seeks to understand nature as a dialectical process.

Scientific data does not specify their interpretation. As multiple interpretations of the same empirical findings demonstrate, 'theories are underdetermined by data' (Gefaell and Saborido 2022: 19). Again, we could attribute this to the irreducible contingency of nature. To properly understand and foster scientific processes we must acknowledge that scientific research agendas are always constrained by ideologies—that shape how we make sense of its findings. But they do not determine the interpretations or undermine them by default. We might speculate whether it is the lack of theoretical fluency has led science to stagnation. Montévil proposes that we combat ideology in science by fostering 'theoretical fluency', the

> ability to recognize that any scientific statement depends on theoretical assumptions and an underlying epistemological framework. Theoretical fluency also requires acknowledging that a change in framework may be required either for empirical reasons or as a result of intrinsic contradictions of a theoretical framework or contradictions with other, established, and relevant theoretical perspectives. (Montévil 2022: 51)

If this capacity is not cultivated, we end up in dogmatism, warns Montévil. I would add that contradictions are not necessarily epistemological but have objective reality, since they concern the generative unity of opposites. Hegel claims that every concrete concept contains contradiction because it is heterogenous, containing different and entangled determinations (Ficara 2020). He demonstrates why contradictions are necessary, and why dialectics demands a specific attitude towards contradiction. It does not simply seek to eradicate them but to identify productive contradictions, which spurs the development of self-determination forward (Bordignon 2021). In this perspective, contradiction is another word for reciprocal action (Foster 2022).

Hegel even says that grasping the contradiction that negativity is as much positive as negative, 'the recognition of the logical principle that negation is equally positive, or that what is self-contradictory does not resolve itself into a nullity' is '[t]he one thing needed to *achieve scientific progress*' (Hegel 2010: 33, emphasis original). While I cannot justify this claim now, we have found that, in most attempts at describing concrete phenomena, negation is part of their construction, as seen in the organism-environment interplay. Hegel also boldly claims that 'a time will come when this science [astonomy] will require for its satisfaction the philosophical Notion' (Hegel 2004: 83). While this has not borne fruit, as Pinkard (2024) notes, there is something to this idea. And while it has not been the case for astronomy, principles similar to those of Hegel have been taken up within science—not only currently but also in the development of sciences such as ecology. The role of philosophy is of course not to replace empirical research but to guide it in a conscious manner.[4]

If we are going to understand the becoming of nature and how it enables life, we cannot reduce it to a static other but treat life as the 'horizon from which something like nature and spirit makes sense' (Illetterati 2023: 206). This suggests why the historicity at the heart of biology should be extended to all sciences. Instead of a futile search for ontological guarantees, it seems we should aim to maintain and make use of this lack instead of filling it with specific content. Some argue that we should view science itself through organic metaphors to understand how it is entangled with other processes—how it is a social practice constrained by the society in which it finds itself, infused with its values (El-Hani and Reis 2021; Bernal 1939).

The call for scientific pluralism is a call for different explanatory strategies based on the level under scrutiny and thus contrasts with reductive strategies. It builds on an organicist conception of science, and demands ideological struggle:

> A particular ideological chain becomes a site of struggle, not only when people try to displace, rupture or contest it by supplanting it with some wholly new alternative set of terms, but also when they interrupt the ideological field and try to transform its meaning by changing or re-articulating its associations, for example, from the negative to the positive. Often, ideological struggle actually consists of attempting to win some new set of meanings for an existing term or category, of dis-articulating it from its place in a signifying structure. (Hall 1985: 112)

In my view, the organisational approach does precisely this: It reclaims organicism by taking up its forgotten components, such as levels of organisation, and reinterprets them in the process. The task is not simply to supply new terms but to interrupt and de-canalise the flow of science. Advocating for a constructivist interpretation of adaptation could be an example of this. Hall suggests that ideologies can be repurposed from the inside. I have tried to show that this is already happening from within science itself and sketched how philosophy can learn from this. Likewise, repeating the MS is re-activating the hidden archive of virtual possibilities, to correct the prevailing discourse. Perhaps, then, instead of extension or expansion, we should speak of negation.

Certainly, a research strategy that is not context-sensitive is incapable of understanding natural processes. Trying to reduce specific objects to generic ones is misleading if the objects we seek to understand are specific; such an approach no longer maps onto the world it aims to grasp. The assumed ground levels are

---

[4] We should note that some have therefore promoted what they call philosophy *in* science, which is to 'use philosophical tools to address scientific problems and to provide scientifically useful proposals' (Pradeu et al. 2021).

not necessarily relevant, as emergent levels are underdetermined by lawful regularities (historicised invariants) (Farnsworth 2022). Contextualised knowledge is needed to understand such phenomena. It is not that regularities are irrelevant but that they are insufficient—incapable of explaining biological phenomena except in abstract terms. A more plural model is needed to get at the specificity of biological objects, and to accommodate differences in context-dependence. A plural model entails 'that context-sensitive knowledge (about specific objects) needs to be more and more produced and considered alongside with decontextualized knowledge (of generic objects) as systems and phenomena increasingly depend on specific historical, contextual, and variable dynamics' (El-Hani and Reis 2021). This principle does not only apply to biology but science in general.

Which Synthesis?

> [T]o use the standard terms never used by Hegel himself, in the final "synthesis" the antithesis is brought to an extreme, fully internalized as constitutive of the entity in question. (Žižek 2020: 179)

This dialectical notion of synthesis seems to apply to the MS: It internalised its opposite, teleology, but disavowed it. It became a constitutive exclusion implicit in their models. The consistency of the synthesis was thus always a mirage. But how does this notion apply to the calls to extend this synthesis? Insofar as a synthesis is accomplished, it is only by way of closing itself off, by hardening into orthodoxy. In this sense, the MS was only seemingly a synthesis. It needed teleology—its apparent antithesis—to stick together.

From what I have said, it seems there can be no final synthesis. Perhaps this much should be acknowledged by a new synthesis. I am therefore sympathetic to Jaeger (2024), who argues that the 'problem is the attempt at synthesis itself' (182), and asks: If evolution generates diversity, why should we strive for a uniform synthesis? He favours a perspectival approach, where explanations are not mutually exclusive but complementary, even if some are more limited than others. We return to his views below.

As we have seen, the MS is already extended to include more levels and processes. But its expansion is restricted if the general outlook and the scientific ideal they strive has not changed. When scientists claim that the MS has been continuously expanded since its infancy, this admission is a cover story: They admit some form of expansion to hide more fundamental problems at the heart of the MS. While it has been changed since its original formulation, the MS is still dominated by a reductive and mechanistic view of nature. Insofar as it has changed, it did so by way of co-opting the challenges to its core and assimilating them to fit its overall thrust. Its core tenets and *modus operandi* remained unaffected, even if it included previously neglected processes into the fold. It integrated them by removing them from their original context, thus blunting their critical sting and repurposing them to its needs.

Because of such attempts at integrating new theories into the fold, trying to accommodate changes in theory and method, the MS became increasingly loose in its structure. I therefore think it is useful to think of it as a philosophy of nature more than a strict research program. But these options are, of course, not mutually exclusive.

From this, it seems too simple to just synthesise new findings into the old synthesis—take care of what was good in the old synthesis and discard its bad parts. This takes for granted that the problem has to do with content, not form. Adding content is easy but changing the way we view the content is more challenging. We must move beyond the additive approach and instead lay the foundations for the emergence of a new scientific approach—a new philosophy of nature—that retroactively changes the meaning of the content and thus the future of research. This follows Lenin's understanding of the scientific process as a repetition of previous stages in a novel way, moving 'in spirals, not in a straight line' (Lenin 1914: 14, quoted by Jablonka and Lamb 2020: 1).

At the level of content, the MS has been accommodating. It is more challenging to make a formal shift in how we relate to our theories—what function they are supposed to play, and which logic they should follow. For example, if we aim for 'objective' theories that minimise the role of the observer, we disavow our implication in what we disclose. By taking ourselves out of the frame, we move in the direction of positivism. But if we are aware of our entanglement with what we study, we might accomplish a self-reflective theory, a theory that refracts its subject matter. Thus, we allow for an open theoretical corpus to be articulated, one that cannot harden without undermining itself. This is the fluency Montévil speaks of.

Such fluency would allow us to address *the scientific blind spot*, as formulated by Frank et al. (n.d.). The claim that this blind spot concerns the experiential dimension of science, and has led science into a state where it cannot account for its own conditions of possibility. While the authors do not make this link themselves, they come very close to propounding a Hegelian (or dialectical materialist) outlook. In this work, I have also tried to outline how this blind spot can be addressed, although I have only been able to sketch how we may overcome it. I have argued that a dialectical and self-reflexive kind of science, in which form and content coincide is the way forward. In this view, theory and practice converge, as theorisation is itself a kind of practice.[5]

We should also take note of the fact that Matteo Mossio claims that organisation is still a blind spot for the EES. Even if it is more attentive to this dimension, it 'does not rely on a characterization of what is an organized system, which rather consists in a particular *level of description* at which phenomena are described and explanations are provided' (Mossio 2024: 7). This, he contends, has implication for the 'explanatory scope and power of evolutionary biology'. Similarly, I have argued that the specific

[5] An example of this would be how we may apply the principles of niche construction, which we have seen are fundamentally dialectical, to scientific practice itself, as is done by Sanches de Oliveira et al. (2023). This also dovetails with the *naturalistic turn* in philosophy of science, which I believe this book is aligned with, although I cannot elaborate much on this within this work but I aim to deal with in forthcoming work (see Rouse 2024). In this view, we should regard science itself as an expression of biological niche construction (Rouse 2024.

kind of causality—dialectical in nature—promoted by the organisational view must be reckoned with, and I have noted how biologists share this view.

The Science of Philosophy

> The advance of science itself creates new philosophical questions. Those who tackle such questions are philosophers, even if they do not acknowledge that name. (Noble 2014: VIII)

The above quotation from Noble takes us back to the introduction, in which I claimed that biologist had been doing philosophy in lieu of philosophers. We see this in Longo and Mossio (2020), who claim that the main problem with the machine metaphor within biology is that it never gave way to a theoretical viewpoint that was distinct from the informational theory that it relied on. The metaphor remained a metaphor but was confused for a theory. (The same could be said for the MS: it confused a philosophy of nature with a scientific paradigm.) They argue that genocentrism is the main culprit for the stagnation in biological theory and that formulating a new framework demands that we jettison this view. In a sense, we need new metaphors, but more importantly, we need metaphors that are supposed to be 'absorbed' by the theoretical framework they elicit. While they might instigate the initial thrust, they are nothing more than stepping-stones (Longo and Mossio 2020). Of course, they can be obstinate, as we have seen. But we should nonetheless aim to excise them or at least be aware of their status. From this perspective, the move from the MS to the EES seems straightforward: It means adopting an approach which emphasises the teleological capacity of the whole organism—abandoning the unidirectional version of causality and the machine metaphor which undergirds it.

In a discussion with Chaix and Longo, Montévil emphasises that the theoretical immaturity of biology can be traced back to the molecular revolution. He claims that biology has, since then, subcontracted 'the theoretical work to other scientists, for example, to computer scientists in bioinformatics, to physicists in biophysics, to mathematicians in applied mathematics, and, to an extent, to philosophers, especially analytic philosophers for logical consistency' (Chaix et al. 2021: no pagination). The problem is that none of these fields does genuine theoretical biology. He takes on analytic philosophy: While 'they typically provide accounts of what biology focuses on and what biologists mean when talking about delicate matters such as functions', they do not 'provide work on scientific theories, integrating a diversity of observations and providing simultaneously a view on living beings and on the way to understand them' (Chaix et al. 2021: no pagination).

This indicates that the role of philosophy is not only to work alongside science to confirm its view but also to question the limits of scientific formalisations, to identify the points at which the sciences transcend themselves—depend on presuppositions that are not justified scientifically. The point is to show that the empirical sciences are never purely empirical, but rational justifications or organisation of data and experience underpinned by unspoken assumptions (Illetterati 2022). Moreover, philosophy depends on experience and science but moves beyond them as it cannot accept their results and assumptions uncritically.[6] It is thus continuous with science, yet its object is not the particular but the totality. Again, science constrains philosophy:

> As I have remarked elsewhere, inasmuch as philosophy is to be science, it cannot borrow its method from a subordinate science, such as mathematics, any more than it can remain satisfied with categorical assurances of inner intuition, or can make use of argumentation based on external reflection. On the contrary, it can only be *the nature of the content* which is responsible for *movement in scientific knowledge,* for it is the content's *own reflection* that first posits and *generates what that content is.* (Hegel 2010: 9)

Here, Hegel touches on the principle of immanent critique, which means that you disclose the contradictions which are inherent to the system. It demands that we follow the inner development of the content. Philosophy is, in Hegel's view, a more radical science than the particular sciences because it aims to grasp the totality in its becoming (Illetterati 2023: 64). Dealing with science, then, philosophers cannot simply use it wherever they please but need to engage in a more systemic manner to get at the theoretical presuppositions guiding the research they engage with. Otherwise, their dealing with science is opportunistic and superficial.

Philosophy should not play the role of a handmaiden that assists science in becoming self-reflexive or critical. I am reminded of the quotation by Whitehead, cited earlier: '[I]f science is not to degenerate into a medley of *ad hoc* hypotheses, it must become philosophical and must enter upon a thorough criticism of its own foundations' (Whitehead 1967: 16–17). We thus not only need to think in terms of the philosophy of science but also of *the science of philosophy*—and how it may inform the other sciences as a critical corrective that does not accept anything as given. While empirical science establishes natural regularities that philosophy have to deal with, only philosophical investigation can unearth the unity of nature—while also preserving the view that nature is to irregular to organise itself into a whole and therefore can never be fully grasped as such. Hence, no 'theory of everything' is attainable.

From Entailment to Enablement

> How can a genuine dialogue [between the humanities and the sciences] take place, one that would both respect the autonomy of each field and redraw its limits and frontiers? (Malabou 2010: 9)

What kind of vision of science does the new evolutionary synthesis suggest? Some attempts at producing a new synthesis follow the footsteps of the MS—not in terms of content but in terms of the view of science it promotes. One example is found in notions that the MS can be overhauled by accumulating scientific data. This not only

---

[6] As Trappes and Leonelli (2025) remarks, 'even a thoroughly naturalist stance does not require the uncritical adoption of all claims and methods from the natural sciences' (16).

follows the predominant quantitative approach to science but also misrepresents the challenge it is faced with. Longo (2018) highlights how

> the genocentric informational/programming views cannot be falsified by experience nor "in theory", because they are not scientific: those views are based on common sense notions of information and program and on the "homunculus" ancient myth, modernized and made literal by encoding it in chromosomes. (95)

This position seems radical but follows what we have said above. We have laboured to define what the MS is and how it may be reformed, as it made use of whatever may salvage its core beliefs. In the end, it comes down to assumptions and metaphors instead of theories. Metaphors are used to hide disagreeable facts. They are taken as the literal truth because they have become linguistically 'dead' (Lakoff and Johnson 2008; Noble 2021). Claiming, as Longo does above, that someone is not even doing science is strict, but is tied to how they are not involved with theories but metaphors. The dominant one is the information metaphor that was adopted uncritically within biology. Common sense understandings of information—as given in discrete bits and contained in a program which is external and unaffected by its material instantiation and may linearly produce its outcome—have sutured theoretical impasses and brought new observables into being. It has fed into causal models, and 'diverted attention from the rich networks of causal and enablement relations, within an organism and an ecosystem' (Longo 2018: 88).

From the organisational perspective, what seems like theoretical closure is also a foreclosure since it makes it impossible to understand evolutionary dynamics (Depew 2017). This applies to the EES as well, insofar as organisation has largely been explained by other processes, instead of being an explanatory device (Moreno and Mossio 2015). This underscores the claim that organisation is a blind spot also for the EES. The EES, then, while an important step towards another kind of biology, must be informed by the organisational perspective and the dialectical approach it implies.

Perhaps the main issue involved in the shift from the MS to the EES is the notion of causality that it employs: 'It would seem that the manner in which biologists think about causality has acted as a metatheoretical conceptual framework to stabilize the dominant scientific paradigm' (Laland et al. 2011: 1516). This reveals why it might not be helpful to think of the shift from the MS to the EES as concerned with content. For instance, the EES maintains the central role of natural selection and DNA inheritance (Herrington and Jablonka 2020). Some cast the shift as one from 'a population-dynamic account to a causal-mechanistic theory' (Pigliucci and Müller 2010: 12). I suppose this means replacing the *deus ex machina* needed to comprehend the route from the microscopic to the macroscopic level, from genotype to phenotype, with an actual theory. However, it seems this shift does little to confront the current view of causality within science. It thus constitutes less of a challenge to the mechanistic worldview than the organisational perspective I have favoured in this study.

Adding new processes to the current synthesis does nothing to change its core; rethinking the causality of these processes, on the other hand, might indicate a more radical alternative with ramifications for science in general. It is not that we do not have other modes of evolutionary change besides selection but a question of how we envision the causality of selection itself, and how it is related to other evolutionary processes. The unidirectionality of evolutionary models has given way to simple cause-effect models, instead of reciprocal and distributed causal processes. Laland et al. (2015) mention constructive development (how organisms respond to and modify internal and external states) and reciprocal causation as two of the ontological assumptions of the EES. I have tried to demonstrate that a dialectical view is implicit in such assumptions.

More recently, Lala et al. (2024) have argued that 'a deeper understanding of evolutionary causation should be viewed as a priority for the field' (142). I have tried to present such an understanding in a way that aligns with their emphasis on reciprocal causation. But I have tied this understanding more explicitly to dialectics than they do.[7] I have done so by reference to Hegel and through the notion of enablement, which I believe could serve as scientific instantiation of what dialectical reciprocity means. From these notions, we get a variegated approach to evolutionary causation, in which it is historically changing and plastic, a heterarchical model of causation (Bechtel and Bich 2024). Hence, while teleology is the truth of mechanism—in the organisational sense that mechanisms must be organised to count as such, just as variation must be organised or facilitated—mechanisms are still operative and at specific junctures could play a large role.[8] Their role also varies between organisms, and changes throughout their lifetime.

By reintegrating development into evolutionary explanations, Lala et al. (2024) believe evolutionary theory could become more predictive. While I do not disagree in principle, I have aimed to show the limitations of any kind of prediction of historical systems.

The more general point is that a dialectical approach always takes the totality into account, and always emphasises the centrality of understanding the notions we apply in an ecological way, as *concrete universal*. Hence, it allows us to formulate an explanation which is not generic, but specific. As such, it accounts for the uniqueness of biological systems, instead of aiming to fit it into a physicalist framework.

Longo's reason for discarding natural selection in favour of enablement is linked with the historical nature of biological systems. Enablement replaces selection as it brings out the enabling role of the plastic context, how variation is canalised historically, and how the constraints on each level shape evolution (Longo 2021a). The point is not only to add organisation, constraints, and contingency into the mix while leaving the mode and level of causation the same: frequency changes in populations caused (or sifted) by selection (Stoltzfus 2012). This would do nothing to undermine the idea of natural selection as the active driver of evolution, which removes the need

[7] Moreover, I take issue with their critique of the notion of 'constraints', which they think is less suitable than the term 'bias'. I think this is based on a one-sided understanding of what is meant by 'constraints', which I have tried to combat throughout this book.

[8] Teleology is also the truth of mechanism in the sense that the purpose of mechanism and physical reality is to enable natural life (Stein 2024).

for other kinds of selection, at different levels, like the one performed by organisms in producing phenotypic variation to conform to the organisation in place (Noble 2021). A more radical move involves discarding the view of evolution as optimalisation and understanding the role of natural selection as excluding the incompatible (Longo and Montévil 2013), which is nonetheless a process of canalisation since negation partakes in constructing what exists.

This perspective is dialectical by underscoring the internal relation between the organism and the environment—between a living being and its configuration space. The principle of optimisation, on the other hand, only works if you already know the space of possibilities. It makes no sense in a changing environment,

> as there is no pre-given space of possibilities where one could give a partial order and define an optimum. So, we should replace selection by enablement: the context makes it possible, history canalizes the variation, the historical traces channel changes, including this amazing chemical trace of history, DNA, the trace of the entire Evolution. DNA and other constraints canalize variation, beginning with the Brownian motions of molecules in the cytosol… then constraints apply at all levels of organization. (Longo 2021b: no pagination)

A niche, for instance, does not cause but enables the survival of the organism. Insofar as we may speak of causes in biology, then, they are *differential*. They modify the space of possibilities to alter enablement relations (Longo and Montévil 2013). Because biological objects are unstable and undergo constant change, they function differently from linear and entailing causes. By contrast, differential causes change the space of possibilities and thus modify the web of enablement relation constituting an organism and its niche. This is a change in what is viable for the organism, and thus makes compatible possibilities that were previously incompatible with the survival of the organism. Biological causes are causal by effectuating a difference that takes the system away from its current enablement relations (Longo and Montévil 2013). We should still speak of causes in the differential sense of making a difference to enablement relations, but we also have to study how the differential cause is enabled in the first place:

> As a matter of fact, one goes to the doctor and rightly asks for the cause of pneumonia—not only what enabled it: find and kill the bacterium, please, which is the cause. Yet, that bacterium has been enabled to grow excessively by a weak lung, a defective immune system or bad habits … so, the therapy should not stop at this differential cause, but investigate enablement as well. (Longo et al. 2012: 1388)

I have tried to elucidate what this more complex view of causality entails through Hegel's notion of a totality of conditions established as necessary for something to emerge after the fact. This totality does not operate in a law-like manner and is not indifferent to what it enables since it is internally related to and depends on it. According to Ng, if we view nature as external, and in strict opposition to the subject, accounting for the possibility of autonomy is impossible. Their relationship is not an either-or dualism or collapse of categories but constant boundary maintenance. Dialectics is not thinking without distinctions, but it concerns internal and dynamic interaction as opposed to external and static ones. In Hegel's words, it demands a relation of freedom, not one of causality (Ng 2020). This points to a kind of causality which is not mechanistic but normative—formalised as enablement relations (Tahar 2022). From this perspective, there is not simply interaction but co-constitution between evolution and development (Malafouris 2015).

The Fourth Perspective

> Hegelian non-naturalistic naturalism is a form of monism (i.e., there is only one world) that aims at neither an ontological nor an epistemological reductio ad unum, which would imply that the different entities originate from some unitary law able to give a linear and continuous structure to reality (be it a teleology of freedom or evolution by natural selection). This kind of monism would not dismember reality into radically heterogeneous spheres, yet it would be able to account for the infinity of orders and differences that are produced within reality—differences that do not imply any duplication or even multiplication of reality. Consequently, the recognition of difference does not necessarily lead to dualism, just as the idea that reality is one does not imply the denial of differences. (Illetterati 2020)

In this work, I have striven to render the perspective laid out above—to make plausible a kind of monism that allows us to grasp the complexity of levels and interactions that we find in nature without succumbing to dualism or idealism. The organisational approach has provided us with the scientific tools to make the scientific implications of this view clearer. It has taken us beyond structuralism, functionalism, and processualism (Jaeger 2024). The first two map onto internalism and externalism: structuralism resembles internalism, focussing on what evolves without explaining the hierarchy they presuppose. Functionalism, on the other hand, is found in the externalism of natural selection, as the only explanation of *why* things evolve. Its main drawback is that it fails 'to accommodate causal interactions between the processes of development, selection, and inheritance' (Jaeger 2024: 167). Finally, the process view makes a crucial contribution by viewing nature as a process. But it has two major weaknesses: It considers the variability generators and their rules to be ahistorical and only deals with regular phenomena. In short, it is not sufficiently historical.

According to Jaeger, the organisational approach is the fourth perspective which ameliorates the pitfalls of the others by reintroducing speculative thinking to complement practical problems. It explains functional conservation through the ongoing maintenance of the organisation and brings out the entanglement between the processes enlisted above; it understands dynamical hierarchies as the coupling of scales and interactions up and down the hierarchy; and finally, it accounts for the radical historicity of biological systems, and how they embody 'self-generated rules' that cannot be specified in advance. As such, the agential view follows from the organisational, as teleological behaviour is essential for the maintenance of organisation.

Discussing the convergence of the agential and organisational view of evolution, Jaeger states that the EES has adopted a cybernetic model of agency which misses the 'dialectical multilevel dynamics underlying biological organisation and the goal-oriented behavior of the organism' (Jaeger 2024: 177). As such, they reduce agency to 'information-processing and feedback-driven goal-seeking' (177). The organisational model is more complex as it encompasses the interaction of multiple levels of organisation, unlike feedback loops, which operate at one level. Jaeger ties this to the distrust of teleology characteristic of modern science, in which the objects of science are defined by principles that lie outside themselves, like laws and forces, and lack the power to effect causal changes. This is evidenced by the neglect of qualitative experience in NCT.

Against the cybernetic model, an agential model accounts for how organisms change their states and organisation and, in doing so, it does not assume explanatory asymmetry, as 'agents both generate and respond to the conditions of their existence' (Jaeger 2024: 178). Hence, the whole does not have precedence over the parts, as they constrain each other mutually. It demands a relational or ecological explanation, in which formal causation is 'not simply imposed on the material flows constitution the organism. Instead, it is continually regenerated, constantly (re)emerging over time through the dialectical dynamic interaction of material processes and the constraints they generate' (Jaeger 2024: 179). This ongoing production of the causal nexus enables and requires a naturalistic teleological explanation.

Conclusion

> You can't know the world unless you're trying to change it. (Prashad 2020)

The above quote indicates the entwinement of theory and practice. Objectivity is produced through a subjective stance that aims to change this objectivity. Similarly, I claim that you may only know a theory by trying to apply it; theory and practice are reciprocally entwined. In this work, I have tried to apply insights from the organisational approach to dialectics and vice versa, to gauge the similarities between them. Perhaps I have gone too far at times, but I believe there is an evident structural similarity that warrants most of my assertions. In any case, only by trying to make use of these theories can I make this claim more legitimate. I do not pretend to have done so conclusively in this work, but I have tried to contribute my part. The mantle has already been picked up by others, cited in this study, and I expect that the theoretical overlaps and tensions (which are not highlighted in this work) will be strengthened in the years to come.

It should not come as a surprise that I concur with the view that '[r]ecent calls for an 'Extended Evolutionary Synthesis' [...] can be unitarily understood together as a call towards a 'dialectical turn' in biology' (Gambarotto and Illetterati 2020: 6). This turn does not just entail a renewed emphasis on reciprocity, agency, organisation, and emergence; it also acts on the level of form, and the status of the synthesis that we are seeking. A dialectical synthesis is never finished, but permanently unstable and ambiguous. The acceptance of precarity demands abandoning the search for ontological guarantees. Similarly, Jablonka and Lamb (2020) warn against the hype surrounding the current calls for an extended synthesis. They say that much of the discussion around the shift is 'unrealistically hopeful, scientifically naïve and sometimes outright misleading' (72). They propose that what is happening is a change in 'thought style', which includes how a collective defines problems of interest, judgements found evident, and the methods it applies. This, the authors note, has similarities to Kuhn's paradigms.

Jablonka and Lamb (2020) also claim that what is underway is a resurgence of organicism. This should be understood, they underscore, as a repetition that modifies what it repeats. If this description is correct, I suggest that the revival of organicism is (1) a return to Hegelian dialectics, and (2) that the organisational approach that I have discussed is the most promising repetition of organicism. Not only does it provide us with a model of how life is constructed and maintained; but it also accounts for the contradictory nature of evolutionary change. Stoltzfus (2017) touches upon this when he says that the MS 'established a cultural identity for evolutionary biology tied to a master theory of evolution' (10). He does not argue against synthetic thinking in general, but he does not want a new synthesis in the sense of a 'campaign to establish a cultural identity that, in the future, will be protected jealously by conformists' (10); he also contends that a master theory that would encompass wide-ranging processes on different levels and timescales would be 'something relatively empty' and thus not a worthwhile pursuit.

As noted, Jaeger similarly says that 'the problem is the attempt at synthesis itself. Evolution is a process that generates diversity. Why not embrace an equally diverse approach to evolutionary explanation?' (2021: 3). Perhaps the task is not articulating general theoretical models, but 'to reconstruct and explain the *actual* history of living beings' (Nuño de la Rosa 2014: 222, my emphasis)?

For Chiesa (2018) a new synthesis risks eradicating ambiguity, like the previous one, by promoting a unitary worldview. As such, it would follow the same principle as the MS and thus eliminate or minimise the importance of the processes that do not fit the majority view. Central proponents of the EES, like Kevin Laland (2018), have argued that plurality must be maintained and that the EES should not strive for the same kind of unity as the MS sought, which was informed by the positivist notion of a 'forced' unification where all sciences should emulate physics. Perhaps it is physics that should learn from biology. At least organic metaphors capture the interplay of sciences more accurately than physical ones.

Based on this, the view that we need a new synthesis seems misguided, at least if this entails that we model the new synthesis on the previous one. Instead of a strict synthesis, what we need is an open-ended orientation, which does not seek simple unification or closure, but instead works as a living totality. Of course, this carries the risk of ambiguity and attempts at hijacking. But insofar as some basic principles are established, such threats should be of minor influence. We find some of these principles in the approaches I have outlined.

Hence, it seems that what is required is not a narrow research program but a more encompassing philosophy of nature, which does not aim to reduce all the different scientific perspectives to one explanatory strategy or presentation of information but allows for plurality (Rouse 2024; Godfrey-Smith 2001). This goes back to my earlier point about the MS itself being more of a philosophy of nature than a strict research program, and that we need a theory which can accommodate multiple research programs. As such, *the organisation of research is more important than unification of knowledge* (Rouse 2024).

So, what does this all say about the relation between science and the humanities? Malabou applies plasticity to science itself, stating that it can only maintain itself by being open to insight from other discourses. If not, they become hardened. In Malabou's view, as Williams (2013) underscores, 'the stabilization of any discipline occurs only in the face of a fundamental capacity for change' (9). The humanities risk being swallowed by the sciences if it does not try to 'think with them'. Without relation to science, these disciplines lose their capacity for change (Malabou 2010). It might happen without their knowledge; they might, for example, adopt a stance that parrots common-sense conceptions of science, forgetting their critical role in challenging simplistic notions of science and identifying unspoken presuppositions within it. It demands a more systematic understanding.

Historicity lies at the heart of biological evolution. It is, according to Longo (2020), 'one of the possible links to the humanities, with no subordination' (93). His views dovetail with the radical naturalism that we find in Hegel's thinking, which aims not to limit the scope of naturalism to the confines deemed acceptable by a narrow conception of science (Illetterati 2020). It is a naturalism 'whose very core is the notion of life' (Illetterati 2023: 188). It expands naturalism by undermining the seemingly insurmountable border erected between nature and culture. It denaturalises nature by accounting for the immanent emergence of subjectivity as something that emerges from, but cannot be reduced to, natural determinants. Instead of mortifying nature, it seeks to grasp our living relation to it (Žižek et al. 2022).

As soon as we make the case that natural science deals with unique and irreversible processes, the distinction between the social and natural sciences becomes plastic. The human and social sciences gain scientific legitimacy without subservience to the natural sciences insofar as we accept the ontological reality of relations (Juarrero 2015). While this view casts the natural sciences as less predictive and concerned with unchanging laws, it opens them up to constructive interaction with other disciplines. Perhaps we should even consider the different disciplines as 'organic parts of the same dialectical whole' (Nuzzo 2024: 98), as both open and closed, just as Hegel's logic? We may achieve unity of science at last, but not the kind that was striven for in the past. Rather, this is a contradictory unity that never attains completeness.—The end of certainty, perhaps, but the beginning of another kind of science:

> The efforts of these organicists indicate that we have not reached the end of science as a fruit of creative human activity. The idea that biology is so complex that computers, which are mere creations of the human mind, will fulfill the role of scientists is erroneous. […] [A] new critical and rigorous biology could become a powerful tool to reconceptualize our world. (Soto and Sonnenschein 2021: no pagination)

These are wide-ranging claims. They argue that a new kind of biology could become the means of reconceptualising the world as such and combat the detrimental consequences of the current outlook, such as 'the environmental problems we are facing today'. This is an integral part of any radical critique of science: the whole of science must be criticised to understand the problems with its instantiations. It is not enough to critique scientific results; one must look at the technological and ideological uses of science, institutional constraints, funding bodies, the pressure to publish, etc. (Zhao 2019). The whole organisation of science is at stake.[9]

Another kind of science seems to necessitate another kind of philosophy. It might be that what is needed is a bastardisation of philosophy, which means that philosophy does not strive for closure, that does not seek to eradicate contradictions to produce identity. Rather, it recognises how every identity is unstable and contaminated by the other (Mohan 2021). At the same time, science and philosophy cannot just be *about* systems but must themselves be systematic, which is to say that all the concepts they apply find their justification through the relation to other concepts and the whole of science. Hence, we should conceive of them as members of an organic system whose meaning cannot be grasped except through their role within the larger organisation. I said above that—at a bare minimum—a more comprehensive theory of evolution requires that we include more processes in its functioning. I have also argued that this is not enough. But more fundamentally, science needs a more comprehensive and systematic philosophical outlook, which is systematic *because* it is permanently incomplete.

Biology may well be the most important science of the twenty-first century. Not because it seeks to explain everything but because it actively delimits its reach, reorganises its boundary conditions, and gets at how it is constrained by its content—entangled with other sciences, discourses, and socio-economic structures. This lack of self-identity was and is its biggest strength. Since it was premised on assumptions and metaphors from other discourses, it was never at peace with itself. And it was acting in the world based on these assumptions, without ontological guarantees. Finally, it seems that the principle of *voir venir* applies to our enquiry. Scientific disciplines are permanently immature and open or reduced to dead metaphors masquerading as theories. We have no choice but to anticipate something that we cannot foresee. Even if there are 'vague forebodings', they are not actualised linearly but in qualitative leaps. We are probably not completely off, as we have identified tendencies; but we cannot state what is possible in advance. We might be at the watershed of a new scientific revolution that places historicity and singularity at the heart of science. It would be foolish to ignore these developments.

Bibliography

Bernal, J. D. 1939. *The Social Function of Science*. London: George Routledge & Sons Ltd.

Bechtel, W., and L. Bich. 2024. Situating Homeostasis in Organisms: Maintaining Organization Through Time. *The Journal of Physiology* 602: 6003–6020. https://doi.org/10.1113/JP286883.

[9] Zhao writes: 'What is to be done can be realistically decided only by accepting an important lesson of critique: that the way to understand the different parts of science, and therefore to *change* them, is to understand the whole of science, particularly its integration in global systems of power and capital' (Zhao 2019: no pagination).

Bordignon, Michela. 2021. Hegel's Logic as a System of Illegitimate Totalities. In *Hegel's Encyclopedic System*, ed. Sebastian Stein and Joshua Wretzel, 116–132. New York, NY: Routledge. https://doi.org/10.4324/9780429022555-9.

Chaix, Victor, Giuseppe Longo, and Maël Montévil. 2021. Science in the Storm 1: GMOs, Agnotology, Theory—Introduction. *Philosophy World Democracy*. https://www.philosophy-world-democracy.org/science-in-the-storm.

Chiesa, Lorenzo. 2018. Lacan with Evo-Devo. In *Inheritance in Psychoanalysis*, ed. Joel Goldbach and James A. Godley, 44–56. New York, NY: SUNY Press.

Depew, David J. 2017. Darwinism in the Twentieth Century: Productive Encounters with Saltation, Acquired Characteristics, and Development. In *The Darwinian Tradition in Context: Research Programs in Evolutionary Biology*, ed. Richard G. Delisle, 61–88. Cham: Springer.

El-Hani, Charbel Niño, and Claudio R. M. Reis. 2021. Research Strategies and Value Outlooks in Scientific Practices: For an Organicist Thinking and a Pluralist Methodology in the Biological Sciences. *Philosophy World Democracy*. https://www.philosophy-world-democracy.org/articles-1/research-strategies-and-value-outlooks-in-scientific-practices.

Farnsworth, Keith D. 2022. How an Information Perspective Helps Overcome the Challenge of Biology to Physics. *Biosystems* 217: 104683. https://doi.org/10.1016/j.biosystems.2022.104683.

Ficara, Elena. 2020. *The Form of Truth: Hegel's Philosophical Logic*. Berlin and Boston: De Gruyter.

Foster, John Bellamy. 2022. The Return of the Dialectics of Nature: The Struggle for Freedom as Necessity. *Historical Materialism* 1–26. https://doi.org/10.1163/1569206X-20222279.

Frank, A., G. Marcelo, and E. Thompson. n.d. *The Blind Spot: Why Science Cannot Ignore Human Experience*. Cambridge, MA: The MIT Press.

Gallagher, Shaun. 2018. Dynamics and Dialectic. *Constructivist Foundations* 14 (1): 114–117. http://constructivist.info/14/1/114.

Gambarotto, Andrea, and Luca Illetterati. 2020. Hegel's Philosophy of Biology? A Programmatic Overview. *Hegel Bulletin* 41 (3): 1–22. https://doi.org/10.1017/hgl.2020.21.

Gefaell, Juan, and Cristian Saborido. 2022. Incommensurability and the Extended Evolutionary Synthesis: Taking Kuhn Seriously. *European Journal for Philosophy of Science* 12 (24): 1–25. https://doi.org/10.1007/s13194-022-00460-2.

Godfrey-Smith, Peter. 2001. On the Status and Explanatory Structure of DST. In *Cycles of Contingency: Developmental Systems and Evolution*, ed. Susan Oyama, P. E. Griffiths, and R. D. Gray, 283–297. Cambridge MA: MIT Press.

Hall, Stuart. 1985. Signification, Representation, Ideology: Althusser and the Post-structuralist Debates. *Critical Studies in Mass Communication* 2 (2). https://doi.org/10.1080/15295038509360070.

Hegel, G. W. F. 2004. *Philosophy of Nature: Part Two of the Encyclopaedia of the Philosophical Sciences*. Translated by A. V. Miller. Oxford, UK: Oxford University Press.

Hegel, G. W. F. 2010. *The Science of Logic*. Cambridge, UK: Cambridge University Press.

Herrington, Emily, and Eva Jablonka. 2020. Creating a "Gestalt Shift" in Evolutionary Science: Roles for Metaphor in the Conceptual Landscape of the Extended Evolutionary Synthesis (EES). *Interdisciplinary Science Reviews* 45 (3): 360–379. https://doi.org/10.1080/03080188.2020.1794383.

Illetterati, Luca. 2020. Nature's Externality: Hegel's Non-Naturalistic Naturalism. *Problemi* 58 (11–12): 51–72.

Illetterati, Luca. 2022. Science as Experience of Freedom: Hegel on the Scientific Nature of Philosophy. In *The Relevance of Hegel's Concept of Philosophy: From Classical German Philosophy to Contemporary Metaphilosophy*, ed. Luca Illetterati and Giovanna Miolli, 55–74. London and New York: Bloomsbury Academic.

Illetterati, Luca. 2023. Beyond a Naturalistic Conception of Nature: Nature and Life in Hegel's Early Writings. In *Nature and Naturalism in Classical German Philosophy*, ed. Luca Corti and Johannes-Georg Schülein, 187–208. New York and London: Routledge.

Jablonka, E., and M. J. Lamb. 2020. *Inheritance Systems and the Extended Evolutionary Synthesis*. Cambridge: Cambridge University Press. https://doi.org/10.1017/9781108685412.

Jaeger, J. 2024. The Fourth Perspective: Evolution and Organismal Agency. In *Organization in Biology*, ed. M. Mossio. Berlin: Springer. https://doi.org/10.1007/978-3-031-38968-9.

Juarrero, A. 2015. What Does the Closure of Context-Sensitive Constraints mean for Determinism, Autonomy, Selfdetermination, and Agency? *Progress in Biophysics and Molecular Biology* 119 (3): 510–521. https://doi.org/10.1016/j.pbiomolbio.2015.08.007.

Lakoff, George, and Mark Johnson. 2008. *Metaphors We Live By*. Chicago, IL: University of Chicago Press.

Lala, K. N., T. Uller, N. Feiner, M. W. Feldman, and S. F. Gilbert. 2024. *Evolution Evolving: The Developmental Origins of Adaptation and Biodiversity*. Princeton, NJ: Princeton University Press.

Laland, Kevin N. 2018. Evolution Unleashed: Is Evolutionary Science Due for a Major Overhaul—Or Is Talk of "Revolution" Misguided? *Aeon Magazine*. https://aeon.co/essays/science-in-flux-is-a-revolution-brewing-in-evolutionary-theory.

Laland, Kevin N., Kim Sterelny, John Odling-Smee, William Hoppitt, and Tobias Uller. 2011. Cause and Effect in Biology Revisited: Is Mayr's Proximate-Ultimate Dichotomy Still Useful? *Science*. https://doi.org/10.1126/science.1210879.

Laland, Kevin N., Tobias Uller, Marcus W. Feldman, Kim Sterelny, Gerd B. Müller, et al. 2015. The Extended Evolutionary Synthesis: Its Structure, Assumptions and Predictions. *Proceedings of the Royal Society B: Biological Sciences*. https://doi.org/10.1098/rspb.2015.1019.

Lenin, V. L. 1914. *The Teachings of Karl Marx*. New York, NY: International Publishers.

Longo, Giuseppe. 2018. Information and Causality: Mathematical Reflections on Cancer Biology. *Organisms: Journal of Biological Sciences* 2 (1): 83–103. https://doi.org/10.13133/2532-5876_3.15.

Longo, Giuseppe. 2020. Naturalizing Physics Or, Embedding Physics in the Historicity and Materiality of the Living. *La Deleuziana—Online Journal Of Philosophy* 132–149.

Longo, Giuseppe. 2021a. Programming Evolution: A Crack in Science. *Organisms: Journal of Biological Sciences* 5 (1): 5–16. https://doi.org/10.13133/2532-5876/17538.

Longo, Giuseppe. 2021b. Science in the Storm 4—Agnotology, Artificial Intelligence, and Democracy. *Philosophy World Democracy*. https://www.philosophy-world-democracy.org/articles-1/science-in-the-storm-part-4.

Longo, Giuseppe, and Maël Montévil. 2013. Extended Criticality, Phase Spaces and Enablement in Biology. *Chaos, Solitons and Fractals* 55. https://doi.org/10.1016/j.chaos.2013.03.008.

Longo, Giuseppe, and Matteo Mossio. 2020. Geocentrism vs Genocentrism: Theories Without Metaphors, Metaphors Without Theories. *Interdisciplinary Science Reviews* 45 (3): https://doi.org/10.1080/03080188.2020.1798588.

Longo, G., M. Montévil, and S. Kauffman. 2012. No Entailing Laws, but Enablement in the Evolution of the Biosphere. *ACM Proceedings of GECCO*: 1379–1392.

Malabou, Catherine. 2010. The Future of the Humanities. In *Theory@buffalo 14: Territory and Thought*, ed. Philip Campanile and Joshua Lam, 8–16.

Malafouris, L. 2015. Metaplasticity and the Primacy of Material Engagement. *Time and Mind* 8 (4): 351–371. https://doi.org/10.1080/1751696X.2015.1111564.

Merchant, Carolyn. 1989. *The Death of Nature: Women, Ecology and the Scientific Revolution*. New York, NY: HarperOne.

Mohan, Shaj. 2021. On the Bastard Family of Deconstruction. *Philosophy World Democracy*. https://www.philosophy-world-democracy.org/other-beginning/on-the-bastard-family-of-deconstruction.

Montévil, Maël. 2022. Conceptual and Theoretical Specifications for Accuracy in Medicine. In *Personalized Medicine in the Making: Philosophical Perspectives from Biology to Healthcare*, ed. Chiara Beneduce and Marta Bertolaso, 47–62. Cham: Springer. https://doi.org/10.1007/978-3-030-74804-3_3.

Moreno, A., and M. Mossio. 2015. *Biological Autonomy*, vol. 12. Dordrecht: Springer Netherlands. https://doi.org/10.1007/978-94-017-9837-2.

Mossio, Matteo. 2024. Introduction: Organization as a Scientific Blind Spot. In *Organization in Biology*, ed. Matteo Mossio, 1–22. Cham: Springer. https://doi.org/10.1007/978-3-031-38968-9_1.

Ng, K. 2020. *Hegel's Concept of Life: Self-Consciousness, Freedom, Logic*. New York, NY: Oxford University Press. https://doi.org/10.1093/oso/9780190947613.001.0001.

Noble, Denis. 2014. Introduction. In *Perspectives on Organisms: Biological Time, Symmetries and Singularities*, VII–X. Berlin and Heidelberg: Springer.

Noble, Denis. 2021. The Illusions of the Modern Synthesis. *Biosemiotics* 14 (1): 5–24. https://doi.org/10.1007/s12304-021-09405-3.

Nuño de la Rosa, Laura. 2014. On the Possible, the Conceivable, and the Actual in Evolutionary Theory. *Biological Theory* 9 (2): 221–228. https://doi.org/10.1007/s13752-014-0173-z.

Nuzzo, Angelica. 2024. The Logic of Nature: Nature as the "Idea in the Form of Otherness". In *Hegel's Philosophy of Nature: A Critical Guide, Cambridge Critical Guides*, ed. Marina F. Bykova, 97–116. Cambridge: Cambridge University Press. https://doi.org/10.1017/9781009023160.007.

Ochoa, C. 2024. Deconstructing the Extended Evolutionary Synthesis: Do We Need a New Theory of Evolution? In *Unity and Disunity in Evolutionary Biology: Deconstructing Darwinism*, ed. R. G. Delisle, M. Esposito, and D. Ceccarelli, 403–439. Cham: Springer International Publishing. https://doi.org/10.1007/978-3-031-42629-2_14.

Pigliucci, Massimo, and Gerd B. Müller. 2010. Elements of an Extended Evolutionary Synthesis. In *Evolution, The Extended Synthesis*, ed. Massimo Pigliucci and Gerd B. Müller, 3–18. Cambridge, MA: MIT Press.

Pinkard, T. 2024. The Prospects for an Idealist Natural Philosophy: Logic and Nature. In *Hegel's Philosophy of Nature: A Critical Guide*, ed. M. F. Bykova, 259–272. Cambridge: Cambridge University Press. https://doi.org/10.1017/9781009023160.018.

Pradeu, Thomas, Maël Lemoine, Mahdi Khelfaoui, and Yves Gingras. 2021. Philosophy in Science: Can Philosophers of Science Permeate through Science and Produce Scientific Knowledge? *The British Journal for the Philosophy of Science* 75 (2): 375–416. https://doi.org/10.1086/715518.

Rouse, Joseph. 2024. Niches and Niche Construction in Biology and Scientific Practice. *European Journal for Philosophy of Science* 14. https://doi.org/10.1007/s13194-024-00604-6.

Sanches de Oliveira, Guilherme, Thomas van Es, and Inês Hipólito. 2023. Scientific Practice as Ecological-Enactive Co-Construction. *Synthese* 202 (1): 4. https://doi.org/10.1007/s11229-023-04215-1.

Soto, Ana M., and Gracen Brilmyer. 2020. On Language, Scientific Metaphor, and Endocrine Disruption: An Interview with Feminist Scientist Ana Soto. *Catalyst: Feminism, Theory, Technoscience* 6 (1): 1–8. https://doi.org/10.28968/cftt.v6i1.33445.

Soto, Ana M., and Carlos Sonnenschein. 2021. The Proletarianization of Biological Thought. *Philosophy World Democracy*. https://www.philosophy-world-democracy.org/articles-1/the-proletarianization-of-biological-thought.

Stein, S. 2024. Nature and Its Limits: Hegel's Idealist Critiques of Physicalism, Naturalism, and Essentialism. In *Hegel's Philosophy of Nature: A Critical Guide*, ed. M. F. Bykova, 36–57. Cambridge, UK: Cambridge University Press. https://doi.org/10.1017/9781009023160.004.

Stoltzfus, Arlin. 2012. Constructive Neutral Evolution: Exploring Evolutionary Theory's Curious Disconnect. *Biology Direct*. https://doi.org/10.1186/1745-6150-7-35.

Stoltzfus, Arlin. 2017. Why We Don't Want Another "Synthesis". *Biology Direct* 12 (1): 23. https://doi.org/10.1186/s13062-017-0194-1.

Tahar, M. 2022. Biological Constraints as Norms in Evolution. *History and Philosophy of the Life Sciences* 44 (9). https://doi.org/10.1007/s40656-022-00483-1.

Trappes, R., and S. Leonelli. 2025. Conceptualising Research Environments using Biological Niche Concepts. *European Journal for Philosophy of Science* 15 (1): 11. https://doi.org/10.1007/s13194-025-00640-w.

Whitehead, Alfred North. 1967. *Science and the Modern World*. New York, NY: The Free Press.
Williams, Tyler. 2013. Plasticity, in Retrospect: Changing the Future of the Humanities. *Diacritics* 41: 6–25. https://doi.org/10.1353/dia.2013.0003.
Wimsatt, William C. 1994. Levels, Robustness, Emergence, and Heterogeneous Dynamics: Finding Partial Organization in Causal Thickets. In *Canadian Journal of Philosophy Supplementary*, ed. Daniel S. Brooks, James DiFrisco, and William C. Wimsatt, vol. 20, 207–274. Cambridge, MA: The MIT Press. https://doi.org/10.1080/00455091.1994.10717400.
Zhao, Helen. 2019. What Is a Radical Analysis of Science? New Directions for Science for the People. *Science for the People* 22 (1). https://magazine.scienceforthepeople.org/vol22-1/what-is-a-radical-analysis-of-science/.
Žižek, Slavoj. 2020. *Hegel in a Wired Brain*. London, UK: Bloomsbury Academic.
Žižek, S., F. Ruda, and A. Hamza. 2022. *Reading Hegel*. Cambridge, UK: Polity.

Correction to: Agency and Organisation: The Dialectics of Nature and Life

Correction to:
R. Sandnes Haukedal, ***Agency and Organisation: The Dialectics of Nature and Life*****, Library of Ethics and Applied Philosophy 44, https://doi.org/10.1007/978-981-95-1924-8**

Unfortunately, the author has missed out a few important corrections including references, an incorrect title (*Organism and Environment: Ecological Development, Niche Construction, and Adaption*" – has been changed to "*Organism and Environment: Ecological Development, Niche Construction, and* ***Adaptation***" on p. 37, 71, 152, 185, 225), and a few grammatical errors in the original published version. These have now been corrected in this revised version.

The updated version of this book can be found at
https://doi.org/10.1007/978-981-95-1924-8

R. Sandnes Haukedal, *Agency and Organisation: The Dialectics of Nature and Life*,
Library of Ethics and Applied Philosophy 44,
https://doi.org/10.1007/978-981-95-1924-8_9

MIX
Papier aus verantwortungsvollen Quellen
Paper from responsible sources
FSC® C105338

If you have any concerns about our products,
you can contact us on
ProductSafety@springernature.com

In case Publisher is established outside the EU,
the EU authorized representative is:
Springer Nature Customer Service Center GmbH
Europaplatz 3, 69115 Heidelberg, Germany

Printed by Libri Plureos GmbH
in Hamburg, Germany